AQA GCSE
ADDITIONAL SCIENCE

B2 1.1 How do cells perform different functions?

B2 1.1
Learning outcomes

Know the names and be able to explain the function of the different parts of animal and plant cells.

In this section we look at the component parts of cells and their role in cell function.

Today we understand that all living things are made up of cells, but this was not always the case. In 1665 a scientist called Robert Hooke used a primitive microscope to discover that samples from a cork tree were made up from repeating structures. He called these structures cells.

Hooke's discovery was only the start of what is now known as cell theory. Although he observed cells in one type of plant, he did not link cells to other species and had no idea what was inside cells.

In 1831 improvements in microscope technology allowed a scientist called Robert Brown to observe nuclei inside a number of different cells. Later in the same decade two German biologists, Theodor Schwann and Matthias Jakob Schleiden, carried out more research and linked cells to all plants and animals. Schwann published these findings without any acknowledgment of Schleiden's or anyone else's contributions. He was correct to suggest that cells are the units that make up all living things. However, he also suggested that new cells formed out of nothing. This was proved to be wrong by Rudolph Virchow in 1858. Virchow proposed the theory, still in place today, that all cells are formed from other cells.

Figure 1.1 Robert Hooke's drawing of the cork cells he could see with his microscope.

Test yourself

1 Cells are often described as the 'building blocks of life'. What does this mean?
2 Robert Hooke observed cells but did not identify anything inside them. What was limiting his research?
3 Why do you think that Robert Brown's discovery was important in developing a better understanding of cells?

What are the functions of different parts of a cell?

Modern microscopes magnify thousands of times and now let us see an incredible amount of detail about the structure of cells — detail that Robert Hooke's microscope simply could not display. We now know that even within cells there are structures that have specific functions. We can identify different parts of cells that each carry out different functions. From our observations we have discovered that all cells have some parts in common, such as a **cell membrane**. We will look at each structure and its function.

Animal cells
Nucleus
The **nucleus** of a cell contains an individual's genetic information as genes on the chromosomes.

Biology 2

Chapter 1 Cells, tissues and organs

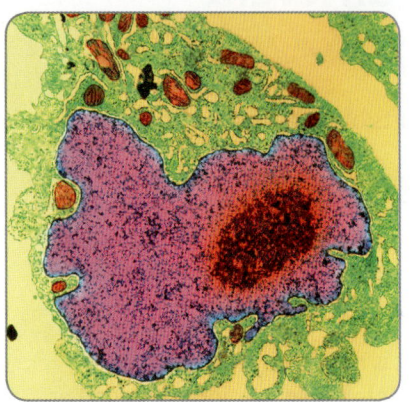
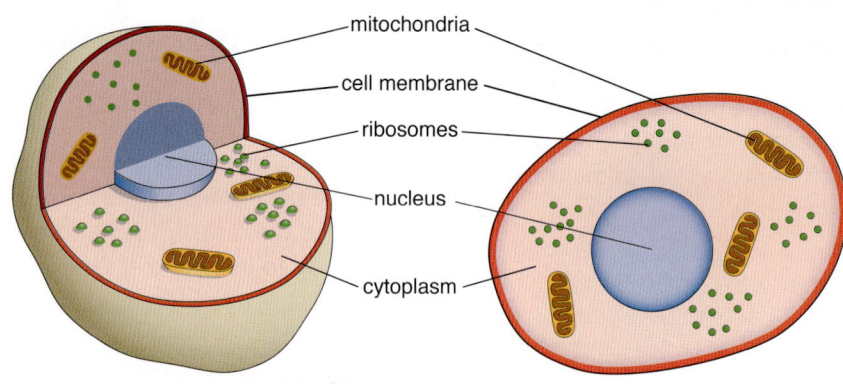

Figure 1.2 An animal cell shown as it is seen with a microscope (magnified × 4800) and as three- and two-dimensional diagrams.

> The **nucleus** controls the cell chemical reactions and cell division.
> The **cell membrane** surrounds the cell and controls the movement of substances into and out of the cell.
> The **cytoplasm** is where the chemical reactions of the cell take place.

The nucleus has two main functions:
- it regulates the cell's activities by controlling the production of enzymes, and
- it controls cell division

How does the nucleus regulate the cell's activities? The chromosomes in the nucleus carry the genetic code for production of all proteins including enzymes. The code is copied and taken to the ribosomes in the **cytoplasm** where the proteins are made.

How does the nucleus control cell division? First, by copying all the chromosomes and then dividing into two; this is followed by the cell cytoplasm dividing.

Cell membrane
The cell membrane surrounds the cytoplasm. The cell membrane controls the movement of many substances in and out of the cell. For example, the glucose used in the mitochondria for respiration enters the cell through the membrane, and insulin produced in pancreas cells leaves these cells through the membrane.

Cytoplasm
The cytoplasm is mostly water and fills the cell membrane. It is where most of the cell's chemical reactions take place. Proteins and enzymes are made in the cytoplasm. Enzymes control all the reactions that take place in the cell. The cytoplasm stores the dissolved raw materials, such as amino acids and glucose, needed for all cellular reactions. These reactions include respiration and protein synthesis.

Some of these enzymes pass out of the cell membrane to control reactions outside the cell. For example, amylase is an enzyme produced by cells in your salivary glands. Amylase passes out of the cells and digests starch into sugar in your mouth and oesophagus. Ideas about enzymes are developed in Chapter 3.

Mitochondria
Your whole body needs energy to keep warm. Cellular energy is used for all chemical reactions, to grow and for muscle activity. Mitochondria release the energy from glucose molecules during a process called respiration. Respiration is a series of chemical reactions that take place on the surface of the channels you can see inside the mitochondrion shown in Figure 1.4. These channels have a large surface area so that more respiration reactions can take place.

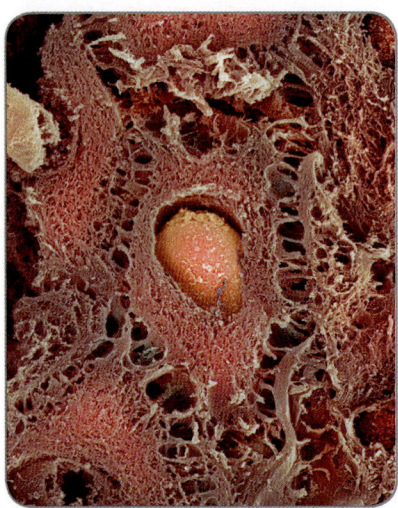

Figure 1.3 The nucleus of a human skin cell.

Figure 1.4 A mitochondrion.

6 AQA GCSE Additional Science

AQA GCSE ADDITIONAL SCIENCE

Nigel Heslop
Steve Witney
Christine Woodward

Graham Hill
Toby Houghton

Philip Allan Updates, an imprint of Hodder Education, an Hachette UK company, Market Place, Deddington, Oxfordshire OX15 0SE

Orders

Bookpoint Ltd, 130 Milton Park, Abingdon, Oxfordshire OX14 4SB
tel: 01235 827827
fax: 01235 400401
e-mail: education@bookpoint.co.uk

Lines are open 9.00 a.m.–5.00 p.m., Monday to Saturday, with a 24-hour message answering service. You can also order through the Philip Allan Updates website: www.philipallan.co.uk

© Nigel Heslop, Graham Hill, Toby Houghton, Steve Witney, Christine Woodward 2011

ISBN 978-1-4441-2077-6

First printed 2011
Impression number 5 4 3 2 1
Year 2016 2015 2014 2013 2012 2011

All rights reserved; no part of this publication may be reproduced, stored in a retrieval system, or transmitted, in any other form or by any means, electronic, mechanical, photocopying, recording or otherwise without either the prior written permission of Philip Allan Updates or a licence permitting restricted copying in the United Kingdom issued by the Copyright Licensing Agency Ltd, Saffron House, 6–10 Kirby Street, London EC1N 8TS.

Cover photo: Anterovium/Fotolia

Other photos are reproduced by permission of the following:

p4 *tl* Power and Syred/SPL, *tc* Susumu Nishinaga/SPL, *tr* Dr Torsten Wittman/SPL, *bl* Manfred Kage, Peter Arnold Inc. SPL, *br* Steve Gschmeissner/SPL; **p5** Omikron/SPL; **p6** *tl* Dr Gopal Murti/SPL, *cl* Steve Gschmeissner/SPL; **p8** Dr Jeremy Burgess/SPL; **p19** Jeff Carroll/AgstockUSA/SPL; **p20** Biophoto Associates/SPL; **p26** laurent dambies/Fotolia; **p27** *tr* Dennis Cox/Alamy; Krot/Fotolia, *bl* Djordje Korovljevic/Fotolia; Wadsworth Controls; **p32** *cl* Chris Howes/Wild Places Photography/Alamy, *cr* Philippe Psaila/SPL; **p38** Mark Boulton/Alamy; **p39** Chaos/Wikipedia; **p43** Topfoto; **p47** Phototake Inc/Alamy; **p49** Darryl Sleith/Fotolia; **p51** *bl* Ingram, *bc* Ingram, *br* Monkey Business/Fotolia; **p53** Vladimir Wrangel/Fotolia; **p55** *tr* Yuri Arcurs/Fotolia, *cl* John Fryer/Alamy; **p58** Melissa Schakle/Fotolia; **p59** James-King-Holmes/SPL; **p61** Adrian T Sumner/SPL; **p65** *tl* Ingram, *cr* CNRI/SPL; **p70** CNRI/SPL; SPL; **p71** Professor Miodrag Stojkovic/SPL; **p72** Pascal Goetgheluck/SPL; **p75** *cr* Robin Weaver/Alamy, *bl* Lucy Cole; **p76** *tl* Sinclair Stammers/SPL, *cl* Tom McHugh/SPL, *bl* Herve Conge, ISM/SPL, *br* Reuters/Corbis; **p77** Christian Jegou Publiphoto Diffusion/SPL; **p78** Christine Woodward; **p83** Deborah Benbrook/Fotolia; **p84** Maximilien Brice, CERN/SPL; **p93** Charles D. Winters/SPL; **p99** *c* Hodder, *bl* Varina Patel/Fotolia; **p101** normond/Fotolia; **p102** *t* Glenn Young/Fotolia, *bl* Jaroslaw Adamczyk/Fotolia; **p104** *tl* Leonard Lessin/Peter Arnold/Still Pictures, *cl* Alamy, *b* Tyler Boyes/Fotolia; **p105** *tl* RTimages/Fotolia, *br* Martinan/Fotolia; **p108** Tormentor/Fotolia; **p109** Pascal Goetgheluck/SPL; **p112** Nicholas Piccillo/Fotolia; **p115** Franck Boston/Fotolia; **p116** Nigel Heslop; **p117** Wellcome Images; **p119** lacabetyar/Fotolia; **p120** Charles D. Winters/SPL; **p123** hjpix/Fotolia; **p127** Cavendish Laboratory, University of Cambridge; **p130** Lorna Ainger; **p134** *tl* Gunter Marx/Alamy, *b* Ronald Hudson/Fotolia; **p137** Andrew Lambert Photography/SPL; **p138** superclic/Alamy; **p140** Alexander Raths/Fotolia; **p141** CBS/Everett/Rex Features; **p142** Tek Image/SPL; **p143** Sandia National Laboratories/SPL; **p148** Kevin Foy/Alamy; **p149** *tl* Vely/Fotolia, *tr* Sonya Etchison/Fotolia, *b* Michael Flippo/Fotolia; **p154** Martyn F. Chillmaid; **p156** *tl* Jacek Chabraszewski/Fotolia, *b* Nigel Heslop; **p157** CandyBoxPhoto/Fotolia; **p160** Tweecat/Fotolia; **p161** AJ Photo/SPL; **p163** *t* Martyn F. Chillmaid/SPL, *c* Nigel Heslop, *cl* Wikipedia; **p166** M.Brodie/Alamy; **p167** Alan Sirulnikoff/SPL; **p169** Corel; **p171** Gustoimages/SPL; **p173** Martyn F. Chillmaid/SPL; **p174** *tl* Nigel Heslop, *bl* Nigel Heslop; **p176** Andrew Lambert Photography/SPL; **p178** James Holmes, Hays Chemicals/SPL; **p182** ParisPhoto/Fotolia; **p190** Buzz Pictures/Alamy; **p196** *tl* PureStock X, *br* Ifstewart/Fotolia; **p197** Hideo Kurihara/Alamy; **p198** Getty Images; **p202** PureStock X; **p205** Nigel Gladwell/Alamy; **p210** fStop/Alamy; **p211** *cl* Anthony Hall/Fotolia, *cr* dicktraven/Fotolia; **p213** Art Directors & Trip/Alamy, **p214** *tl* Lou-Foto/Alamy, *c* Pakhnyushchyy/Fotolia; **p218** Steve Bloom Images/Alamy; **p220** *tl* NASA; **p221** Gordo/Fotolia; **p223** billy stock/Alamy; **p224** j. gruau/Fotolia; **p225** Hank Morgan/SPL; **p227** *bl* Iosif Szasz-Fabian/Fotolia, *br* photosoup/Fotolia; **p228** *bl* Devyatkin/Fotolia, *br* GeoM/Fotolia; **p231** Monkey Business/Fotolia; **p233** demarco/Fotolia; **p234** Peratech; **p239** Sheila Terry/SPL; **p241** Flake/Alamy; **p242** Sheila Terry/SPL; **p243** *bl* Sergey Peterman/Fotolia, *br* pavelr/Fotolia; **p247** *b* TopFoto, *t* NASA; **p248** Graham Tomlin/Fotolia; **p252** *cr* Alaska Stock LLC/Alamy, *cl* Oulette & Theroux, Publiphoto Diffusion/SPL, *bl* Dario Sabljak/Fotolia; **p257** *cl* ISM/SPL, *b* Cordelia Molloy/SPL; **p259** Martyn F. Chillmaid/SPL; **p260** History/Alamy

Printed in Italy

Hachette UK's policy is to use papers that are natural, renewable and recyclable products and made from wood grown in sustainable forests. The logging and manufacturing processes are expected to conform to the environmental regulations of the country of origin.

Contents

Getting the most from this book 2

Biology 2

Chapter 1 Cells, tissues and organs 4
Chapter 2 Photosynthesis, organisms and communities 19
Chapter 3 Enzymes: function, industrial uses and their role in exercise 38
Chapter 4 How do our bodies pass on characteristics? 58
Exam questions 81

Chemistry 2

Chapter 5 Bonding in reactions 84
Chapter 6 Mass, moles and yield 123
Chapter 7 Controlling rates of chemical reactions and energy transfer 148
Chapter 8 Using ions in solutions 167
Exam questions 186

Physics 2

Chapter 9 Forces and their effects 190
Chapter 10 Work, energy, power and momentum 210
Chapter 11 Using electricity 223
Chapter 12 The atom and radioactivity 247
Exam questions 268

Index 272
Periodic table 283

Getting the most from this book

Setting the scene
These sections at the start of each chapter introduce the topics that are going to be covered and encourage you to think about what you are going to learn. They also explain how you can develop your **practical** and **ICT** skills through the activities in the chapter.

Welcome to the *AQA GCSE Additional Science Student's Book*. This book covers all the content of the units Biology 2, Chemistry 2 and Physics 2 for the new AQA specification, as well as the 'How Science Works' elements. You have to sit three written papers for GCSE Additional Science plus the Controlled Assessment. The material tested in each written paper matches the content of the three units of this book. The following features are used throughout this book.

Learning outcomes
Each section starts with a list of **learning outcomes** that help you to target your thinking. Look back at these outcomes once you have read a section to check that you have understood each of them before moving on.

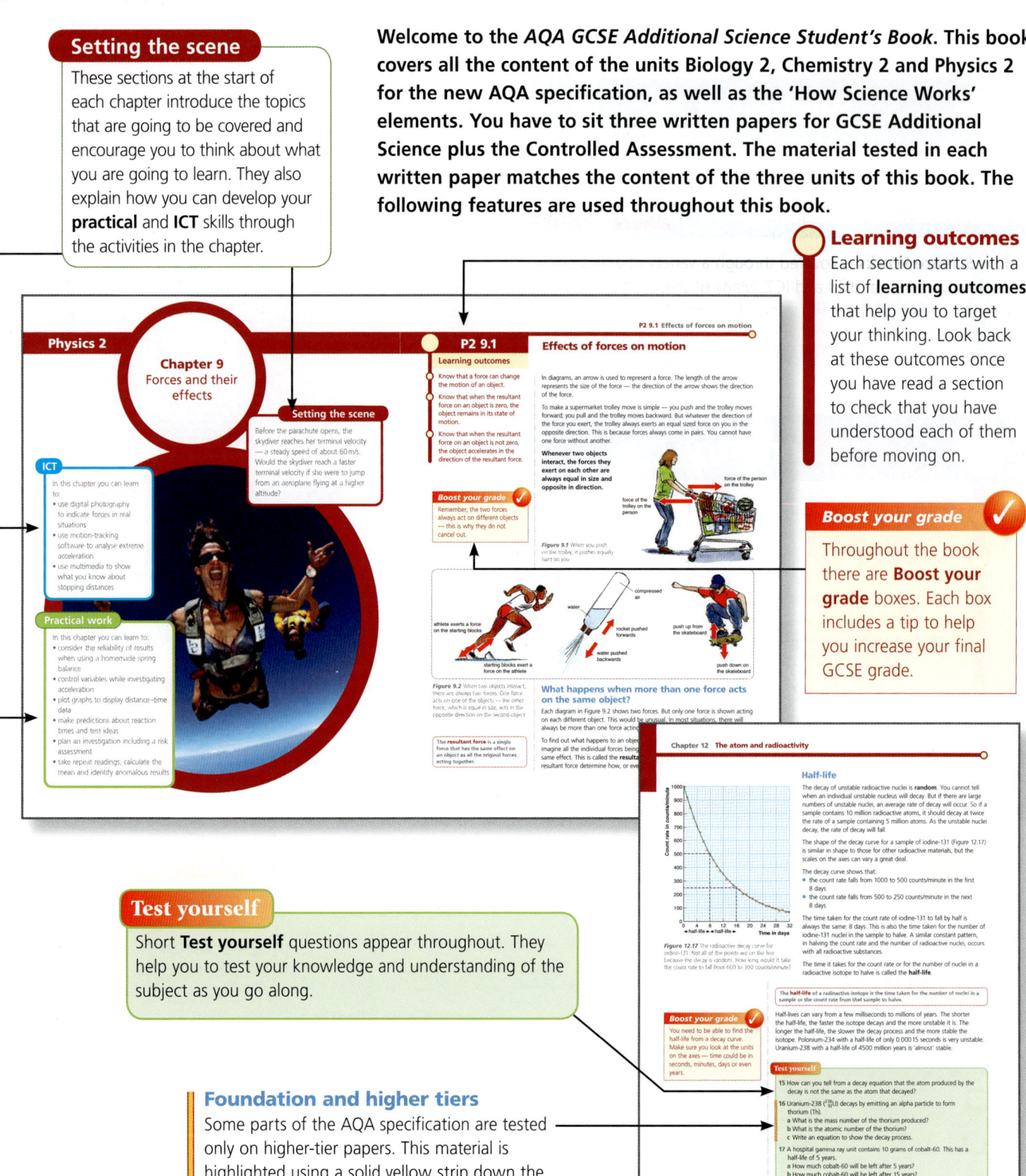

Boost your grade
Throughout the book there are **Boost your grade** boxes. Each box includes a tip to help you increase your final GCSE grade.

Test yourself
Short **Test yourself** questions appear throughout. They help you to test your knowledge and understanding of the subject as you go along.

Foundation and higher tiers
Some parts of the AQA specification are tested only on higher-tier papers. This material is highlighted using a solid yellow strip down the side of the text.

2 AQA GCSE Additional Science

Getting the most from this book

Key words are highlighted in the text and definitions are given in the margin to help you pick out and learn these important terms and concepts.

How Science Works is covered through a variety of activities that involve practical skills and ICT. Many of these activities are designed to be completed in practical lessons. Others present you with data and take you through new and important aspects of How Science Works. Some of the activities involve real-life applications and implications of science.

Homework questions

At the end of each chapter there are **Homework** questions that help you to make sure you have understood the content in the chapter.

Exam corner

The **Exam corner** sections at the end of each chapter contain typical answers (some good and some less good) to examination-style questions. You can read what the examiner thinks about each answer and pick up tips on how to improve your examination grade.

Exam questions

At the end of each unit are **Exam questions** that you can use to prepare for your GCSE examinations. These questions are similar in style to those set in the AQA unit exams.

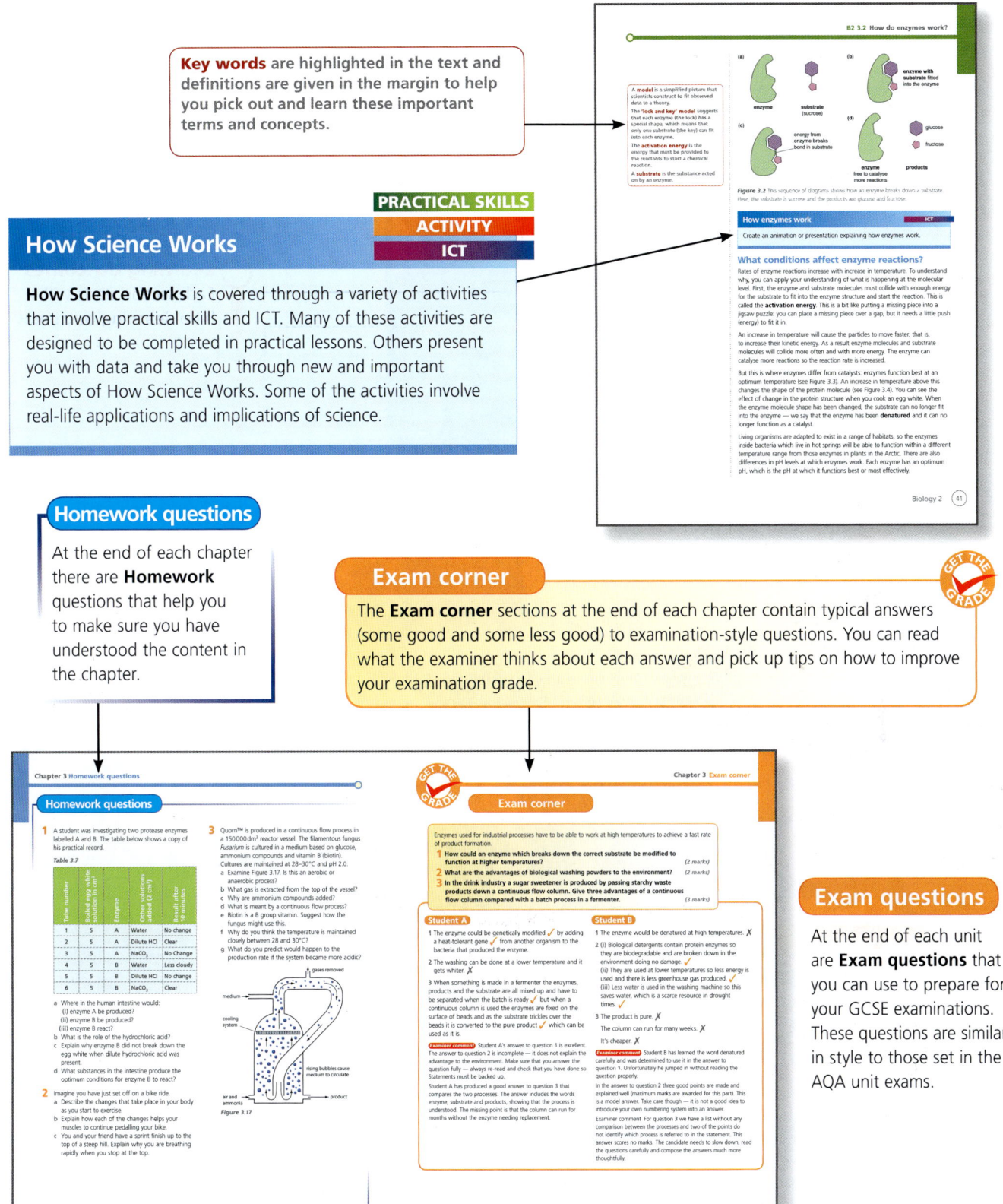

AQA GCSE Additional Science

Biology 2

Chapter 1
Cells, tissues and organs

Setting the scene

Cells are the basis of life. The DNA in cell nuclei controls all the processes in your body that keep you alive and is responsible for reproduction. The cells in the pictures are all different and each is specialised for a specific function. Can you describe the function of the cells in the photos? Roughly how big is each cell? Can you remember the names of any of the structures inside cells?

ICT

In this chapter you can learn to:
- create a resource that shows the similarities and differences between plant, animal, yeast and bacterial cells
- use multimedia to explain diffusion

Practical work

In this chapter you can learn to draw conclusions from the results of a diffusion experiment.

B2 1.2 Plant, algal, bacterial and yeast cells have additional structures

> **Boost your grade** ✓
>
> Some of the diagrams used here are three-dimensional to help you understand that the cell is a three-dimensional structure. You will only be expected to draw or complete two-dimensional diagrams in an examination (as Figure 1.6).

Ribosomes

The genes in the nucleus control production of **proteins** in the cytoplasm. Ribosomes are structures that link together the amino acids that make up proteins. The amino acids are linked in the order specified by the genetic code.

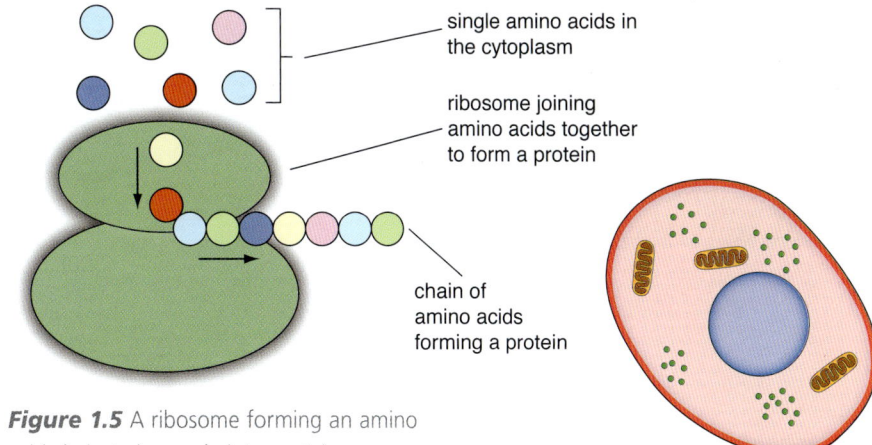

Figure 1.5 A ribosome forming an amino acid chain to be made into protein.

Figure 1.6 Animal cell.

Check Figure 1.6 and make sure that you can find and name all the parts.

> **Mitochondria** are the parts of a cell responsible for releasing energy from glucose by respiration to make cellular energy (to talk about one use 'a mitochondrion' or 'the mitochondrion').
>
> **Ribosomes** join amino acids in the correct order to make a specific protein. Each protein is coded for by a gene.
>
> **Proteins** are long chains of **amino acids** that form the basis of many cell structures such as cell membranes, mitochondria and ribosomes, and body structures such as skin, muscle and hair. They also form chemicals in the body such as enzymes.

> **Test yourself**
>
> 4 Muscles use a lot of energy for contraction. Which sub-cellular parts will you expect to see in large numbers in muscle cells?
>
> 5 Cells of the pancreas produce insulin, which is a protein. Which sub-cellular parts will you expect to see in large numbers in cells of the pancreas?

B2 1.2

Learning outcomes

Know and be able to describe the differences between animal, plant, bacterial and yeast cells.

> **Test yourself**
>
> 6 Why don't animal cells need cell walls?

> **Chloroplasts** use light energy to make glucose and oxygen from carbon dioxide and water by photosynthesis. They are found only in some plant cells but in most algal cells.

Plant, algal, bacterial and yeast cells have additional structures

Plant cells

What components can you find that are not in an animal cell?

Plant cells are larger than animal cells but we still need a microscope to examine them. Plant and algal cells have all the parts of an animal cell described in the last section and, in addition, they have the following.

Cell wall

Plant and algal cells have a rigid cell wall around the cell membrane that strengthens the cell and keeps its shape. It is made from a chemical compound called cellulose. Without cell walls plants would not be able to stay upright.

Chloroplasts

Chloroplasts are structures that contain a chemical called chlorophyll. Chlorophyll absorbs light energy to make food for the plant or alga(e) by a

Biology 2   7

Chapter 1 Cells, tissues and organs

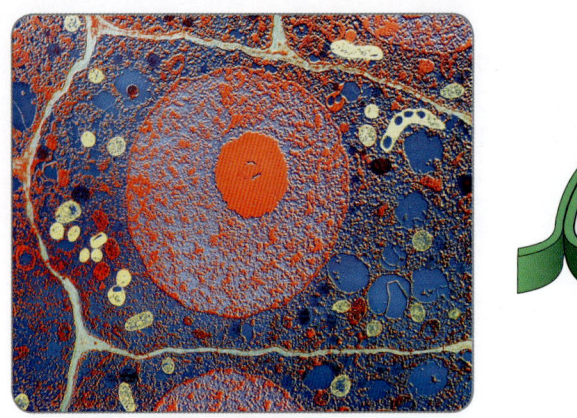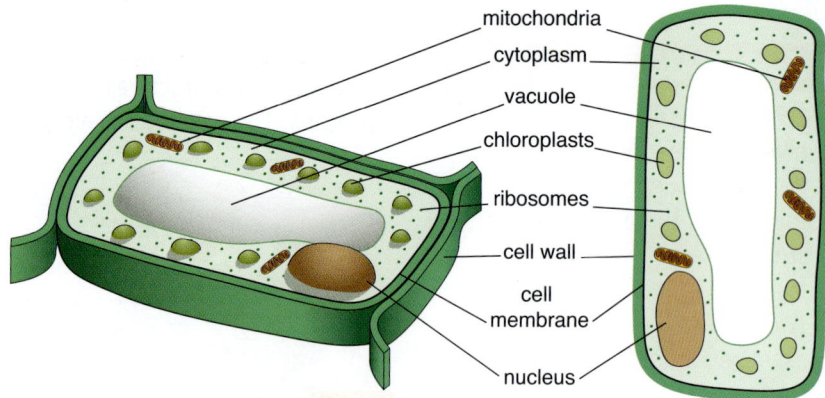

Figure 1.7 A plant cell shown as seen with a microscope (magnified ×2800) and as three- and two-dimensional diagrams.

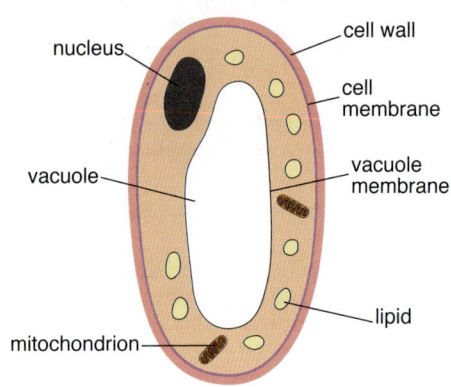

Figure 1.8 A yeast cell.

process called photosynthesis. Photosynthesis is actually a series of chemical reactions that are all controlled by enzymes.

Permanent vacuole

The vacuole in plant and algal cells contains cell sap. Cell sap contains water, mineral ions and some dissolved chemicals such as sugars. The vacuole acts like a reservoir for the cell.

The vacuole presses against the cell wall and helps to support the cell by keeping it rigid.

Yeast cells

Yeast is a single-celled fungus. It is important to humans because it can be used to make bread, beer and wine.

Yeast cells have a nucleus, cell membrane, cytoplasm, mitochondria and ribosomes with the same functions as plant and animal cells. A yeast cell also has:
- a fungal cell wall for support which is *not* made of cellulose, and
- a large vacuole containing dissolved substances such as mineral ions and sugar

Bacterial cells

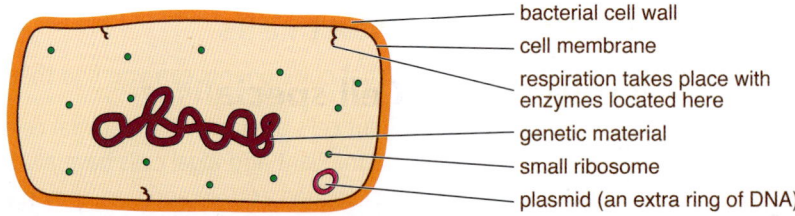

Figure 1.9 A bacterial cell (e.g. *E.coli*).

If you look at an evolutionary tree you will see that **bacteria** evolved before plants and animals. They do not have all the parts that plant and animal cells have.

Cytoplasm

The bacterial cell contains cytoplasm in which the chemical reactions of the cell take place. Small ribosomes are present in the cytoplasm for protein synthesis.

> The word **bacterium** is singular: 'a bacterium'. We usually talk about many bacteria. The word can also be used as an adjective, for example, a bacterial cell.

8   AQA GCSE Additional Science

B2 1.3 How do cells perform specialised functions?

Bacterial cell wall

This gives structural support to the cell. It is *not* made of cellulose.
- There is no nucleus.
- The genetic information of the cell is a long chromosome, which is free in the cytoplasm. The chromosome has the bacterial genes. (In the section on genetic engineering you saw a small ring of DNA called a plasmid — this is extra DNA.)
- There are no mitochondria.
- Respiration takes place using enzymes attached to extensions of the cell membrane.

Test yourself

7 Copy and complete this table

Table 1.1 Differences between plant, animal, yeast and bacterial cells.

	Plant and algal cell	Animal cell	Yeast cell	Bacterial cell
Nucleus	present			
Cell membrane				
Cell wall				Present — not made of cellulose
Mitochondria				
Chloroplasts				
Ribosomes				
Vacuole				

ICT

Create a resource that shows the similarities and differences between plant, animal, yeast and bacterial cells.

B2 1.3

Learning outcomes

- Know the features and identify a range of specialised cells.
- Understand how to relate the structure of a specialised cell to its function.

As cells mature they become different shapes or have different enzymes as they become specialised. We call this process **differentiation**.

How do cells perform specialised functions?

Cell specialisation

We have seen the specialised functions in the sub-cellular parts and now we will look at how some cells have special shapes for particular functions. Remember that an embryo starts as a ball of cells that specialise as the organism develops. We say that the cells **differentiate** so that they can perform different functions. Think of specialised cells you may already know — red and white blood cells, nerve cells, the light receptor cells, specialised cells in the pancreas to produce insulin, a fibroblast cell making new connective tissue (see the photograph at the chapter opening). This process is essential for the body to heal after an injury. These cells are mature and, once specialised, do not change shape or function.

However, there are some cells, called stem cells, which begin as being unspecialised but can produce a range of different cells. Stem cells are found in

Biology 2

Chapter 1 Cells, tissues and organs

> **Test yourself**
>
> 8 Do specialised animal cells such as a muscle cell contain the same sub-cellular parts as skin cells?

embryos, adult bone marrow (for forming the different blood cells) and some other body tissues such as in the liver and even at the bed of your fingernails.

Tissues

Large multicellular organisms need to group similar cells together to form body **tissue**, as the examples in Figures 1.10–1.12 show.

In these diagrams you can see that the cells within the tissue are the same shape.

> A **tissue** is a group of cells with similar structure and function.
> **Epithelium** means membrane. It can also be used as an adjective: epithelial cells.

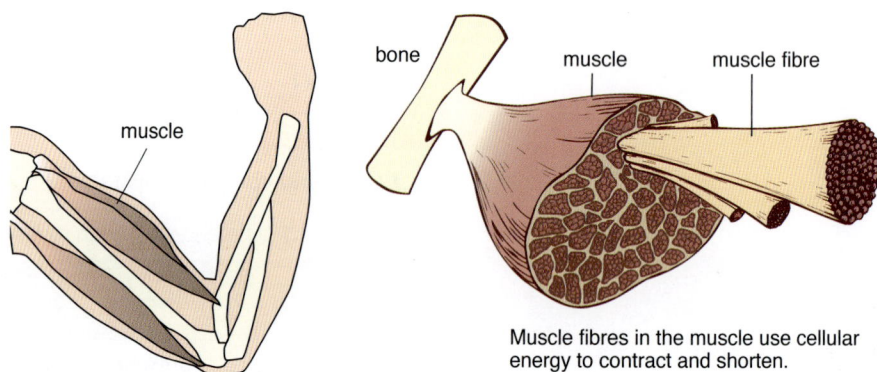

Figure 1.10 Muscle tissue is specialised to contract to bring about movement.

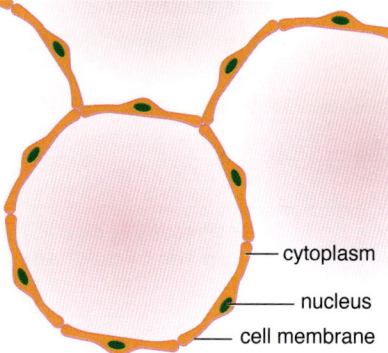

These epithlial cells around an alveolus are thin to allow gaseous exchange.

Figure 1.11 Epithelial tissue covers or lines some parts of the body. The thin **epithelium** lining the lungs allows gases to diffuse through.

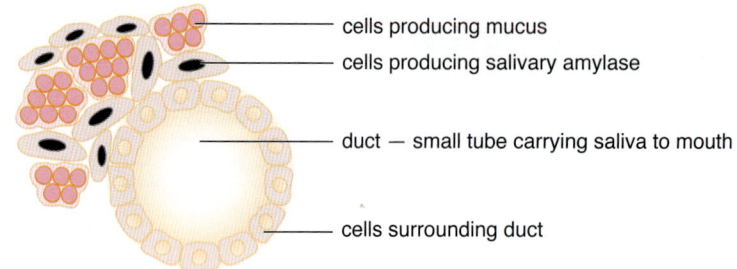

Figure 1.12 Glandular tissue produces substances such as enzymes and hormones.

B2 1.4

Learning outcomes

- Know why organisms require organ systems.
- Know the features and be able to identify a range of specialised organs in plants and animals.
- Understand the linking of organs into systems.

Even more organisation levels to form the whole organism

Organs

Further organisation comes from grouping tissues into organs. An organ contains more than one type of tissue so that the organ has one important function. Think of the body organs you know. The heart is an organ made up of largely muscular tissue to pump the blood but is covered and lined with epithelial tissue.

The stomach is an organ that contains the three tissue types shown in Figures 1.10–1.12.
- The stomach wall contains muscular tissue to churn and mix the contents with digestive juices.
- The juices contain enzymes produced by the glandular tissue.
- Epithelial tissue covers the outside of the stomach and lines the inside.

AQA GCSE Additional Science

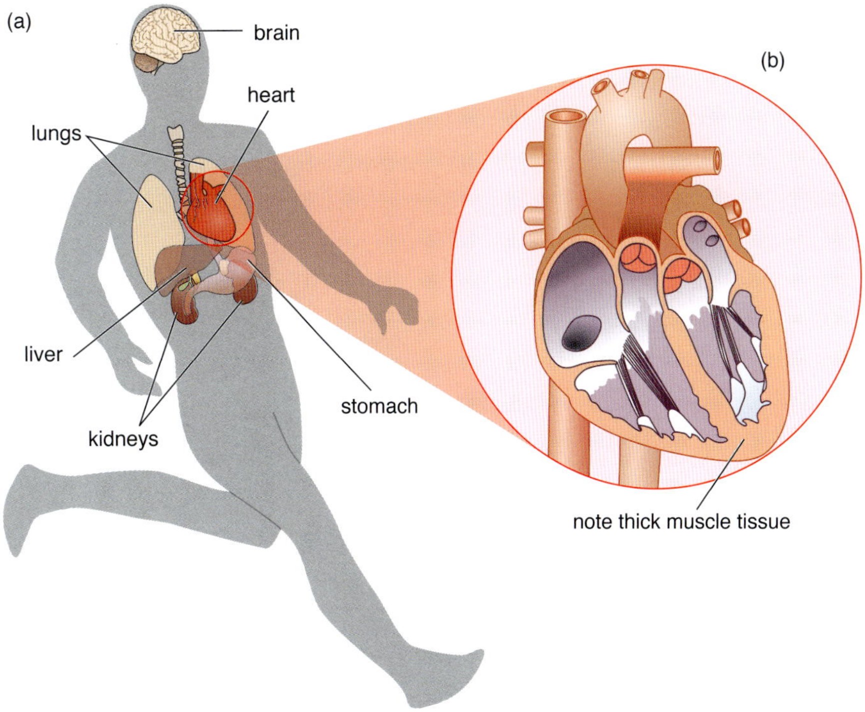

B2 1.4 Even more organisation levels to form the whole organism

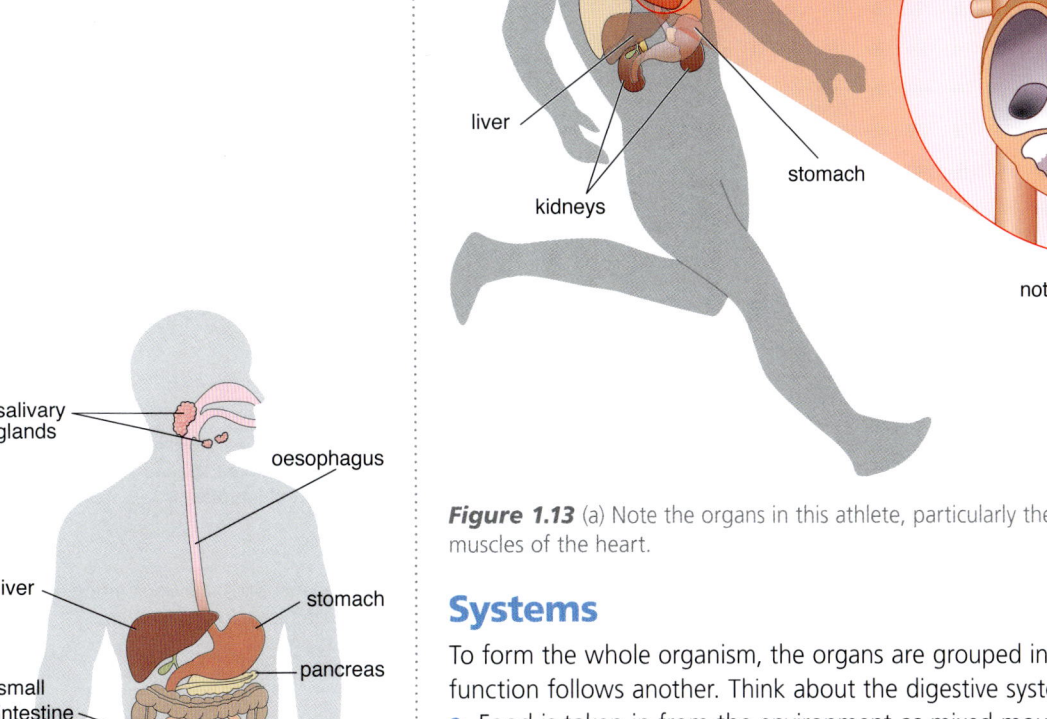

Figure 1.13 (a) Note the organs in this athlete, particularly the heart. (b) Note the thick muscles of the heart.

Figure 1.14 Digestive system.

Systems

To form the whole organism, the organs are grouped into systems so that one function follows another. Think about the digestive system:

- Food is taken in from the environment as mixed mouthfuls and has to be broken down chemically into small molecules that can be used by the body.
- Muscle tissue in the stomach wall mixes the food with the juices.
- Glandular tissue and organs produce the digestive juices containing enzymes that carry out these chemical reactions. The glands that produce the enzymes are the salivary glands, the stomach and the pancreas.
- The liver produces bile, which changes the pH in the small intestine to alkaline.
- The digestion takes place in the stomach and small intestine.
- Having produced small soluble molecules, such as amino acids and sugars, they must be absorbed into the blood further along the small intestine through the thin epithelial lining tissue of the villi (the villi are the tiny finger-like projections into the small intestine, which increase the surface area for absorption).
- This leaves a watery mass of indigestible matter. Water is absorbed in the large intestine and the indigestible matter is formed into faeces. This is stored temporarily in the rectum before being returned to the environment.

The whole organism is made up of all the different organ systems which function as a well-coordinated whole. As doctors know, if one system has a problem the whole body is affected.

> **Test yourself**
>
> 9 List the following starting with the smallest:
>
> mitochondrion, human chromosome, cheek cell, heart, plant cell, bacterial cell, digestive system

Biology 2 11

Chapter 1 Cells, tissues and organs

Plant tissues and organs

Multicellular plants have specialised tissues and organs. The roots anchor the plant and absorb nutrients and water, the stem is for transport and the leaf is an organ specialised for photosynthesis. The arrangement of tissues in the leaf is shown in Figure 1.15.

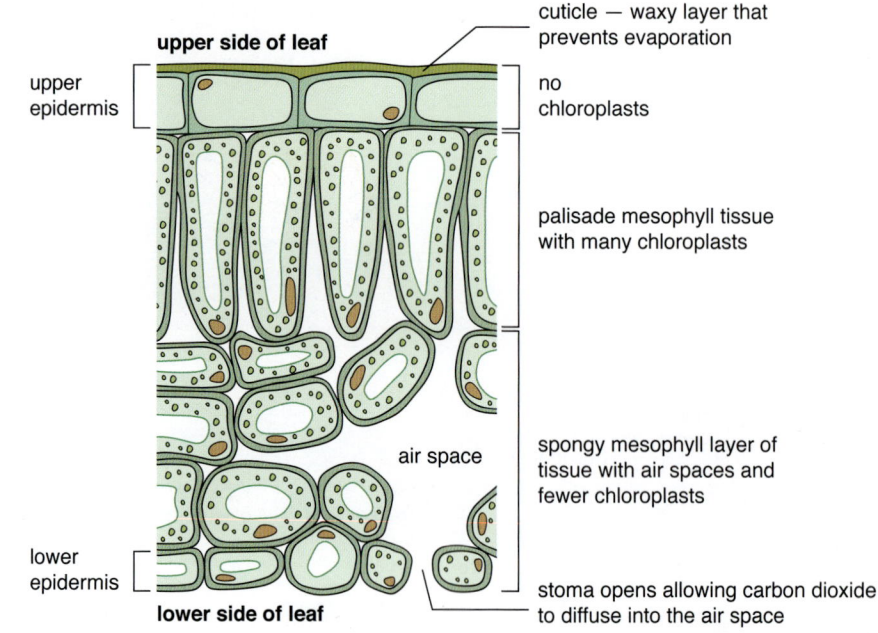

Figure 1.15 The organisation of tissues in the leaf, which is the plant organ for photosynthesis.

> **Test yourself**
>
> 10 If the stem is the organ for transport in the plant, what two types of tissue will you find there?
>
> 11 Complete the statements and PRINT the correct level of organisation beside each statement. The first is completed for you.
>
> The basic unit of all living things: CELL
>
> Contains similar _____ cells performing the same function: _____
>
> Made up of two or more _____ with related functions: _____
>
> Structures with related functions are linked into: _____
>
> 12 Where in a plant would you expect to find cells with the most chloroplasts? Give a reason for your answer.
>
> 13 Where in a plant cell would you find (a) DNA, (b) chlorophyll, (c) cellulose, (d) enzymes for respiration and (e) mineral ions in solution?

B2 1.5 How do dissolved substances move into and out of cells?

B2 1.5
Learning outcomes

- Explain how substances move into and out of cells by diffusion.
- Understand how this applies to examples in animals and plants.

How do dissolved substances move into and out of cells?

So far this chapter has referred to a range of processes, such as respiration and photosynthesis, which take place inside cells. These processes take place as a series of chemical reactions and therefore need chemicals that can get into a cell. They also produce chemicals that need to leave the cell. In this section we will consider how these substances move into or out of a cell.

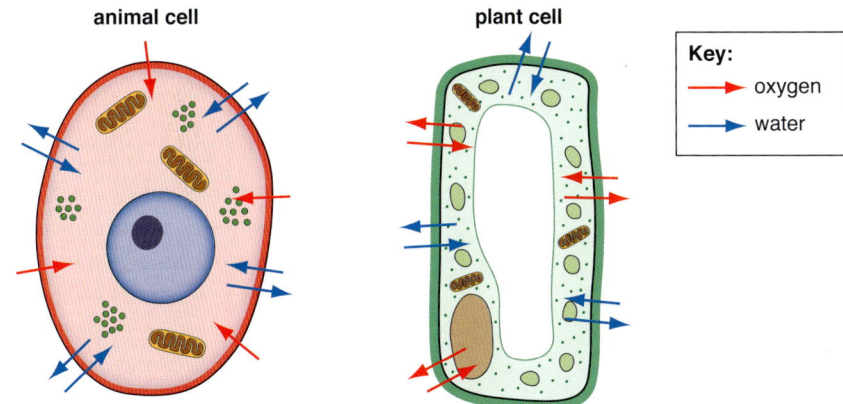

- oxygen required for respiration
- oxygen required for respiration
- oxygen produced by photosynthesis

Figure 1.16 The movement of oxygen and water into and out of animal and plant cells.

Diffusion

If you spray air freshener in one corner of a room it is not long before it can be smelt throughout the room. This happens because the particles that make up the air freshener move around. To start with there are lots of particles in the space where you sprayed the air freshener, but quickly they spread out to where there are fewer particles. They move from a region where there is a higher concentration of particles to a region where there is a lower concentration. This process is called **diffusion**. Imagine if your class first huddled in the corner of the room and then ran about until they were evenly spread around the room. This gives a 'picture' or a model of how the particles behave when diffusion takes place.

Particles of a substance that is dissolved in a liquid will also diffuse. If you put a drop of ink in a beaker of water you can see the ink gradually spread throughout the water. This happens because the particles in the ink are diffusing amongst the water particles. They are moving from where there is a higher **concentration** of 'ink' particles.

How does this apply to living cells? Oxygen will dissolve in water and so can diffuse when in solution. This is how it moves into cells when it is used for respiration. Since oxygen is used up quickly in the mitochondria for respiration, there will always be a lower concentration of oxygen inside the cell than in the liquid outside the cell. We say there is a concentration gradient. This ensures that the oxygen will always diffuse into the cell. The greater the difference in concentration, the faster the rate of diffusion. Diffusion takes place because of the concentration gradient — it does not use energy.

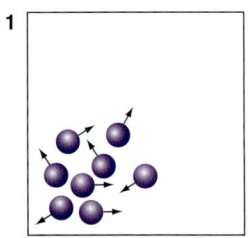

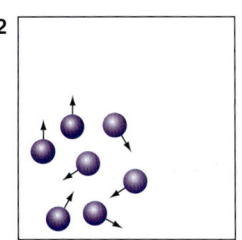

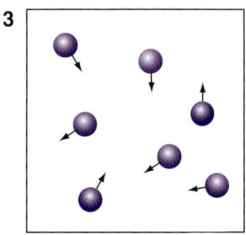

Figure 1.17 Molecules in a gas spread out by diffusion.

Biology 2 13

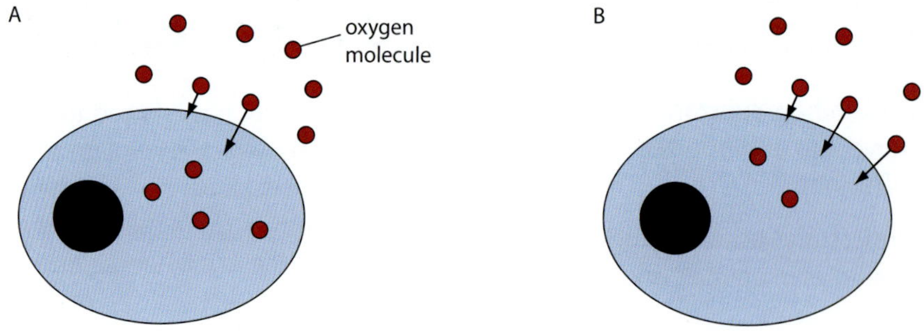

Chapter 1 Cells, tissues and organs

Figure 1.18 Count the number of molecules of oxygen in both cells. Into which cell will the oxygen diffuse faster?

Oxygen diffusion in the lungs

Blood transports oxygen to living cells but the blood first needs to be loaded with oxygen and this can only happen in the lungs. Diffusion will be faster if there is a large surface area for diffusion. In the lungs the surface area is increased by large numbers of alveoli. The concentration of oxygen in the lungs is kept high by breathing in air that contains 21% oxygen. The oxygen dissolves in the moisture of the lungs, diffuses through the thin epithelium and into the blood capillaries, diffuses into the red blood cells and is carried away as the blood is pumped along. Because the blood is always taking the oxygen away there is always a concentration gradient. Body cells produce carbon dioxide by respiration. This is carried back to the lungs in the blood. It diffuses in the opposite direction from the oxygen.

> **Test yourself**
>
> 14 What gas must diffuse into active muscle cells? What gas diffuses out of active muscle cells?

> **A simple demonstration of diffusion** — ACTIVITY
>
> Put a drop of strong-smelling oil into a balloon, blow the balloon up and tie it. Can you smell the oil from outside the balloon? If so, explain how it got from the inside of the balloon to the outside through the membrane of the balloon. How is this similar to what happens in the lungs?

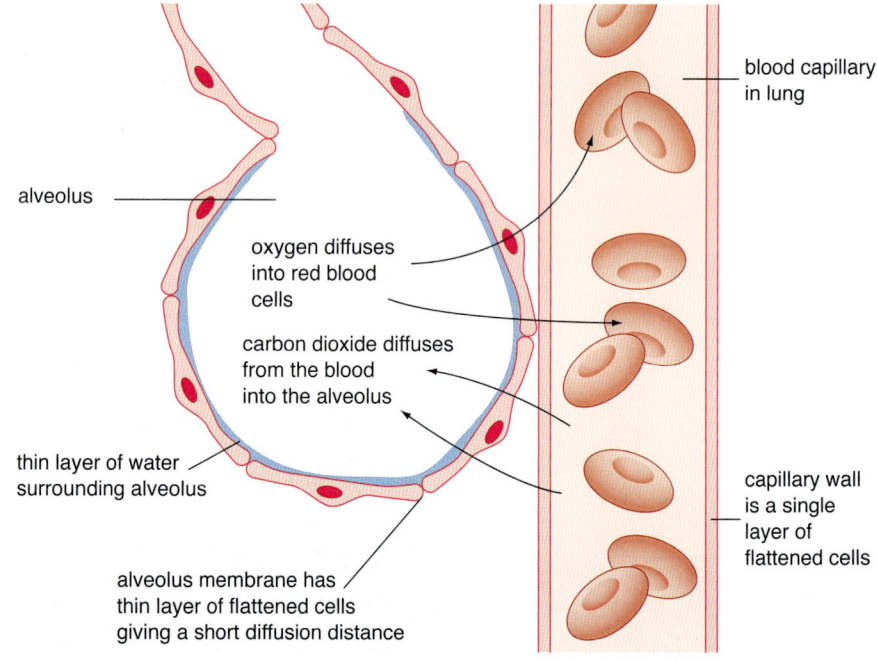

Figure 1.19 Diffusion of gases between an alveolus and a blood capillary in the lung.

Of course, oxygen, or anything else trying to get into a cell, has to pass through the cell membrane. The membrane will not let through larger molecules such as starch. Therefore the cell membrane is referred to as a **partially permeable membrane**. Small water-soluble molecules such as glucose can diffuse quickly.

> **Concentration** is the amount of one substance in a given volume of another.
>
> **Diffusion** is the movement of particles in a gas or any dissolved substance from a region of higher concentration to a region of lower concentration.
>
> A **partially permeable membrane** is a membrane, such as a cell membrane, that will allow only some substances to pass through it.

14 AQA GCSE Additional Science

B2 1.5 How do dissolved substances move into and out of cells?

Diffusion

**HOW SCIENCE WORKS
PRACTICAL SKILLS**

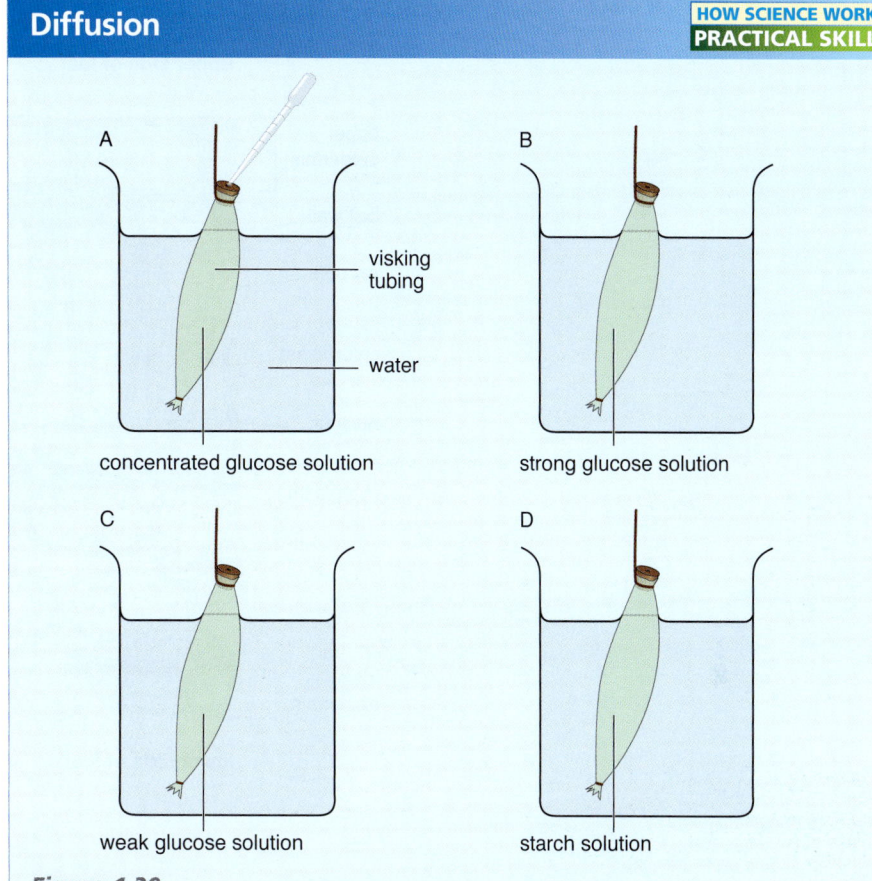

Figure 1.20

Set up some visking tubing as shown in Figure 1.20.
- Make up 50 cm³ of glucose solution with as much glucose as you can dissolve.
- Use a pipette to add this to visking tube A.
- Dilute the remaining glucose with an equal volume of water.
- Add this to visking tube B.
- Further dilute the remaining glucose solution to give a dilute solution.
- Add this to visking tube C.
- Into visking tube D add starch solution.
- Carefully wash the outsides of the visking tube because there must be no spilt glucose solution.
- Hang the tubes to equal depths in beakers of water.
- Draw a neat results table.
- At 5 minute intervals test the water in beakers A, B and C for glucose, and D for starch.

In which beaker did you first detect glucose?

Compare your results with the rest of the class.

What can you conclude from this experiment?

Did you detect any starch in D?

Can you explain this result?

Use ICT to present your results and findings.

Biology 2

Chapter 1 Cells, tissues and organs

Does diffusion take place in leaves?

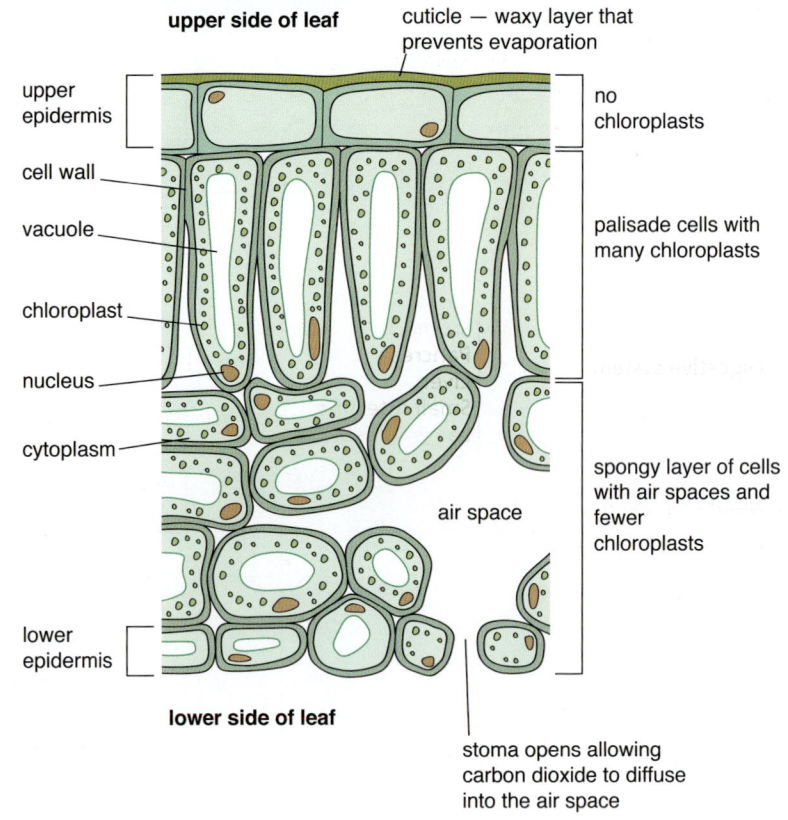

Figure 1.21 A section of a leaf

Examine Figure 1.21.

Assume the leaf is photosynthesising. Carbon dioxide gas is required as the raw material for photosynthesis. The concentration of carbon dioxide in the air is 0.37%. What do you think the concentration of carbon dioxide is inside a photosynthesising leaf? It will be zero. Therefore, even though the concentration of carbon dioxide in the air is small, there will still be a concentration gradient between the air and the leaf cells inside the leaf. Carbon dioxide will diffuse into the leaf through the open stomata. One product of photosynthesis is oxygen. Think about concentration gradients inside and outside of the leaf and work out what will happen to the oxygen produced on a bright sunny day.

Summarising the conditions for diffusion

Diffusion takes place:
- more rapidly in gases than in liquids
- faster for small water-soluble molecules
- always from a high concentration to a lower concentration

The rate is proportional to the concentration gradient.

Increasing the surface area will increase the rate of diffusion.

Test yourself

15 Describe the process of diffusion and explain why it happens. Use a 'muscle cell' and 'oxygen' in your example.

16 Describe two ways in which the lungs are adapted to increase the efficiency of diffusion.

17 Explain the term 'concentration gradient'. Why is it important in diffusion?

Homework questions

1 The human body has organ systems in which the organs carry out related functions. Copy and complete the table below. The first example has been given to you.

System	Organs
Digestive system	Mouth Oesophagus Stomach Pancreas Liver Small intestine Large intestine Rectum
Nervous system	
Transport system	
Urinary system	
Female reproductive system	

2 a List the structures that are found in both plant and yeast cells.
b List the difference between plant and yeast cells.

3 Read the following passage.

Blood transports oxygen to living cells but the blood first needs to be loaded with oxygen in the lungs. Diffusion will be faster if there is a large surface area. In the lungs the surface area is increased by large numbers of alveoli. The concentration of oxygen in the lungs is kept high by breathing in air with 21% oxygen. The oxygen dissolves in the moisture of the lungs, diffuses through the thin epithelium and into the blood capillaries, diffuses into the red blood cells and is carried away as the blood is pumped along. Because the blood is always taking the oxygen away there is always a concentration gradient.

In the paragraph above there are seven factors that make the lungs an efficient exchange surface. List them.

4 You have three visking tubes each sitting in a beaker. The tubes and the beakers all contain solutions of glucose. The table below shows the concentrations of glucose in each case.

	% concentration of glucose in visking tube	% concentration of glucose in beaker	How will the glucose move? from–to	Explanation
A	20	5		
B	5	15		
C	10	10		

a Copy and complete the table.
b Will the diffusion rate be fastest in A, B or C? Why?

5 The pancreas is the organ that produces the hormone insulin in specialised cells. The intestine and the blood system are linked.
a Describe the route the amino acids needed to make insulin travel from the small intestine to the pancreas.
b How is the insulin transported to the liver?
c What sub-cellular organelle would you expect to find in large numbers in the cells producing insulin?
d Where is the code for the production of insulin located?
e By what process is cellular energy made in these cells?

Chapter 1 Exam corner

Exam corner

1 A leaf is well adapted as the organ of photosynthesis. Explain this. *(4 marks)*
2 List the differences between plant cells and bacterial cells. *(3 marks)*

Student A

1 For photosynthesis a leaf needs to trap light energy. It has special cells called palisade cells which have lots of chloroplasts ✓ which capture the light energy. It needs carbon dioxide and this diffuses in through open stomata in the bottom of the leaf and between the spongy cells. ✓ It needs water and this is transported in the phloem. ✗ Because it is using the carbon dioxide to make sugar there is always a shortage of carbon dioxide so it always diffuses in from the higher concentration of the air outside. ✓

2 Cell wall, no mitochondria, no nucleus. ✗

Examiner comment You need to think through a question such as this before you start writing. It is a good idea to jot down the key words that you want to use.

Student A's answer to question **1** contains no reference to tissues — it only refers to cells. Did she miss the word 'organ' in the question? In question **2**, if you are asked a question about differences you must make it clear which cell you are referring to, i.e. 'A plant cell has … but a bacterial cell has …' Always use this format and you can't go wrong.

Student B

1 The leaf organ has layers of tissues. ✓ The top layer is the palisade mesophyll and the cells in this layer all have many chloroplasts to trap light. ✓ The bottom layer tissue is spongy so carbon dioxide can diffuse quickly ✓ through the spaces. The water is transported to the leaf in the xylem. ✓ The light energy is used to combine the carbon dioxide and water to make glucose. Carbon dioxide is used up so quickly that there is none in the leaf so it diffuses into the leaf quickly. ✓

2 The plant cell has a cellulose cell wall but the bacterial cell wall is not made of cellulose. ✓

The plant cell has mitochondria but there are none in bacteria. ✓

The plant cell has a nucleus but there is none in the bacteria — its chromosome is loose in the cytoplasm. ✓

Examiner comment Student B's answer to question **1** shows that she understands that an organ is composed of tissues and she explains this well. The point that the carbon dioxide is used up etc. is good but would have been stronger if it had referred to concentration differences or concentration gradient. The answer to question **2** is excellent — just what examiners are looking for.

18 AQA GCSE Additional Science

Biology 2

Chapter 2
Photosynthesis, organisms and communities

ICT
In this chapter you can learn to:
- create a podcast for vegetable growers with advice on increasing yield
- use a social networking platform to discuss ideas with a wide range of students

Setting the scene
Why do you think these tomatoes are grown in a greenhouse rather than outside? What conditions might the commercial grower be trying to control inside the greenhouse?

Practical work
In this chapter you can learn to:
- make observations of cut flower stems and use them to explain how water travels through the plant
- identify which variables to control in a photosynthesis experiment and identify patterns in the results
- take measurements in a survey of land, calculate means and look for patterns in results
- evaluate the data from a grass survey
- work safely when investigating soil samples

Chapter 2   Photosynthesis, organisms and communities

B2 2.1   Photosynthesis

Learning outcomes

Understand the process of photosynthesis.

> **Photosynthesis** is the process in green plants which captures light energy and converts it to chemical energy used to make carbohydrates. The carbohydrates are used to form biomass. The word photosynthesis can be split into two parts: 'photo' meaning 'light' and 'synthesis' meaning 'to make'.

Green plants use light to convert the simple molecules carbon dioxide and water into larger glucose molecules using enzymes in chloroplasts. This process is called **photosynthesis**.

We can break down the summary equation for photosynthesis into three parts, as shown below.

Raw materials	Conditions	Products
Carbon dioxide + water $6CO_2 + 6H_2O$	Light energy $\longrightarrow$	Glucose + oxygen $C_6H_{12}O_6 + 6O_2$

How is the light captured?

Some plant cells have chloroplasts to absorb light energy. Figure 2.1, a photograph taken at high magnification with an electron microscope, shows that the chloroplast has many membranes inside it. These membranes contain chlorophyll, which makes the leaves appear green. When light reaches the chlorophyll, enzymes in the chloroplasts use this energy to catalyse reactions that produce glucose. Glucose is quickly converted to starch that can be stored in the leaf, and oxygen is released as a waste product.

How is the raw material, carbon dioxide, obtained?

During the day, the leaf is photosynthesising and using carbon dioxide. The **stomata** (small pores) are open and the carbon dioxide diffuses into the leaf from the air. Diffusion always takes place from a region of higher concentration to one of lower concentration. Although the concentration of carbon dioxide in the air is on average only about 0.037% (or 370 ppm), when the leaf is

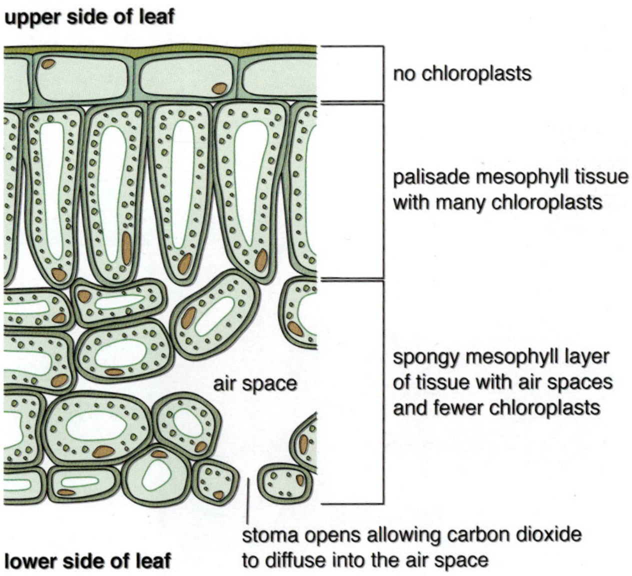

> **Test yourself**
>
> 1 The box above shows that a condition for photosynthesis is light energy. What must be present in the leaf cells to trap this light energy?

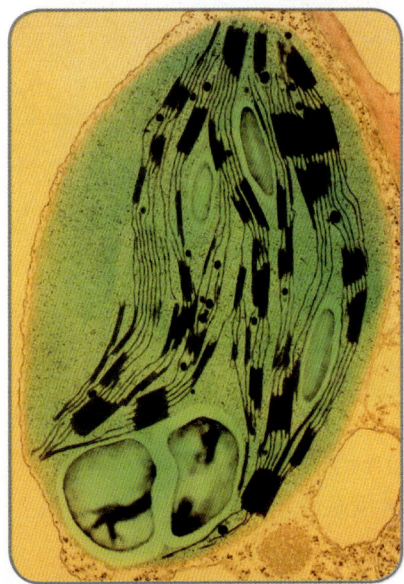

Figure 2.1 The structure of a chloroplast (magnification about ×15 000).

upper side of leaf

- no chloroplasts
- palisade mesophyll tissue with many chloroplasts
- spongy mesophyll layer of tissue with air spaces and fewer chloroplasts
- stoma opens allowing carbon dioxide to diffuse into the air space

air space

lower side of leaf

Figure 2.2 In this leaf section note the different shapes of the cells in the tissue layers.

B2 2.2 What factors affect the rate of photosynthesis?

photosynthesising it will be using up the carbon dioxide inside the leaf and so maintaining the concentration difference needed for diffusion.

The carbon dioxide concentration of the atmosphere can be written as a percentage or as ppm (parts per million). Atmospheric carbon dioxide levels are changing, and in 2006 a record high level of 381 ppm was recorded.

How is the raw material, water, obtained?

Roots take up water from the soil. The uptake is maximised because the root surface area is increased by many tiny root hairs. The water is then transported up the xylem vessels in the stem to the leaves. Oxygen is a valuable by-product from the reaction and diffuses out of the stomata into the atmosphere.

> **Visible water transport** PRACTICAL SKILLS
>
> You can see which parts of the stem take water to different parts of the leaf by cutting a white carnation flower stem lengthways into several strips. Put each of these strips into a different coloured beaker of dye. This will create a 'tie-dyed' flower. Photograph your results.

Test yourself

2. What materials are needed for photosynthesis?
3. What is the energy input for photosynthesis?
4. A variegated plant has many white leaves.
   a. What essential condition for photosynthesis is missing from these leaves?
   b. Explain why this plant would not grow as rapidly as a green-leaved plant.
5. What are the products of photosynthesis?

B2 2.2

Learning outcomes

Know the factors that affect the rate of photosynthesis.

What factors affect the rate of photosynthesis?

A reaction rate is the quantity of product produced or reactant used per unit time. You could compare this to a model with which you are familiar, such as the money you earn for washing cars on a Saturday job. For example, if you get paid £2 per wash and you wash two cars an hour, your hourly rate is £4.00 per hour. If you wash five cars an hour, your hourly rate is £10 per hour. The more cars you can wash in an hour, the greater your hourly rate of pay.

You can measure the rate of photosynthesis by finding the volume of oxygen produced per hour using a range of input values.

To find out how the input of carbon dioxide, light intensity and temperature limit the rate of photosynthesis, we need to measure the **rate of photosynthesis**.

From the equation for photosynthesis given in Section 2.1 you can see that carbon dioxide is one of the raw materials. Although leaf cells respire 24 hours a day, carbon dioxide cannot leave the leaf at night, since the stomata are closed (light intensity controls when the stomata open). By daybreak there will be a build-up of carbon dioxide in the leaf from the respiration that has taken place overnight. Initially light intensity will be a **limiting factor** for photosynthesis, but as the day gets brighter there will be more rapid photosynthesis, using up the carbon dioxide.

Water is essential for support of the leaf cells; if there is a shortage of water, the stomata will close to reduce water loss and photosynthesis will stop. This may happen around midday on a hot sunny day when there is a high water loss from the leaves or during drought conditions when the soil is dry.

Test yourself

6. Why does the concentration of carbon dioxide in the atmosphere limit the rate of photosynthesis?
7. What factor limits the rate of photosynthesis at dawn?

> The **rate of photosynthesis** is the speed at which the photosynthesis reaction takes place. It can be measured as the rate of formation of oxygen.
> A **limiting factor** is the factor that controls the rate of the product formation.

Biology 2

Chapter 2 Photosynthesis, organisms and communities

How does carbon dioxide concentration affect the rate of photosynthesis?

Carbon dioxide forms a small percentage of atmospheric gases and can be a limiting factor when the light intensity is high.

Analysing the effect of carbon dioxide concentration on the rate of photosynthesis

HOW SCIENCE WORKS
PRACTICAL SKILLS

A group of students decided to investigate the effect of carbon dioxide concentration on the rate of photosynthesis. They decided to calculate the rate of photosynthesis by measuring the volume of oxygen produced from an aquatic plant, *Elodea* (see Figure 2.3).

To investigate the effect of carbon dioxide concentration, the students made a **series dilution** of sodium hydrogencarbonate solution. They started with a 5.0% solution. They followed these instructions to make solutions with different concentrations.

- Take six 1000 cm³ beakers. Label them A to F.
- Fill beaker A with 5% solution.
- Use a measuring cylinder to add 900 cm³ of water into beaker B.
- Using a second measuring cylinder transfer 100 cm³ sodium hydrogencarbonate solution from A to beaker B.
- Stir beaker B with a glass rod to mix. This gives a 0.5% sodium hydrogencarbonate solution.
- Make up a second beaker B as you require 2000 cm³.
- Use the 0.5% solution to make up the solutions in the proportions shown in Table 2.1.

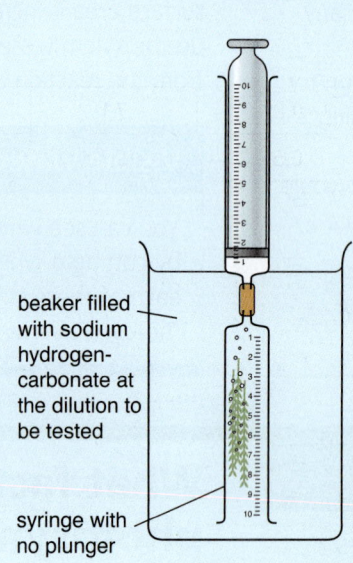

1. Connect the lower syringe to the empty upper syringe with a rubber connection.
2. Put three sprigs of *Elodea* in the lower syringe with the cut end upwards; there should be no air in this syringe.
3. Set in constant light for 10 minutes to equilibrate.
4. To start the experiment, draw any gas up into the top syringe and record the volume.
5. Leave the apparatus in constant light for at least 2 hours.
6. Draw up any gas produced in the lower syringe. Record the volume produced for each sodium hydrogencarbonate concentration.

Figure 2.3 The apparatus and method used to measure rate of photosynthesis.

Table 2.1 A summary of the series dilution.

Beaker	Starting concentration of NaHCO₃	Volume of NaHCO₃ in cm³	Volume of H₂O in cm³	Final concentration
B	5%	100	900	0.5%
C	0.5%	800	200	
D	0.5%	600	400	
E	0.5%	400	600	
F	0.5%	200	800	

❶ Copy and complete the final column of Table 2.1.

The students used the apparatus and method shown in Figure 2.3. They knew that they should only change one input variable in each investigation. The input variable was the range of concentrations of sodium hydrogencarbonate solution. They knew that the light intensity and temperature should be kept constant.

❷ Which factor were the students controlling?

❸ This experiment was completed outside on a hot June day. What factors may not have been controlled?

To develop the experiment further the students decided to calculate the rate of photosynthesis in each experiment as the rate per gram of dry mass per hour. They measured the total dry mass of the *Elodea* used

AQA GCSE Additional Science

B2 2.2 What factors affect the rate of photosynthesis?

each time by placing it in a folded filter paper in a drying cupboard set at 100°C. The samples were dried and weighed every hour until the mass of dried *Elodea* remained constant for two successive weighings. Table 2.2 shows the set of results they obtained.

Table 2.2

Concentration of NaHCO$_3$ as %	Starting volume in cm^3	Final volume in cm^3	Volume of oxygen in cm^3	Mass of *Elodea* in g	Time in h	Volume of oxygen in cm^3/g/h
0.5	0.2	9.6	9.4	1.341	2	
0.4	0.4	8.6	8.2	1.399	2	
0.3	0.3	7.1	6.8	1.439	2	
0.2	0.5	5.9	5.4	1.510	2	
0.1	0.4	4.8	4.4	1.522	2	

4. Copy the table and complete the calculations. Use the following equation to complete the final column:

$$\frac{\text{volume of oxygen in cm}^3}{\text{mass of } Elodea \text{ in g}} \times \frac{1}{\text{time (h)}}$$

5. Use the results in the table to construct a graph with volume of oxygen in cm^3/g/h on the y-axis and concentration of sodium hydrogencarbonate solution on the x-axis.

6. Suggest a suitable title for this graph, which links the output variable with the input variable.

7. Describe the pattern shown by your graph.

8. Suggest how you would modify this experiment to investigate the effect of light intensity on the rate of photosynthesis.

> A **series dilution** is a range of concentrations. A minimum of five dilutions from concentrated to weak would give the input variable for the carbon dioxide concentration. The students planned the dilutions to give sensibly spaced points on the x-axis of their graph.

How does light intensity affect the rate of photosynthesis?

First, work through this simple model to make sure you understand what **light intensity** means. Switch on a torch, hold it close to a wall and shine the beam at the wall — you will get a small circle of light. As you move the torch further from the wall the circle of light gets larger and dimmer. The light energy given out by the torch is the same, but the light energy falling on each square centimetre of the wall (the light intensity) is less when the torch is further away.

Light intensity is proportional to $1/d^2$, where d is the distance from the light source. This relationship shows that as the distance from a light source is increased, the light intensity reduces rapidly (Figure 2.4). You could use this graph of light intensity against distance to plan an experiment to find how light intensity affects the rate of photosynthesis. Unfortunately, normal light bulbs are inefficient and give off heat as well as light. Could this introduce another variable when the lamp is too close to the apparatus? Energy-efficient fluorescent bulbs or tubes give off less heat.

Light intensity generally increases towards midday and then decreases towards evening, but it can change quite quickly on a cloudy day. Light is the energy input and will have a significant effect on the rate of photosynthesis. The light energy capture stage of photosynthesis depends on the light intensity and is independent of temperature.

> **Light intensity** is the amount of light (energy) per unit area of the surface. Light intensity decreases rapidly with increasing distance from the light source.

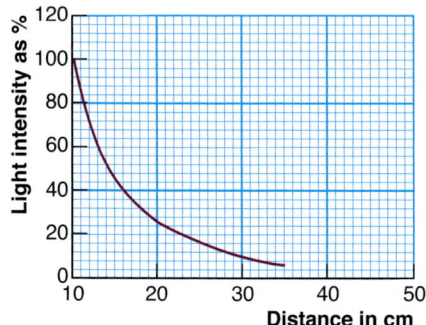

Figure 2.4 For this graph the light intensity was taken as 100% at 10 cm distance. Look at the shape of the graph. What happens to the light energy as the light source is moved away?

Biology 2 — 23

Chapter 2 Photosynthesis, organisms and communities

How does temperature affect the rate of photosynthesis?

After the light has been captured, enzymes in the chloroplasts catalyse the chemical reactions that convert carbon dioxide to glucose. Therefore, you would predict that the rate of glucose production would increase with increase in temperature up to the optimum temperature of the enzymes. The rate of photosynthesis increases on a warm day since the enzyme reactions are temperature-dependent. The production of glucose and hence starch will be slower on a cold day.

Figure 2.5 shows that at higher temperatures the rate of photosynthesis is faster, for a fixed concentration of carbon dioxide. Above the optimum temperature, photosynthesis slows down.

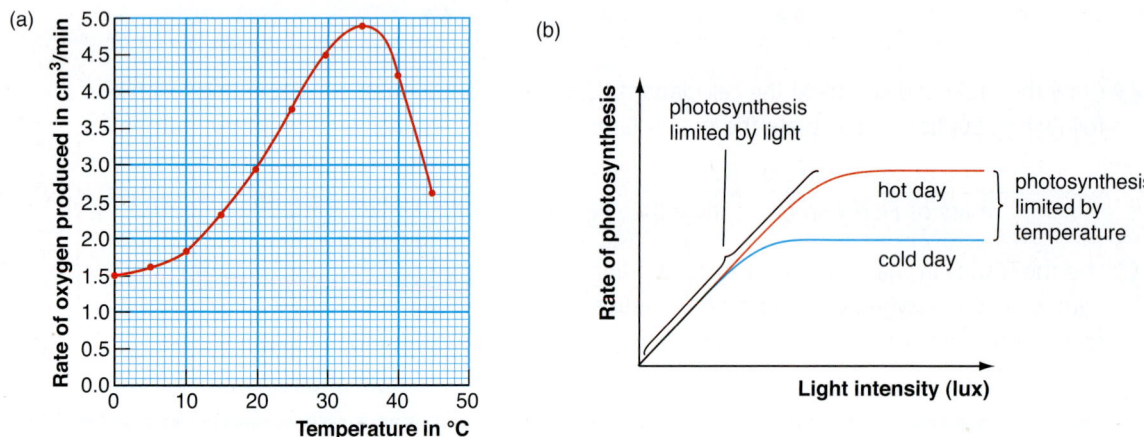

Figure 2.5 (a) Effect of temperature on the rate of photosynthesis at a fixed light intensity and fixed carbon dioxide concentration. (b) Effect of increasing light intensity on the rate of photosynthesis.

Having worked through this section, you can see that the rate of photosynthesis at any time is a result of the interaction of several factors. Any one factor of low temperature, low light intensity or low level of carbon dioxide could limit the rate of photosynthesis.

Boost your grade ✓

When asked 'What factors will increase the rate of photosynthesis?' too many candidates answer 'light'. Two points to note here: (1) Does the candidate mean light colour or light intensity? (2) The question includes the word 'increase'. The correct answer should be: 'an increase in light intensity'. Apply this to the other factors that should be in the answer.

Test yourself

8 What three factors reduce the rate of photosynthesis?

9 Explain what effect the fish in a fish tank will have on the rate of photosynthesis first thing in the morning if the air pump is not running?

24 AQA GCSE Additional Science

B2 2.3 How do plants use glucose?

Learning outcome

Know how plants use the products of photosynthesis.

Glucose is the end product of photosythesis. Figure 2.6 shows how glucose is used in plants.

After glucose is produced, it is converted to starch in the leaf. We can test for starch to find out whether a leaf has been photosynthesising (Figure 2.7). Materials can only be carried to other parts of the plant when dissolved in water, so the insoluble starch is converted to soluble sugar for transport throughout the plant to growing points, fruits, storage organs and seeds.

All parts of the plant need a supply of glucose for respiration. The growing shoots of the plant need glucose for respiration, to release chemical energy for making new cells. Just as you need minerals to keep healthy, so do plants. Nitrates are required for making amino acids and therefore for building proteins to make the cytoplasm and enzymes in the new cells. Magnesium ions are needed to produce chlorophyll molecules. The mineral ions are taken up from the soil into the roots. If plants are unable to produce sufficient protein, they will grow slowly and remain small. If they are unable to obtain enough magnesium they will not be able to make chlorophyll, so their lower leaves will become yellow and not photosynthesise efficiently.

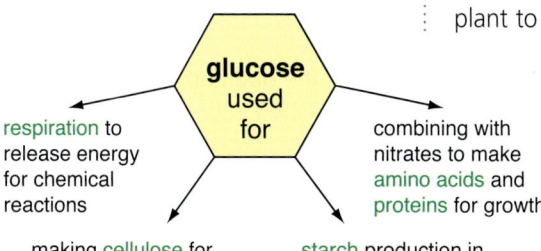

Figure 2.6 How the plant makes use of the glucose produced.

Nutrient stores in seeds are important as they support the embryo growth before photosynthesis starts. Many seeds have lipid or oil as the main storage product. These lipid stores are made from the glucose produced during photosynthesis. Lipids are the most compact form of energy and so this material is found in small and lightweight seeds. We take advantage of the larger seed stores as a source of oil for cooking e.g. peanut oil (also known as groundnut oil), corn oil, sunflower oil, and also the fruit of the olive tree, which gives olive oil. Other seeds, such as wheat and barley, have starch as the main storage material — we use these grains for flour. Soya beans have protein as the main storage product.

Boost your grade ✓

Remember sequences for experiments, Figure 2.7 becomes D D S F T — dip, decolorise, soften, flatten, test. Once you remember the sequence you can fill in the details.

1 Dip
Take a leaf from a plant standing in sunlight and dip in boiling water for 10 seconds to denature enzymes and make the membrane permeable.

THEN TURN OFF THE BUNSEN because alcohol vapour is inflammable.

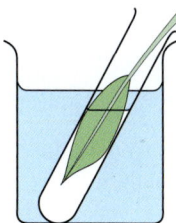

2 Decolorise
Put a tube of alcohol into the water bath and allow it to warm up. Add the leaf and shake gently until the alcohol becomes green.

3 Soften
Soften the leaf by rinsing in the hot water.

4 Flatten
Flatten the leaf underside up on a white tile and add a few drops of iodine. Where starch is present it will give a blue-black colour.

This method can be used with a leaf which has been partly covered with a stencil or photonegative, or with a variegated leaf.

Figure 2.7 Testing a leaf for starch.

Biology 2 25

Chapter 2 Photosynthesis, organisms and communities

Test yourself

10 A plant was put into a black plastic bag and left for 24 hours. The leaves were then tested for starch. Explain why you would not expect to find starch in the leaves even during the day.

11 Where are the enzymes located that control the conversion of carbon dioxide to glucose?

12 When testing a leaf for starch it is dipped into boiling water to stop enzyme reactions.
  a Suggest which enzyme reactions these are.
  b Why does the alcohol become green?
  c If parts of the leaf become blue-black when tested with iodine solution, it indicates that starch is present. What could you conclude if the iodine solution stayed brown?

13 Draw sketch graphs to compare the rate of oxygen produced on a bright winter day and on a bright summer day. Assume the light intensity is the same on both days.

14 A student grew a tomato plant in sand. After a month he compared it with his friend's that had been grown in potting compost. Both plants had been watered regularly but the plant in sand was shorter, had much smaller leaves and the lower leaves were yellow. Suggest reasons for these differences.

B2 2.4 What factors limit the rate of photosynthesis in the environment?

Learning outcomes

Know the factors affecting photosynthesis in a range of habitats and agricultural situations.

We have seen that under laboratory conditions light intensity, carbon dioxide concentration and temperature can all affect the rate of photosynthesis. In the environment all these factors interact, as you can see in the following activity.

From laboratory to outdoors ACTIVITY

Use the information gained from laboratory examples to consider photosynthesis in the different environments shown below.

A field crop on a hot, sunny day (Figure 2.8)

1 Will light intensity be a limiting factor?

2 By 9 a.m. the rate of photosynthesis has reached a constant level and does not increase even though the light intensity is increasing. What could be the limiting factor?

3 At night the stomata are closed and the plant continues to respire, resulting in a build-up of carbon dioxide inside the leaf. What effect would you expect this to have on the rate of photosynthesis when it first gets light in the morning?

Figure 2.8

26 AQA GCSE Additional Science

B2 2.4 What factors limit the rate of photosynthesis in the environment?

A forest in Iceland (Figure 2.9)

It has been said that you can't get lost in a forest in Iceland because the birch trees only grow to about 1.5 m high (see Figure 2.9). Iceland has periods of 24 hours darkness in winter and 24 hours light in summer, although the light intensity is not the same during those 24 hours.

4 What factors limit the growth of the trees?

5 Is carbon dioxide concentration likely to be a limiting factor?

Figure 2.9

A pond (Figure 2.10)

Ducks use the pond and their waste has resulted in extra nitrates (plant food) in the water. As a result there is dense weed growth and many snails feeding on the weeds.

6 What do you think will happen to the carbon dioxide concentration in the water overnight? Explain your answer.

7 During mid-morning, streams of tiny bubbles can be seen rising to the surface. What gas do you think these bubbles contain, and where does it come from?

8 What do you think happens to the carbon dioxide content of the water during the day? Explain your answer.

Figure 2.10

Figure 2.11 A large commercial greenhouse.

Controlling the greenhouse environment

The environment in commercial greenhouses is closely monitored using sensors that measure temperature, carbon dioxide concentration and light intensity. If any of these factors change, computer controlled systems artificially restore the factor to optimum so that the plants receive ideal conditions of light, temperature and carbon dioxide. For example if a drop in carbon dioxide is detected a valve automatically opens to increase the concentration and closes again when the optimum concentration is reached. This saves adding it all the time and wasting the gas. Lights will switch on until the sun rises and the light intensity is high enough and then switch off until late afternoon when the light intensity drops.

Figure 2.12 A computer control panel that controls conditions in a greenhouse.

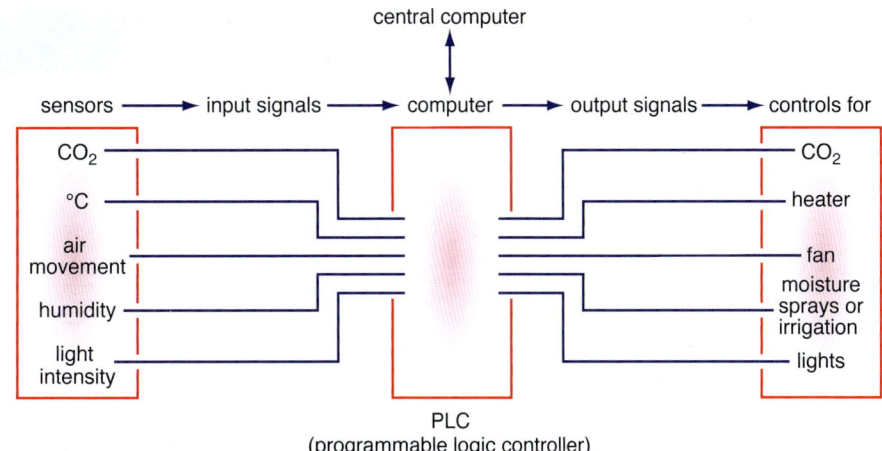

Figure 2.13 A control system for the greenhouse. Does this resemble any human system?

Biology 2

Chapter 2 Photosynthesis, organisms and communities

Greenhouse crop production
HOW SCIENCE WORKS ACTIVITY

For maximum profit, greenhouse crops must have a high yield and good quality. Producers also need to raise the crops to maturity in the minimum possible time. You saw in the activity in Section 2.2 that carbon dioxide could be a limiting factor for photosynthesis. Many experiments have been carried out to find the optimum light intensity and carbon dioxide concentrations for different plants.

Use the graph in Figure 2.14 to answer the following questions.

In a closed greenhouse in winter with artificial lighting and no ventilation, the levels of carbon dioxide can drop during the day to 200 ppm. (The rate of photosynthesis at 370 ppm carbon dioxide (atmospheric level) is taken as 100%. This is shown in red on the graph in Figure 2.14.)

1. Use the graph to find the effect on the rate of photosynthesis of reducing carbon dioxide levels from 370 ppm to 200 ppm.

Increased carbon dioxide concentration results in larger leaves, which increases the size and quality of lettuces. Colorado State University found that increasing carbon dioxide levels to 550 ppm meant that they could fit only 16 instead of 22 lettuces into a box.

2. Use your knowledge of photosynthesis to explain why the lettuces were bigger.

3. From the graph, by what percentage was the rate of photosynthesis increased by changing from 370 to 550 ppm carbon dioxide?

If the carbon dioxide concentration is raised to 1000 ppm, for salad crops such as tomatoes, peppers and cucumbers, they flower earlier producing more flowers and more fruits.

4. What shape is the curve above a concentration of 1000 ppm?

5. Suggest why the grower does not usually increase the concentration above 1000 ppm.

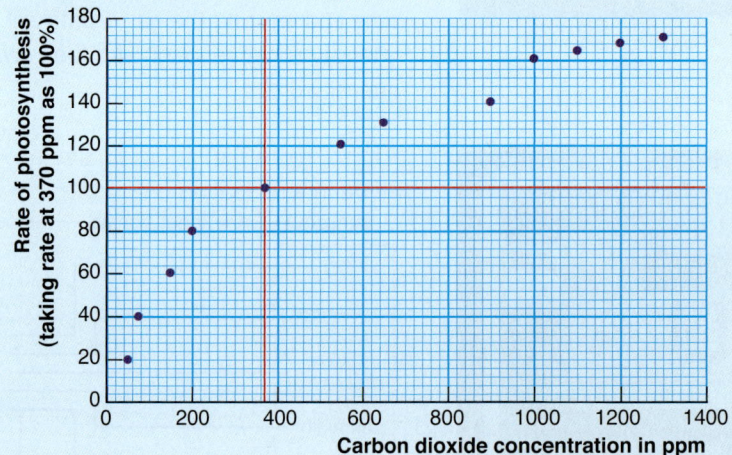

Figure 2.14

Victorian greenhouses were heated by manure. The decomposition produced sufficient heat for fruits such as peaches to be produced early in the season. The rotting manure also raised carbon dioxide levels.

B2 2.5 Sampling methods to find relationships within a community

In modern greenhouses the carbon dioxide concentration can be raised by using a heater burning oil. This will increase the temperature as well as carbon dioxide levels. However, there are problems:
- the water produced during combustion increases the growth of mildew
- if impure fuel is burnt, sulfur dioxide and oxides of nitrogen are given off; these cause acid rain-type damage to leaves, thus reducing photosynthesis

The best, but not the cheapest, way of supplementing the carbon dioxide is to use storage cylinders of liquid carbon dioxide and computer control to regulate carbon dioxide gas concentration within the greenhouse.

6 Sensors linked to a computer measure the carbon dioxide concentration. Suggest why the computer is programmed to activate a valve that increases the carbon dioxide concentration as the light intensity increases.

Test yourself

15 Why would a grower want to artificially control the conditions in his greenhouse?

16 A flower grower wishes to have pot plants ready for Mothering Sunday but they are beginning to open too early. What factor would he control? What other advantage would this give to the grower?

Advice for vegetable growers **ICT**

Create a podcast or multimedia resource for vegetable growers, explaining all the ways to increase the yield of their crops.

B2 2.5

Learning outcome

Understand how physical factors of the environment may affect the communities present.

All the plants and animals living in one habitat form the **community**.

Sampling methods to find relationships within a community

Animals and plants live in environments to which they are well suited because of specific adaptations. So far, this chapter has investigated how environmental conditions affect the rate of photosynthesis and production of plant biomass. The plant biomass available will affect the variety and number of herbivores, and this in turn regulates carnivores. Here we investigate the **communities** of plants and animals and the physical factors that may affect them.

Obtaining quantitative data

You are given the task of finding the average number of daisy plants per square metre in a recreation ground. The field is too big to crawl all over it on your hands and knees counting every daisy, so you need to use a sampling method.

You could use random sampling with a quadrat to compare areas which have different mowing frequencies, e.g. the area by the tennis courts is only mowed monthly but the rest is mowed weekly in the summer. Is the number of daisies per square metre the same in each area? A second use would be to sample for daisy plants before and 3 months after treatment with weedkiller to see if the treatment has reduced the number.

Whilst carrying out the activity on daisy plants, suppose you observed that there appear to be more daisies growing along a path which runs from corner to corner across the recreation ground than in the rest of the field. You could test this observation by counting along a transect line which crosses the path.

Biology 2

Chapter 2 Photosynthesis, organisms and communities

Random sampling with quadrats

**HOW SCIENCE WORKS
PRACTICAL SKILLS**

Task: find the average number of daisy plants per m² in a recreation ground that is regularly mowed.

This method is used when the distribution appears to be even over the area. So this method is suitable for use on the recreation ground.

You are given a quadrat, which is square with each side 50 cm long (Figure 2.15).

❶ What is the area of your quadrat in m²?

   area = length × width = 0.5 m × 0.5 m = _____ m²

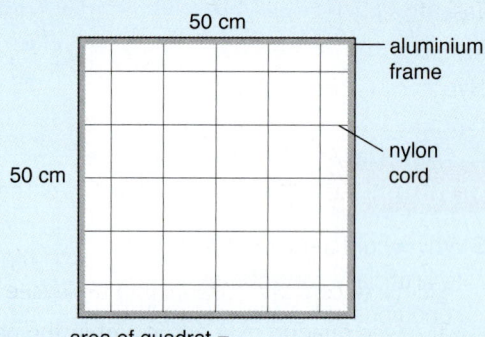

area of quadrat =

Figure 2.15 Calculate the area of this quadrat in m².

Lay out two measuring tapes along two sides of the field with the zero points together at the corner.

❷ Calculate the area of the field in m² from Figure 2.16.

Ideally you would sample sample 10% of the field area.

The ideal number of quadrat samples is $\frac{10\% \text{ of the area}}{\text{area of quadrat}}$.

Ideal sample size = _____ quadrats.

(The field is large, so you may reduce the number of samples to 100.)

❸ Collect a sheet of random numbers (or generate them on your calculator.)

❹ Use the random numbers to place the quadrat. If your first number pair is 12, 18, walk to 12 m along one side and send your partner to 18 m on the other side. Walk out from the tape (in a straight line) until you meet. Place the quadrat here.

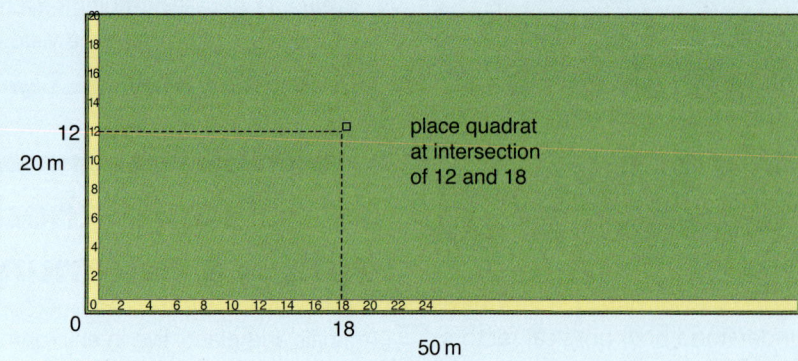

Figure 2.16 Place the quadrat according to the random number.

❺ Count the number of daisy plants, not flowers, in the quadrat.

Problem: some daisies are half-in and half-out of the quadrat.

Rule to solve problem: count all the daisies half-in on two sides only, e.g. bottom and left sides (see Figure 2.17).

Record the final number in a results table like the one shown below.

Date	Names
Quadrat number	Number of daisies
1	6
2	4

Figure 2.17

❻ Repeat until you have enough results.

❼ Add up the total number of daisy plants = _____.

❽ Calculate the area sampled = number of quadrats × area of each quadrat = _____ m².

❾ Calculate the number of daisy plants per m² = _____.

30 AQA GCSE Additional Science

B2 2.5 Sampling methods to find relationships within a community

Counting along a belt transect

**HOW SCIENCE WORKS
PRACTICAL SKILLS**

Task: to find if the distribution of daisies changes across the path.

Hypothesis: there are more daisy plants per m² along the path than on the areas beside the path.

A belt transect can be used when
- the vegetation is short
- the plants counted can be identified as individual plants

1. Place a measuring tape across the path extending 2 m each side of the path.
2. Place the quadrat at the zero point and count the number of daisy plants. Record your results in a table (see Table 2.3 for an example).
3. Turn the quadrat over and count in the next 0.5 m area. Continue until you reach the end of your transect.
4. Repeat in four other transects across the path.
5. Use the sample of results in Table 2.3 to make a histogram of the average number of daisy plants across the path.
   a Calculate the **mean** value for all five transect lines at each distance along the transect line, and add to the results table. (The mean is the average. So for distance 0 add the results together 0 + 1 + 2 + 1 + 2 = 6 and divide by the number of transects, which is 5.)
   b Draw a histogram of the results with distance along the transect along the bottom.

From the histogram:
c At what distance along the transect line was the path?
d Was your hypothesis correct?
e What else can you observe about the distribution of daisy plants from the results?

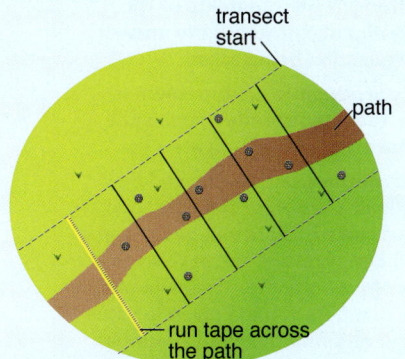

Figure 2.18 Place the transect tape across the path like this.

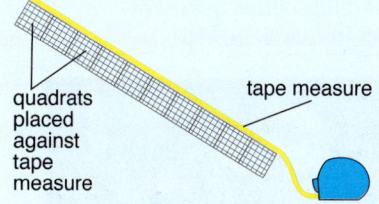

Figure 2.19 You will end up with a row of touching transect results, at five places across the path.

Table 2.3 Number of daisy plants in each quadrat along five transect lines.

Transect number	Distance along transect line (m)											
	0	0.5	1.0	1.5	2.0	2.5	3.0	3.5	4.0	4.5	5.0	5.5
1	0	3	3	0	2	5	3	4	2	1	2	0
2	1	2	1	2	2	6	2	5	2	1	2	1
3	2	2	1	2	3	4	2	6	1	0	1	1
4	1	1	1	1	4	3	0	5	1	1	0	0
5	2	2	2	0	2	5	3	5	2	1	0	0
Total	6											
Mean	1.2											

These results are **reproducible** enough to show a general trend and you have a **valid** result. If the experiment was repeated by someone else, they would find a similar pattern and conclude that there were more daisy plants in the path area. What factors could account for the difference in distribution? The soil is compressed by walking so it will drain more slowly. There will be fewer air spaces between the soil particles so the roots

Biology 2

Chapter 2 Photosynthesis, organisms and communities

will have less oxygen for respiration. If you examine a daisy plant you will see that the leaves are arranged in a rosette shape with the growing point close to the ground so it is not damaged by normal mowing, or walking. Because the path has some bare areas it might get warmer in the sun. So the distribution of the daisy plants could be affected by availability of water or oxygen, temperature (abiotic factors — see Section 2.6) or people walking along the path (biotic factor).

Validity is the suitability of the investigative procedure to answer the question being asked.

A **reproducible** measurement is one that consistently remains the same after several repeats.

The **median** value is the middle value in a set of results that have been sorted. There will be the same number of results above this value as below. To find the median, place the numbers you are given in value order and find the middle number.

The **mean** is the average of the results. Add the results and divide by the number of results taken.

The **mode** is the number that appears most often in a list of results (the modal value).

Sampling animals

The usual method of sampling water organisms is kick-sampling, using a D-shaped net. The catch can be identified and then separated into trophic levels to draw up a food chain. Certain important species such as mayfly larvae are only found until May when they complete their life cycle and fly off as adults. So in many locations you are not likely to collect sufficient information to draw up a pyramid of biomass. The species present can be compared from different areas of one stream or from still and flowing water. Whenever you collect animals don't forget to record the producers present (plants) so that you can study the whole community. Also record abiotic factors such as light or flow rate of the stream. (Once you have completed the survey you should return the organisms to the water.)

Collecting invertebrates in a meadow can be great fun. They can be sampled using a sweep net across the vegetation, collected from the net using a pooter (Figure 2.21) and identified and separated into trophic levels.

To relate the invertebrates to a specific quadrat, the organisms can be collected directly with a pooter (Figure 2.22). Larger animals can be identified by signs — rabbit dropping are a give-away, but you might also find footprints.

Figure 2.20 Using a sweep net.

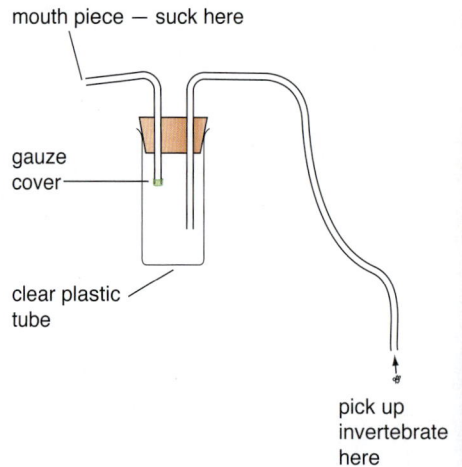

Figure 2.21 A pooter used to collect insects.

Figure 2.22 Collecting the sample with a pooter.

An investigation can be carried out to compare the invertebrates on different types of trees. Are there more invertebrates associated with cultivated trees, such as flowering cherries, or indigenous wild species, such as hawthorn? This would be carried out using a beating tray and collecting the organisms with a pooter. Simply hold the beating tray under a branch and shake or tap the branch for a fixed period of time (20 s). Collect the insects from the tray using the pooter (there should be no damage to the tree).

AQA GCSE Additional Science

B2 2.5 Sampling methods to find relationships within a community

What should I do with the results?

**HOW SCIENCE WORKS
PRACTICAL SKILLS**

Read the following carefully and critically.

How does the average height of grass vary along a transect line?

Student's method: At 0.5 m intervals the student measured one piece of grass to the nearest centimetre. This was repeated along five transect lines. Table 2.4 shows the results. The student knew that the numerical results had to be analysed so found the **modal** value for each distance along the transect. It gave a pattern.

Table 2.4 shows a set of results of heights of the grass along the transects (to the nearest centimetre).

Table 2.4 Heights of the grass along the transects.

Transect number	Distance along transect											
	0	0.5	1.0	1.5	2.0	2.5	3.0	3.5	4.0	4.5	5.0	5.5
1	7	7	6	7	4	3	2	3	5	5	6	7
2	8	6	7	5	4	2	3	3	6	6	5	7
3	7	6	7	6	5	3	2	2	6	7	7	7
4	7	7	6	5	4	3	2	3	5	7	8	8
5	6	6	7	5	4	3	2	3	6	7	8	8
Modal value	7	6	7	5	4	3	2	3	6	7	8	7

Working as a small group, use the following list of questions to design a more reproducible experiment and a better method of handling the results.

Points to consider: Was this a suitable procedure that would give a valid result? Was there any bias in choosing which piece of grass to measure? Were enough results taken? Was it appropriate to take the modal value? Was the measurement to the nearest centimetre accurate enough? If friends repeated the experiment along the same transect line would they come up with the same set of results? Is this experiment reproducible?

The conclusion must be that this method was not valid and did not give reproducible results because the sample size was too small: too few blades of grass were measured at each distance.

How could you improve the method? It would have been better to measure more stem lengths. How could this be done?

How could you analyse the results? Was the modal value appropriate? Discuss whether to use the mean height of your sample or the **median** value. Using the median value helps to remove the effect of anomalous results, although if a result is 'way-out' you may decide to remove it before taking the mean.

Use your school's VLE forum or a social networking platform to discuss and answer these questions. Alternatively share your results using collaborative software such as Google docs and analyse them.

Biology 2

Chapter 2 Photosynthesis, organisms and communities

B2 2.6 The relationships between organisms and their physical environments

Learning outcomes

Understand how physical factors of the environment may affect the communities present.

Abiotic factors are the 'non-living' physical factors of the environment that are external influences upon the community. They include climate — rainfall, light, wind; soil factors — such as particle size which affects drainage and air pockets, and humus; the location — does the tree trunk or slope face the sun, how steep is the slope or is it a microhabitat below a rock.

A **population** is the number of one species living in a habitat.

Competition and predation are examples of biotic factors that control **population** numbers. When we look at the list of non-living factors it becomes obvious that it is not possible to study any one factor in isolation. Organisms such as woodlice which dry out easily will be found in moist places such as under a rock or heap of vegetation and these are usually also dark. Similarly, plants with thin cuticles will not be found on dry, sunny slopes.

Investigating abiotic factors

HOW SCIENCE WORKS — PRACTICAL SKILLS

You might consider taking the daisy plant investigation a bit further to see whether water or humus content were possible factors causing the pattern of distribution. This can be done by sampling the soil at points along the transect. If you do not have access to a soil auger, soil samples can be collected using a small empty food can. Sample at 1 m intervals along the transect. Push the can into the ground and, using a small trowel, collect a vertical sample. Place this in a labelled, sealed plastic bag for analysis in the laboratory. (SAFETY: be aware of sharp edges, and place a thick cloth on the can before pushing it down with your hand or foot.)

To find the water content of the soil in the laboratory:

① For each sample remove the fresh plant material, stones and any invertebrates. Weigh out approximately 25 g soil into an evaporating dish. Record the exact mass (g) to 2 decimal places. Clearly label all samples.

② Place the sample in a drying oven at 100°C. Prepare a results table for all samples, leaving space for the results that you calculate, i.e. percentage water and percentage humus.

③ (SAFETY: the dish will now be hot.) Re-weigh, until the mass is constant (that is, you have two readings of the same mass). Record this new mass. Keep the labelled sample.

④ Calculate the percentage of water in the soil sample:

starting mass of fresh soil − mass of dry soil = mass of water

$$\% \text{ water in sample} = \frac{\text{mass of water}}{\text{starting mass of soil}} \times 100 = \_\_\_\_\%$$

Add to results table.

To find the percentage of humus:

⑤ Put the dried soil into a crucible and heat over a Bunsen burner until red hot and stops smoking (the smoke is the burning humus). SAFETY: wear safety glasses and be aware that the crucible will remain hot after you have finished heating.

⑥ Allow to cool and re-weigh. The drop in mass is equivalent to the mass of dried humus present before burning.

⑦ Calculate the percentage of humus in the sample:

$$\% \text{ humus} = \frac{\text{mass of dried humus}}{\text{mass of dried soil}} \times 100 = \_\_\_\_\%$$

Add to results table.

Test yourself

17 List the abiotic factors that might affect a plant growing on the sunny side of a wall.

18 Why might you find more single-celled algae on one side of a tree trunk than the other?

B2 2.6 The relationships between organisms and their physical environments

> Analyse your results to see if either water content of the soil or humus content of the soil could have contributed to the pattern of distribution of the daisy plants. Do you think any other factor affected the distribution?
>
> Present your experimental results in an interesting way using ICT.

Other abiotic factors that could be investigated are:
- Dissolved oxygen in water: does this vary in a pond as photosynthesis takes place through the day? This can be recorded using a dissolved oxygen sensor attached to a datalogger and recorded over a 24-hour period.
  Take care with this delicate, expensive equipment.
- Light intensity: does this vary along a transect through a wood? Does the shade from the trees limit the vegetation growing on the ground? Record the light intensity with a light sensor, making sure that the sensor is vertical for each reading.
- Nutrients: garden centres sell simple dip strips for analysing soil.
- Temperature: does this affect germination of seeds?

Chapter 2 Homework questions

Homework questions

1. Draw a sketch graph to show how the rate of photosynthesis changes with increasing light intensity on a hot and cold day.
   a. What are the raw materials for photosynthesis?
   b. How does the leaf obtain these raw materials?
   c. What are the products of photosynthesis?
   d. Why do trees in the Arctic Circle grow so slowly?

2. a. How would you investigate whether the invertebrates in an apple tree and an oak tree were the same species?
   b. How would you try to make this a fair test?
   c. What might affect your results?

3. A greenhouse crop grower wants to maximise his profits from selling cucumbers. Write a paragraph explaining what factors he should consider.

4. a. A chloroplast is a part of the plant cell which absorbs light energy. How is the light energy absorbed?
   b. Why is the distribution of starch uneven in a variegated leaf?
   c. Growth rings on a tree show the biomass produced in a particular year. What effect would the following have on the width of the rings: (i) a severe drought, (ii) severe defoliation by caterpillars (they ate all the leaves), (iii) a mild but rainy spring followed by a warm summer?

5. Some students noticed that the grass under a tree was less dense and shorter than the grass in the open field. They decided to investigate the effect that the tree was having on the grass growth in the following ways:
   - Take an interrupted belt transect on the north, east, south and west sides of the tree for 5 m.
   - Take soil samples along the transect line in the first and third quadrats from the trunk.

   They made observations at lunchtime about activity near the tree.

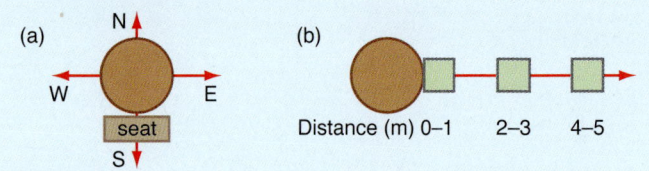

Figure 2.23 (a) The position of the seat on the south side of the tree; (b) the method for placing the quadrats.

Hypothesis to test: The shade from the tree reduces the grass growing beneath the tree.

The results showing the % plant cover were as follows:

	N			S		
Quadrat distance from tree (m)	0–1	2–3	4–5	0–1	2–3	4–5
% cover: bare soil	60	15	0	80	60	0
moss	40	25	20	20	20	20
grass	0	60	80	0	20	80
daisy	0	0	0	0	0	0

	E			W		
Quadrat distance from tree (m)	0–1	2–3	4–5	0–1	2–3	4–5
% cover: bare soil	90	100	75	90	50	5
moss	10	0	5	5	20	0
grass	0	0	5	5	25	95
daisy	0	0	10	0	5	0

Soil sample results showed that the % humus and water were highest on the north side. The south side had little humus or water present. East and west had similar values but lower than the north.

a. Explain how the soil samples would have been (i) collected and (ii) treated to find the results.
b. Which side of the tree had the highest percentage of moss cover?
c. On which side of the tree was there grass cover closest to the tree?

Observations showed that there was always a group of students on the seat at lunchtime.

d. State two effects the use of the seat would have on the soil composition.
e. Explain how trampling could have affected the distribution of (i) grass and (ii) moss.
f. Go back to the original hypothesis. Do the results support this hypothesis? Could you suggest an alternative?
g. What changes to the method would you make to test your new hypothesis?

AQA GCSE Additional Science

Chapter 2 Exam corner

Exam corner

A student was investigating the uptake of carbon dioxide by water plants at different light intensities. The bicarbonate indicator solution changes from red to yellow as the carbon dioxide is used up. The six tubes were set up as follows: each had a sprig of waterweed 8 cm long folded to fit into the tube, they were filled with the indicator solution and sealed with a cap. Five tubes were placed at different distances from a fluorescent tube light box. Tube 6 was tightly wrapped in foil.

Table 2.6

Tube number	Distance from lamp (cm)	Initial colour of indicator	Colour after 5 min	Colour after 10 min	Colour after 20 min	Colour after 25 min
1	2	red	orange	yellow	yellow	yellow
2	4	red	red	orange	yellow	yellow
3	6	red	red	red	red	orange
4	8	red	red	red	red	orange
5	10	red	red	red	red	red
6	covered in foil	red	red	darker red	darker red	darker red

a Explain why tube 1 became yellow first. (2 marks)
b Why did tube 6 become darker? (3 marks)

Student A

a Tube 1 changed first because it was closer to the lamp. ✗
b Tube 6 didn't change because it was covered in foil. ✗

Examiner comment The question relates to photosynthesis so this should appear in the answer to part (a).

Student A appears not to know that carbon dioxide is a raw material for photosynthesis and just answers in terms of the variable that has been changed in both part (a) and part (b). Look how easy it is to throw marks away!

Student B

a Tube 1 changed first because this tube was closer to the lamp and therefore received a higher light intensity which means it gets more light energy. ✓

b Tube 6 became darker because as it was covered it could not photosynthesise ✓ without light energy. ✓ The weed was respiring ✓ and giving off carbon dioxide so the solution went darker.

Examiner comment In part (a) Student B relates the distance to the light energy but does not refer to the rate of photosynthesis, which will be highest in tube 1 and therefore the carbon dioxide uptake is quickest in tube 1.

(b) This is a good answer — both photosynthesis and respiration are stated in the answer. The colour change is noted and related to respiration producing carbon dioxide.

Biology 2

Biology 2

Chapter 3
Enzymes: function, industrial uses and their role in exercise

ICT

In this chapter you can learn to:
- create an animation of the lock and key model for how enzymes work
- use audio and visual resources to show the many uses of enzymes in industry

Setting the scene

Why is biological washing powder called biological when it is full of chemicals?

Practical work

In this chapter you can learn to:
- plan and carry out an investigation into how different conditions affect enzyme reactions
- use graphs to display the results of an investigation into the effect of exercise

B2 3.1

Learning outcomes

- Know that proteins act as structural components of muscles, hormones, antibodies and enzymes.
- Know that enzymes are proteins which act as biological catalysts that increase the rate of reactions in living organisms.
- Understand that protein molecules and enzymes are made of chains of amino acids coiled together.

Proteins are specialised biological molecules

Ribosomes assemble amino acids to form proteins. The order of assembly is dictated by the DNA code in the nucleus. Proteins have many different functions.

Building protein for growth

Growing cells need to build new structures inside the cell. To do this they make proteins. Proteins are components of cell membranes and the membranes of the sub-cellular parts such as mitochondria. Amino acids can be joined together to form structural proteins and enzymes. Enzymes catalyse the reactions that make proteins inside the cell. This is called protein synthesis — different enzymes control which proteins can be built. Different cells have different functions and so there are thousands of different proteins.

A protein that you will know well is keratin, which makes up hair, feathers and also hooves. It is a structural protein. Keratin is not soluble. Muscle tissue is composed of a protein that is specialised to contract.

Hormones are soluble proteins that are transported in the blood plasma to target organs. Think back to genetic engineering and the manufacture of growth hormone. The growth hormone molecule has about 200 amino acids. The chain of amino acids is folded into a specific three-dimensional shape so that the hormone can bind onto receptor sites of the target organ membranes.

Antibodies also have a surface shape. Antibodies are specific to the antigen of a pathogen. This means that the antibodies can attach to the pathogen surface and inactivate it.

Enzymes are biological catalysts. They are also three-dimensional molecules. Here we look at them in greater detail.

> **Test yourself**
>
> 1 Name two structural and three soluble globular (three-dimensional) proteins.
> 2 What gives three-dimensional proteins their specific shape?

What are enzymes and what are their functions?

Enzymes are called biological catalysts because they control most of the reactions that take place in living organisms. All catalysts change the rate of chemical reactions but are not themselves changed by the reaction. Enzymes catalyse essential reactions such as respiration and photosynthesis that take place inside cells. The enzyme is not used up so it can catalyse the same reaction many times.

The function of enzymes is to enable reactions to take place at lower temperatures. Without enzymes, at low temperatures, such as the temperature of the human body, the chemical reactions inside cells would take place slowly.

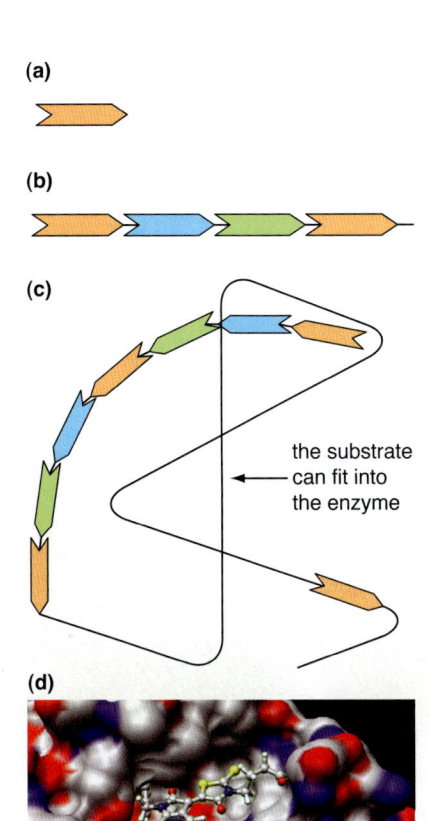

Figure 3.1 (a) A single amino acid; (b) a long chain of amino acids joined together; (c) for three-dimensional proteins the chain folds and forms bonds across the chain between amino acids to make it more stable; (d) a three-dimensional model of an enzyme molecule with the substrate in position.

Biology 2

Chapter 3 Enzymes: function, industrial uses and their role in exercise

In the food and biotechnology industries enzymes are used in many processes so that the reaction temperature can be kept low. This saves energy used to heat the container and so saves money. Enzymes also increase the reaction rate so that large amounts of product can be made quickly.

Enzymes are biological molecules that catalyse reactions in living organisms. Enzymes are referred to as biological catalysts.

Enzymes differ from chemical catalysts because they are protein molecules. Therefore, they have some properties of both catalysts and proteins. As catalysts, enzymes change the rate of reactions and as proteins they have a complex three-dimensional structure that allows other molecules to fit into the enzyme.

Test yourself

3. What is the benefit of enzyme-controlled reactions inside living cells?
4. Give three advantages of using enzymes in biotechnology applications.
5. Give two similarities and two differences between enzymes and other catalysts.

B2 3.2 How do enzymes work?

Learning outcomes

- Know that enzymes have an optimum temperature and optimum pH.
- Explain why the shape of an enzyme is vital for its function.
- Explain why heat affects enzyme function.

During digestion or breakdown reactions, each enzyme works on one particular substrate giving certain products. Different enzymes work on different food types. Protease enzymes speed up digestion of protein and amylase breaks down the substrate amylose (commonly called starch) to give glucose.

Why does an enzyme only work on one substrate? Let us look at the structure of enzymes in more detail. You will need to think about the three-dimensional structure of the molecule. Remember that enzymes are protein molecules.

Amino acids are the building blocks of proteins. They are joined end to end to form a long chain. This molecular chain is stabilised when it is folded into a three-dimensional shape and cross-bonds form between the amino acids (rather as a girl might twist up her hair and put in slides). All enzymes have this same basic structure, but they have a different sequence of amino acids to form different protein molecules. The order in which amino acids are joined together is controlled by the DNA code in the nucleus.

Each tightly folded protein molecule or enzyme has ridges and grooves, giving it a unique shape. The idea of the substrate fitting into the enzyme is called 'the **lock and key**' model for enzyme action. Just as two keys can look similar but will not work in the same lock, so similar substrates cannot be broken down by the same enzyme. This is why enzymes are described as specific. Once the correct substrate is attached to the enzyme molecule, a reaction at the surface active site (ridge or groove) of the enzyme takes place and a chemical bond is broken in the substrate.

The sequence of a *breakdown* reaction is:
- the enzyme and substrate collide
- the substrate attaches to a special ridge or groove on the enzyme surface
- the enzyme breaks the bond in the substrate
- products are released

Test yourself

6. Work out the sequence of the reaction to build up a single product from two substrate molecules — for example, to build up a molecule such as sucrose. The structure of sucrose is shown in Figure 3.2.

B2 3.2 How do enzymes work?

> A **model** is a simplified picture that scientists construct to fit observed data to a theory.
>
> The **'lock and key' model** suggests that each enzyme (the lock) has a special shape, which means that only one substrate (the key) can fit into each enzyme.
>
> The **activation energy** is the energy that must be provided to the reactants to start a chemical reaction.
>
> A **substrate** is the substance acted on by an enzyme.

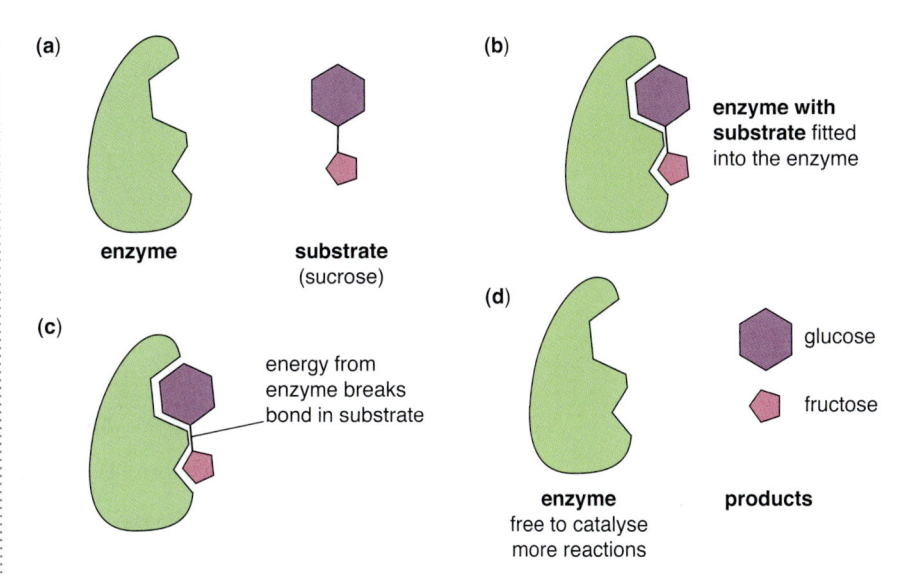

Figure 3.2 This sequence of diagrams shows how an enzyme breaks down a substrate. Here, the substrate is sucrose and the products are glucose and fructose.

How enzymes work — ICT

Create an animation or presentation explaining how enzymes work.

What conditions affect enzyme reactions?

Rates of enzyme reactions increase with increase in temperature. To understand why, you can apply your understanding of what is happening at the molecular level. First, the enzyme and substrate molecules must collide with enough energy for the substrate to fit into the enzyme structure and start the reaction. This is called the **activation energy**. This is a bit like putting a missing piece into a jigsaw puzzle: you can place a missing piece over a gap, but it needs a little push (energy) to fit it in.

An increase in temperature will cause the particles to move faster, that is, to increase their kinetic energy. As a result enzyme molecules and substrate molecules will collide more often and with more energy. The enzyme can catalyse more reactions so the reaction rate is increased.

But this is where enzymes differ from catalysts: enzymes function best at an optimum temperature (see Figure 3.3). An increase in temperature above this changes the shape of the protein molecule (see Figure 3.4). You can see the effect of change in the protein structure when you cook an egg white. When the enzyme molecule shape has been changed, the substrate can no longer fit into the enzyme — we say that the enzyme has been **denatured** and it can no longer function as a catalyst.

Living organisms are adapted to exist in a range of habitats, so the enzymes inside bacteria which live in hot springs will be able to function within a different temperature range from those enzymes in plants in the Arctic. There are also differences in pH levels at which enzymes work. Each enzyme has an optimum pH, which is the pH at which it functions best or most effectively.

Biology 2

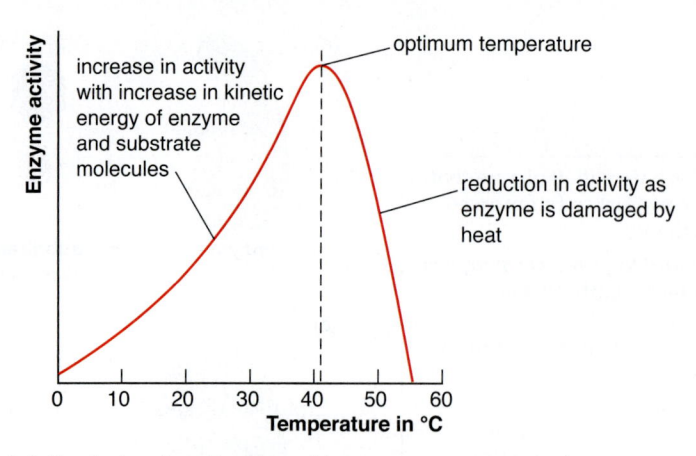

Chapter 3 Enzymes: function, industrial uses and their role in exercise

Figure 3.3 Graph showing the effect of heat on an enzyme reaction.

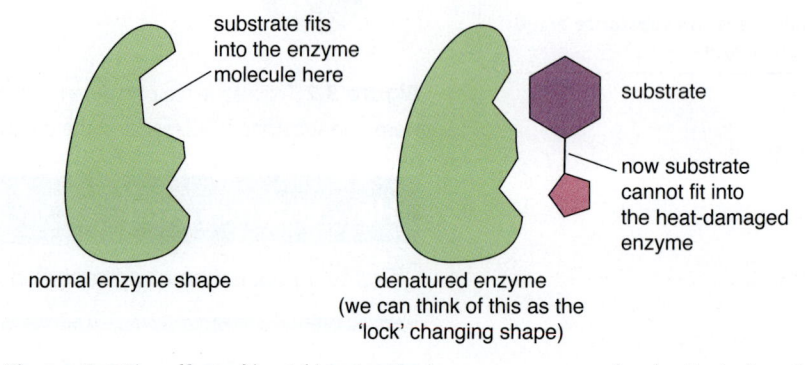

Figure 3.4 The effect of heat (denaturation) on an enzyme molecule. Heat alters the bonds and changes the three-dimensional shape.

When temperature destroys the three-dimensional shape of a protein molecule or enzyme, it is **denatured**.

Effect of temperature and pH

HOW SCIENCE WORKS — PRACTICAL SKILLS

Plan and carry out an experiment to show how one of these variables (temperature or pH) affects the action of enzymes.

Test yourself

7. Why are enzyme reactions slow at low temperatures?
8. Why does the rate of product formation increase as the temperature increases between 5°C and 30°C in an enzyme reaction?
9. A student was carrying out an experiment using the enzyme protease on boiled egg white. He found that the white solution became colourless for all temperatures up to 45°C, but above this temperature the solution remained cloudy. Can you explain this?
10. What term is given to the temperature at which enzymes function most efficiently?
11. The mouth is slightly alkaline but the stomach has a low pH. Why do you think salivary amylase stops working when the food it is mixed with enters the stomach?
12. Why can protease break down protein but not starch?

B2 3.2 How do enzymes work?

Digestion in action — ACTIVITY

How could you act out as a drama the process of digestion from a food being eaten and broken down to being absorbed in the small intestine and passing into the blood?

Who was Emil Fischer? — HOW SCIENCE WORKS ACTIVITY

Emil Fischer was born in Germany in 1852. His father, who said that Emil was too stupid to be a businessman and had better be a student, sent Emil to the University of Bonn in 1871 to study chemistry. Emil Fischer won a Nobel prize in 1902. This was the first prize awarded for organic chemistry. He was famous for rigorous teaching and making links between science and industry. Six of his students later went on to become Nobel prize winners.

Emil Fischer first suggested the 'lock and key' model in 1894. The amazing thing about Fischer's model was that he suggested this before scientists had discovered that enzymes were protein molecules.

Figure 3.5 Emil Fischer.

The 'lock and key' model is used today to develop computer simulations that investigate possible new drugs. The computer programs design and test thousands of different molecules with different shapes, to see how they could 'fix' onto receptors on the surface of target disease cells.

1. Write and illustrate a newspaper report about Fischer's great award. Find out about him by searching for his name and biography. How long after his discoveries was he rewarded?

2. Find the answers to the following questions using books or an internet search.
   a In Emil Fischer's model, what does the lock represent?
   b After the 'lock and key' model was suggested, how long was it before the nature of enzymes was worked out, and how long before a three-dimensional model of a protein was first made?

3. Draw a diagram to show the 'lock and key' model as a lock and key, and a more modern diagram showing an enzyme and substrate.

Biology 2

Chapter 3 Enzymes: function, industrial uses and their role in exercise

B2 3.3

Learning outcomes

- Understand and describe the action of digestive enzymes that catalyse the breakdown of different foods.
- Know where the enzymes are produced and where they function.

Boost your grade ✓

Cracking the name code for enzymes

The ending -ase tells you that the substance is an enzyme. The first part of the name tells you the substrate. For example, **protease** is the enzyme that breaks down protein, and **amylase** breaks down amylose (starch). Think of some compounds with names ending in -ose. What type of substance are these?

Are enzymes used outside cells in the body?

Digestive enzymes catalyse the breakdown of food as it passes down the intestine from the mouth. The digestive system can be thought of as a tube through the body carrying food.

Digestive enzymes are produced by specialised cells in **glands**, such as the salivary glands or the pancreas, or by cells in the lining of the stomach and intestine. These enzymes pass out of the cells and are released into the intestine.

The purpose of digestion is to break down large insoluble molecules into small soluble molecules. For example, amylase breaks down the substrate amylose (starch) to give glucose. Smaller molecules are needed because they are soluble and can pass (diffuse) from the small intestine into the blood plasma for transport to body cells. Smaller molecules, like glucose and amino acids, are also in a form that cells can use.

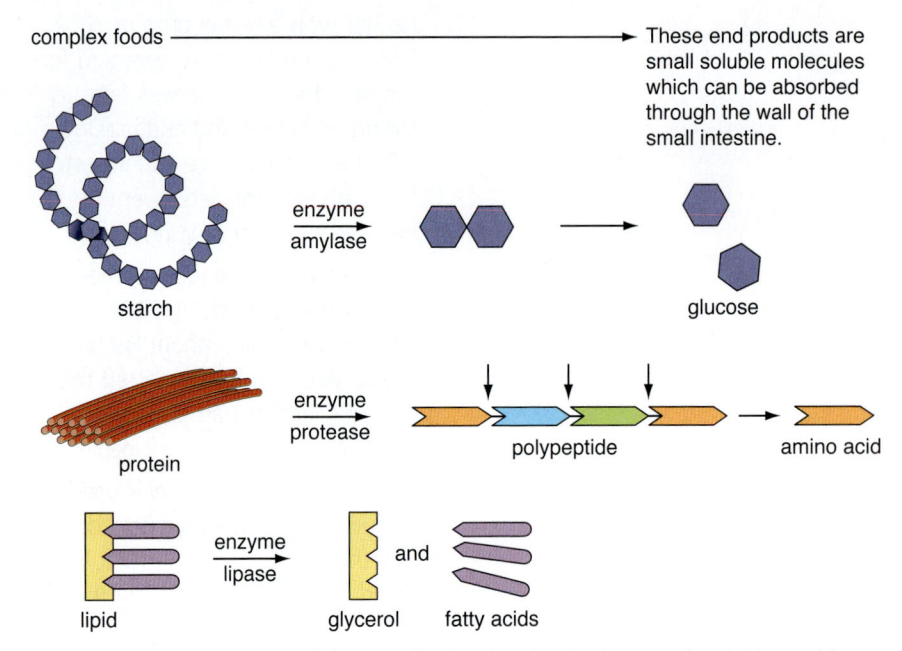

Figure 3.6 The breakdown of the complex food molecules into small, soluble, usable molecules.

Protease enzymes that break down proteins are produced by cells in the stomach lining, so protein digestion begins in the stomach. However, different proteases are produced by the pancreas and secreted into the small intestine where proteins continue to be broken down into amino acids. The stomach protease and the pancreatic protease function at different pH levels, as you can see in Table 3.1.

Other substances are released that help the enzymes to function by providing the optimum pH. Hydrochloric acid is released into the stomach to provide the low pH required by the protein-digesting enzyme called protease. **Bile** produced in the liver cells is stored in the gall bladder and added to the small intestine via a small duct (tube). Bile neutralises the acidic food, leaving the stomach to provide the higher pH needed by enzymes that work in the small intestine.

B2 3.3 Are enzymes used outside cells in the body?

A **gland** is an organ that produces and releases (secretes) a substance used by other cells.

Bile is a substance produced by the liver and released during digestion into the small intestine to provide alkaline conditions. Bile also reduces the size of fatty droplets, making them easier to digest.

Amylase is the digestive enzyme that breaks down starch into glucose (sugar).

Lipase is the digestive enzyme that breaks down lipids (fats and oils) into fatty acids and glycerol.

Protease is the digestive enzyme that breaks down proteins into amino acids.

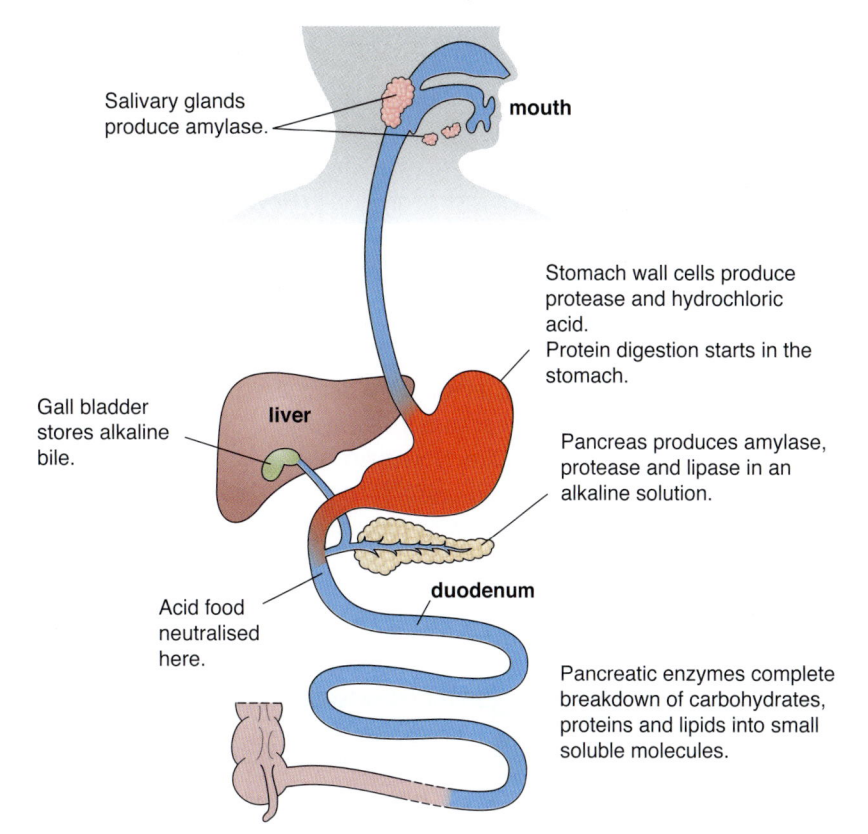

Figure 3.7 Digestive enzymes control reactions that take place in the digestive system.

Fats are broken down by the enzyme **lipase**, which is produced by the pancreas and small intestine. The products are fatty acids.

Locate the digestive enzymes in Figure 3.7. What do the colours red and blue indicate?

Table 3.1 Enzymes of the digestive system.

Enzyme	Site of production	Site of action	pH	Substrate	Products
Salivary amylase	Salivary glands	Mouth	8	Starch	Sugars
Pancreatic amylase	Pancreas	Small intestine	8	Starch	Sugars
Protease	Cells in stomach wall	Stomach	2 (acid produced by the stomach acidifies food)	Protein	Polypeptides and amino acids
Protease	Pancreas	Small intestine	8 (bile from the liver neutralises acid food in the duodenum)	Protein and polypeptides	Polypeptides and amino acids
Protease	Small intestine wall	Small intestine	8	Peptides	Amino acids
Lipase	Pancreas and cells in small intestine wall	Small intestine	8	Lipids	Fatty acids and glycerol

Biology 2

Chapter 3 Enzymes: function, industrial uses and their role in exercise

Test yourself

13 a The food leaving the mouth is slightly alkaline. How is the pH lowered in the stomach?
  b What is the optimum pH for the enzyme amylase?
  c The food leaving the stomach is strongly acidic. How is the pH raised?
  d What is the optimum pH for all the pancreatic enzymes?
  e What is the optimum pH for stomach protease?

14 At the start of the chapter, in question 3, 'What is the benefit of enzyme-controlled reactions inside living cells?' you considered temperature. Now think about the rate of energy production and give a second answer. Suggest how slow uncatalysed reactions would affect life.

15 Copy Table 3.2 and complete the four blanks.

Table 3.2

Enzyme	Large complex molecules	Small soluble end products of digestion
Amylase	Starch	(a) _____
Protease	(b) _____	Amino acids
(c) _____	Lipids (fats and oils)	(d) _____ and glycerol

16 a Name the two places where amylase is produced.
  b Name the two sites of protein digestion.
  c Describe where fats and oils are digested.

17 a Where is bile produced?
  b Where is bile stored?
  c What is the function of bile?

B2 3.4 Industrial production and use of enzymes

Learning outcomes

- Understand and describe how enzymes produced by microorganisms are used in detergents to remove food stains.
- Understand and describe applications of enzymes used in industry including pre-digested baby food and products containing sugar or fructose syrup.
- Evaluate the advantages and disadvantages of using enzymes in the home and in industry.

The list of industrial applications for enzymes is amazing and includes food, wine and beer production, making fructose sweeteners for soft drinks, laundry detergents, producing the 'stone-washed' effect in jeans and manufacturing pharmaceuticals. The food industry uses many enzymes, which can be 'tailor-made' to give an exact product for the development of a new food.

Like all living organisms, bacteria and fungi produce and secrete enzymes. The enzymes pass out of their cells into the environment. The enzyme products, small soluble molecules, are then taken into the organism's cytoplasm. Microorganisms are collected from different locations around the world and the enzymes they produce are tested. Biotechnologists look for enzymes that could be used for particular processes, and that work at the temperature of the industrial plant.

As bacteria and fungi normally live in cool environments, their enzymes usually function at low temperatures. However, if an enzyme is required in industry to work at high temperature or extremes of pH, then enzymes have to be found from microorganisms living in those environmental conditions. Enzymes that act on the right substrates but at the wrong temperature may need to be modified by gene transfer.

To produce the enzymes, the microorganisms are cultured in a fermentation process. At the end of the batch process, the **fermenter** is emptied and the enzymes are collected from the solution in which the bacteria have grown.

Many industrial enzymes come from soil microorganisms. One of the best sources of industrial proteases used in laundry detergents is *Bacillus*, a common soil bacterium.

B2 3.4 Industrial production and use of enzymes

> A **fermenter** is a large steel vessel used for biochemical reactions. Sensors monitor the conditions inside. The sensors send information to a computer, which then controls input valves to maintain the temperature, pH, nutrient and oxygen levels at the optimum values.

Enzymes in the food industry

An early application of enzymes in the food industry was the production of sweet syrups by breaking down starch. This process can be done chemically by boiling starch with acid, but the reactions using the enzyme method give a pure and reliable product, with no other by-products. Another advantage is that energy costs are lower as the enzyme process can be carried out at a lower temperature.

Glucose syrup is widely used as an ingredient and food additive. Most glucose syrup is produced from maize (corn) using enzymes. Maize processing results in a large amount of starch waste products. Using a common and cheap (waste) material as the raw material for glucose syrup is cost effective.

Many **carbohydrase** enzymes are used in sequence to produce a variety of syrups of different sweetness. For example:
- maize starch is treated with one form of amylase to convert it to a thick starch paste
- this paste is then reacted with different amylase enzymes to form sugars such as maltose or glucose. The reaction can be stopped here, or
- the final stage is the conversion of glucose syrup to fructose syrup, using the enzyme **isomerase**

In the early 1970s a continuous flow system (see Figure 3.9) was developed in which glucose solution was added through the top and fructose solution was delivered from the bottom. The continuous process can run for about 6 months non-stop. As enzymes are proteins, their structure eventually breaks down and the enzyme needs replacing. The old enzymes are biodegradable in the environment so do not result in toxic waste. Fructose syrup is much sweeter than sucrose or glucose syrup so it can be used in smaller quantities as a sweetener. This saves food manufacturers money, but another result was that in the 1970s fructose syrup began to be used in slimming foods to give sweetness with fewer calories.

Although there are clear economic advantages of using enzymes in industrial processes, there are some disadvantages:
- development and production of new enzymes is costly
- enzyme action is highly specific, so they have only one use
- enzymes require exact conditions of temperature and pH in which to work
- as enzymes are proteins, they can cause allergic reactions, and so must be encapsulated to minimise the risk of skin contact or inhalation
- some people may object to genetically modified (GM) enzymes being used to produce food

Figure 3.8 An industrial fermenter for producing enzymes. The raw material is converted by the enzymes as it flows slowly over beads with enzymes trapped on the surface (see Figure 3.9).

> **Carbohydrase** enzymes break down complex carbohydrate molecules such as starch into simple sugars.
>
> **Isomerase** is an enzyme that converts glucose into fructose.

Test yourself

18 List the economic advantages of using enzymes in the food industry.

19 a In a continuous flow column, what is the advantage of having the enzymes immobilised on beads? (See Figure 3.9.)

  b In a batch process, the raw materials and the enzymes are reacted together in a reactor vessel until an economic concentration of products has been formed. The remaining raw materials and the enzymes are then separated from the products. What are the advantages of 'continuous flow' rather than 'batch' processes?

Biology 2   47

Chapter 3 Enzymes: function, industrial uses and their role in exercise

Protease enzymes break down proteins.

(a) enzyme (e) is attached to gel beads

(b) raw material in (e.g. glucose syrup)

column packed with beads with enzymes trapped on surface

Because the enzyme is immobilised (fixed) to the beads it remains in the column.

pure product out (e.g. fructose syrup)

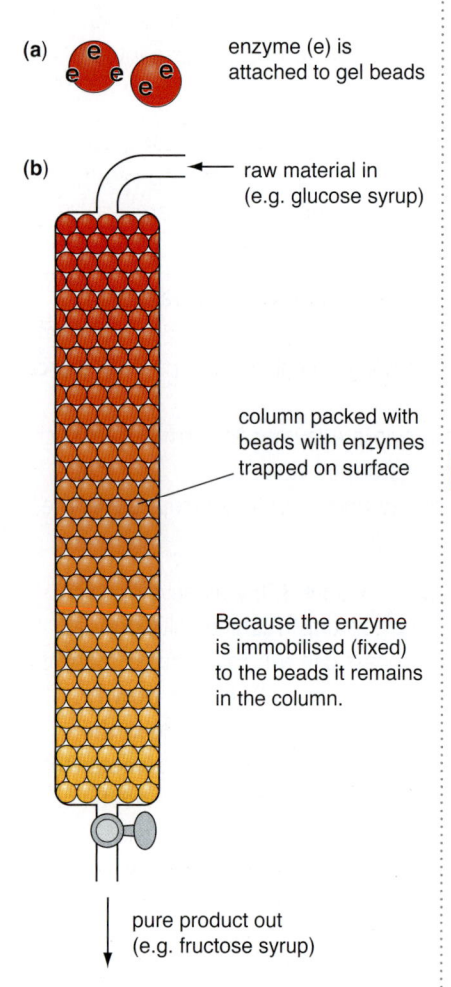

Figure 3.9 In a continuous flow column the substrate (glucose) is added through the top and reacts as it flows slowly over beads, which have the enzyme attached to the surface. The product can be collected from the bottom of the column.

Other food products made from enzymes

You would not expect much in common between food for babies and food for body builders, but there is. Some baby food is pre-digested using **proteases** so that the baby can use the amino acids without digestion. Baby-milk powders are manufactured from cows' milk and are treated with enzymes to break down the proteins. This has the advantage that they are less likely to cause allergic reactions. The protein supplements taken by body builders are also pre-digested into amino acids, which can be quickly taken up for building and repairing muscles if consumed immediately after training.

Soft-centred chocolates start with the centre as a hard sugar paste, which can be easily handled for chocolate coating. The addition of invertase enzyme converts the sucrose to glucose and fructose syrup within 1 to 2 days. In other words, by the time the chocolates reach the shop the hard centre has become soft. This has other advantages in that fructose retains water within the centre and prevents the sugar crystallising out.

Test yourself

20 Complete Table 3.3 showing the use of enzymes in food production.

Table 3.3

Food product	Enzyme used in manufacture	Substrate	Product	Advantage of the product
Baby food	Protease		Amino acid	Readily available amino acids for baby growth
Slimming products	Invertase	Glucose syrup		Sweetens without adding calories
Soft drinks	Carbohydrase	(Maize) starch		
Baby food		Fats	Fatty acids	No digestion needed

Laundry detergents

Washing used to be carried out at high temperatures with a great deal of agitation to break down dirt. This was costly in terms of electricity used to heat the water. Modern detergents use enzymes that function efficiently at low temperatures, reducing fuel costs.

Food or biological stains on clothes can include fats and oils from fried items, butter or oily dressings, proteins from eggs, grass or blood and starches from sauces made with flour. Biological washing powders have the following enzymes added: lipase to break down fats, proteases to digest proteins and carbohydrases to remove starch stains. Similar enzymes are used in dishwasher detergents.

Although using 'biological detergents' improves stain removal and reduces costs, so much washing detergent is used in industrial production, in the hotel and catering industry and in the home that environmental factors must also be

B2 3.4 Industrial production and use of enzymes

considered. Biological detergents are biodegradable, because enzymes are proteins and break down naturally in the environment. They also have other environmental advantages in terms of energy efficiency and water efficiency.

Table 3.4 The requirements for a good washing powder.

What do the homemaker, laundry and hotel industry want from a detergent?	What are the environmental requirements for a detergent?
A good clean wash	No foam in water-treatment works
Low temperatures to reduce the cost of heating the water	Reduced amount of energy used, to reduce air pollution and greenhouse gas emissions from electricity generation
Less water used, as many houses and businesses now have water meters	Reduced amount of water used, to avoid water shortages particularly in the south of England
Less agitation to reduce wear on fabrics	Biodegradable materials
All stains removed	
Fabrics to feel soft after washing	
Shorter washing time	

Figure 3.10 What stains might be on a toddler's bib? How could these be removed in a cool wash?

Test yourself

21 Figure 3.10 shows a typical toddler's bib. Explain how biological washing powder will solve the problem for each stain, including the egg, butter and tomato sauce.

22 What are the environmental advantages of using enzymes in detergents?

23 The instructions on the box state that biological washing powders are not effective when used at temperatures above 40°C.
   a Explain why this is so.
   b If you had dropped fruit and cream down the front of your best shirt, what would be the advantage of using biological washing powder?

Powerful molecules — ICT

Produce a multimedia resource using ICT on 'The uses of enzymes'.

Stone-washed jeans — ACTIVITY

Denim fabric is made by weaving. The cross-threads are unbleached cotton and the warp (vertical threads) are dyed with indigo. Indigo is a stable dye and strong hypochlorite bleach (high pH and high temperature) is needed to remove it chemically.

Stone-washed jeans were originally produced literally with stones, in an enormous machine that shook 150 kg of pumice stone with 150 pairs of jeans for up to 6 hours. The stones damaged the cellulose fibres and released the indigo dye. This process was damaging to the seams of the jeans, the machines and the operators because of the pumice dust in the air, and it took a lot of water to rinse off all the grit particles from the jeans.

The modern method uses an enzyme called DeniMax, which produces the stone-wash effect faster without damage to the fabric and uses a minimum quantity of water.

The new enzyme process removes the need for pumice stones, the final look is the same and the jeans last longer.

1 List four problems with the old stone-wash system.

2 A new enzyme has been developed which is specific to indigo and which works under such mild conditions that it can also be used on stretch denim. The result of using the enzyme is to reduce the depth of the colour and produce an aged effect. What is meant by the term 'the enzyme is specific to indigo'?

3 What are the environmental advantages of the enzyme process used in the manufacturing of jeans?

Chapter 3 Enzymes: function, industrial uses and their role in exercise

B2 3.5

Learning outcomes

- Know that enzymes catalyse the reactions in processes such as respiration, photosynthesis and protein synthesis.
- Explain aerobic respiration and explain how energy is released by chemical reactions inside mitochondria.

Aerobic respiration occurs when oxygen reacts with glucose and releases energy. Carbon dioxide and water are the products of the reaction.

Test yourself

24 Name two large molecules made in plant cells by joining together smaller molecules.

25 In which part of the cell is most of the energy in respiration released?

26 Combustion is one reaction but respiration takes place as a series of small reactions. What points are the same about these reactions and how do they differ?

Protein synthesis is the building of protein molecules from a chain of amino acid molecules, using enzymes to catalyse the reaction.

What reactions do enzymes control inside cells?

Most chemical reactions in the cell take place in the cytoplasm. Different enzymes catalyse each of the chemical reactions in the cell. Enzymes working inside cells have a constant temperature and pH.

Respiration to release energy

Making larger molecules from smaller molecules needs energy. All cells in living organisms release energy from food by respiring. **Aerobic respiration** is a reaction in cells that uses glucose and oxygen to release energy. Aerobic respiration occurs as a series of reactions, each controlled by an enzyme.

The overall reactions in aerobic respiration are summarised by the equation:

glucose + oxygen → carbon dioxide + water (+ energy)
$C_6H_{12}O_6 + 6O_2 \rightarrow 6CO_2 + 6H_2O$ (+ energy)

First, glucose is broken down in the cytoplasm to form smaller molecules that can enter the mitochondria. In the mitochondria a sequence of reactions takes place. In the final stage hydrogen atoms removed from the glucose are combined with oxygen to form water and energy is released.

So why is respiration a series of reactions and not a single reaction? If you oxidise glucose by setting fire to it you end up with the same overall reaction and the same products, but a great deal of energy is released rapidly producing heat — we call this combustion. This type of reaction within cells would damage the proteins. Instead, in respiration the energy is released bit by bit at each reaction, in small units that have sufficient energy to power the other reactions in the cell. You could compare this to arriving at a holiday destination with all your money as large-value notes — it's best if you have smaller values and coins for buying drinks and chocolates.

What do cells and organisms do with these small units of energy?

Remember that energy is needed for building up small molecules into large molecules, such as amino acids into large protein molecules; this is known as **protein synthesis**. Other chemical reactions in the cytoplasm also need energy. For example, plants need energy to add nitrates to the sugars produced by photosynthesis and form amino acids. Plants also build up glucose molecules into cellulose for their cell walls.

Photosynthesis itself is a chemical reaction, and without plant enzymes this reaction could not take place. In photosynthesis the enzyme systems in chlorophyll convert light energy into chemical energy in the form of glucose. The majority of food webs are based upon this reaction.

Energy released by respiration is used by warm-blooded animals to maintain their body temperature, and for muscle fibre contractions so that the animals can move.

B2 3.6 What is exercise?

> **Test yourself**
>
> 27 Examine the following statements and identify which are correct.
>   a Plants respire at night and photosynthesise in daylight.
>   b Plants respire for 24 hours a day.
>   c Plants photosynthesise for 24 hours a day.
>   d For plants to grow, the rate of photosynthesis must be greater than the rate of respiration over the year.
>   e For growing plants the rate of photosynthesis equals the rate of respiration over the year.
>   f Plants can photosynthesise for longer each day in spring than in winter.
>
> 28 a Suggest three ways in which animals make use of the energy released in respiration.
>   b Give two ways in which plants use the energy released in respiration.

B2 3.6 What is exercise?

Learning outcomes

- Know that muscles use the energy released during respiration to contract.
- Understand how energy is released from glucose in muscle cells by aerobic and anaerobic respiration.

Exercise improves your body in a number of different ways. It can make your heart more efficient, it can strengthen your muscles, it can stop you becoming obese and can improve your flexibility and balance. There are many different ways to exercise, but all types of exercise have one thing in common. They involve moving parts of your body and this means making your muscles work. When you move, your muscles have to contract and pull on bones in your skeleton.

A muscle needs energy whenever it contracts. This energy is released during **respiration** from nutrient molecules derived from the food we eat. Respiration takes place in almost all the cells in your body. It is particularly important in muscle cells.

Figure 3.11 Young children should be active for 1 hour per day; older people can snatch a gym session for aerobic or strength workouts. Pilates exercises improve coordination, balance and core muscles.

Biology 2

Chapter 3 Enzymes: function, industrial uses and their role in exercise

> **Respiration** is a series of chemical reactions that release energy from glucose and other nutrients. It takes place in the cells of living organisms.
> **Aerobic respiration** requires oxygen to release energy from nutrients.
> **Anaerobic respiration** releases energy from nutrients without oxygen.

In this chapter, two types of respiration are discussed: **aerobic respiration** and **anaerobic respiration**. Aerobic respiration is the most efficient type of respiration that uses oxygen to release the energy from nutrients. The most important nutrient used in respiration is glucose, a type of sugar. Anaerobic respiration is less efficient, but it does not need oxygen to proceed. Both types of respiration are really a series of chemical reactions, although each one can be summarised by a single equation.

Aerobic respiration:
glucose + oxygen → carbon dioxide + water (plus the release of energy)

Anaerobic respiration:
glucose → lactic acid (plus the release of a small quantity of energy)

The glucose and oxygen used during respiration have to be transported to the muscle cells for respiration to take place. At the same time, carbon dioxide produced during respiration has to be removed from the cells and excreted. Your body has a transport system that it relies on to do these jobs. We will study this in the next section.

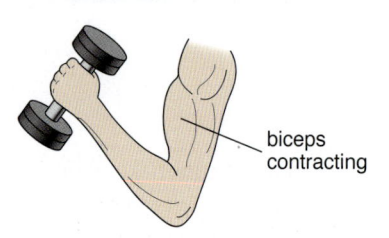

Figure 3.12 The biceps muscle has to contract to pull the arm upwards.

> **Test yourself**
>
> 29 For each picture in Figure 3.11, write down the muscles that are contracting during the exercise shown.
> 30 What substances do the muscles need to obtain energy?
> 31 How do these substances get into the body?

B2 3.7

Learning outcomes

- Understand and describe how exercise changes the heart rate, breathing and arteries.
- Know how these changes affect the supply and removal of substances to and from the muscles.
- Know how muscles use glycogen during exercise.

How does your body respond when you take exercise?

So far in this chapter you have learnt that respiration releases the energy that allows your muscles to contract when you are exercising. You have also learnt how the glucose and oxygen needed for respiration are transported by the circulatory system to your body cells. When you exercise, your rate of respiration increases to provide your muscles with enough energy for the extra exertion. This has an immediate effect on the systems involved in providing glucose and oxygen to your muscles.

Look at the runner in Figure 3.13. His breathing system is providing sufficient oxygen to maintain his body as he waits on the starting line. As soon as the race starts, his muscles will begin contracting more quickly. This needs more energy, so his rate of respiration has to increase. This in turn requires more oxygen and more glucose. His circulatory system will need to work at a rate sufficient to transport this oxygen and the glucose needed to his muscles and other organs.

- When the race starts, the runner's breathing must respond so that more oxygen is absorbed into the blood in his lungs. The number of times he breathes in and out each minute (his breathing rate) will increase. He will also take in more air every time he inhales. On average, the lungs can increase the amount of air inhaled by about eight times when someone starts exercising.

AQA GCSE Additional Science

B2 3.7 How does your body respond when you take exercise?

Figure 3.13 When this 400 m runner starts the race, his muscles will need more glucose and oxygen.

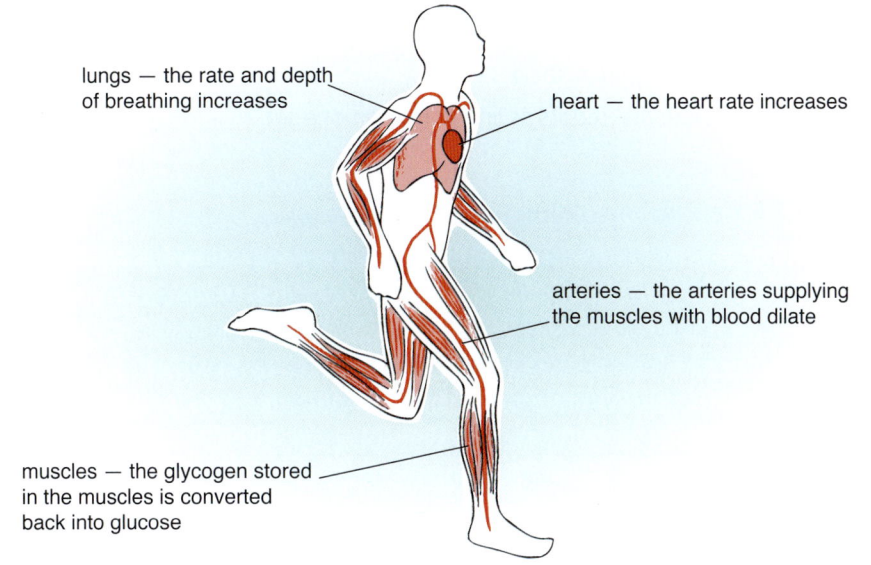

Figure 3.14 When you exercise, different parts of your body respond to increase the rate at which energy is released in your muscles.

- lungs — the rate and depth of breathing increases
- heart — the heart rate increases
- arteries — the arteries supplying the muscles with blood dilate
- muscles — the glycogen stored in the muscles is converted back into glucose

- His heart rate will also increase when he starts running. This will transport oxygen and glucose to the muscles more quickly.
- To allow more blood to reach the muscles, the arteries supplying the muscles dilate (become wider).
- As the rate of respiration increases, more carbon dioxide is produced. The increased blood flow and increased breathing rate deal with this by removing the extra carbon dioxide as it is produced.

Glycogen: the glucose store

Although the blood is being pumped to the muscles more quickly, sufficient glucose may not be absorbed from the small intestine in the first place. When this happens, the muscles use some of their stored **glycogen**. Muscles can store any excess glucose they receive as glycogen. Then, when respiration increases, the muscles convert this glycogen back to glucose so that energy can be released from it. This will help a runner during a race, but it can take 2 days to replace the glycogen that is used up.

Test yourself

32 How do you think smoking would affect the body's ability to respond to exercise?

Glycogen is an insoluble polymer formed from glucose and stored in muscles. It can be converted back to glucose and used during vigorous exercise.

Measuring the effects of exercise on the body

HOW SCIENCE WORKS — PRACTICAL SKILLS

Sara and Kerry decided to investigate the effect of exercise on their heart rate and breathing rate. They used data loggers to measure their heart and breathing rates at rest and during exercise. Their heart rate was measured in beats per minute and their breathing rate in breaths per minute. Their results are shown in Table 3.5.

1. Why did they measure their heart and breathing rates at rest?
2. a One of Sara's results does not fit the overall pattern. Which result is this?
   b What could have caused this anomalous result?
3. The first two breathing rates are the same for both girls. Why do you think their breathing rates did not increase when they started walking?
4. Describe and explain the general pattern in each student's results using the scientific ideas covered in this chapter.

When Sara and Kerry had stopped running, they measured their heart rate every minute until it returned to the normal heart rate at rest. The time it takes to do this is called your recovery time. Recovery time gives an indication of fitness — shorter recovery times are linked

Biology 2

Chapter 3 Enzymes: function, industrial uses and their role in exercise

to higher levels of fitness. Their results for this part of the investigation are shown in Table 3.6.

5 Plot both students' results as line graphs on one set of axes.

6 a What was the recovery time for each student?
b Which student does this suggest is the fitter?
c What other factors should be considered before drawing a firm conclusion from these data?

Table 3.5 Sara and Kerry's results.

	Sara		Kerry	
	Heart rate in beats per minute	Breathing rate in breaths per minute	Heart rate in beats per minute	Breathing rate in breaths per minute
Resting	68	12	72	12
Walking	63	12	86	12
Running on the flat	112	16	120	14
Running up hill	126	19	134	17

Table 3.6 Sara and Kerry's results for recovery time.

Time in minutes	Sara	Kerry
0	126	134
1	115	128
2	107	121
3	95	112
4	83	105
5	74	96
6	68	88
7	68	80
8	68	72
9	68	72

B2 3.8 What are the effects of vigorous exercise?

Learning outcomes

- Explain how muscles respond to long periods of vigorous exercise.
- Explain how the body 'repays' an oxygen debt.
- Understand how an oxygen debt builds up in muscles during anaerobic respiration owing to the formation of lactic acid.
- Understand how to interpret data about the effects of exercise on the body.

Our muscles can keep working provided they have a sufficient supply of oxygen and glucose. Even when the supply of glucose is limited, our muscles convert glycogen into glucose in order to maintain respiration. However, during long periods of vigorous exercise this situation can change.

The person in Figure 3.15 is exercising hard. If the exercise continues for a long time his muscles will start to suffer from a condition called **muscle fatigue**. This occurs when the blood cannot supply the muscles with enough oxygen. When this happens, the muscles start to respire anaerobically.

Lactic acid and oxygen debt

During anaerobic respiration, energy is released from glucose without oxygen. However, anaerobic respiration releases far less energy per gram of glucose than aerobic respiration. This is because anaerobic respiration only partly breaks down glucose, producing a chemical called **lactic acid**. As lactic acid builds up in our muscles, it prevents them contracting properly. This is when muscle fatigue sets in. Just hold your arms straight out in front of you for a long time if you want to experience muscle fatigue!

> **Muscle fatigue** occurs during long periods of vigorous exercise and causes the muscles to work less efficiently.

54 AQA GCSE Additional Science

B2 3.8 What are the effects of vigorous exercise?

Lactic acid builds up in muscles during anaerobic respiration.

An **oxygen debt** is the amount of oxygen needed to oxidise the lactic acid that has built up in muscles during anaerobic respiration.

Figure 3.15 Exercising vigorously for long periods of time can put your muscles under great stress.

During a sprint, athletes expend so much energy so quickly that they must rely on anaerobic respiration. Their muscles have to work so hard that the blood simply cannot supply their muscles with oxygen quickly enough. This is possible for the duration of the race, but at the end of the race their muscles will contain a lot of lactic acid. This must be broken down otherwise it will start to damage their muscles. Lactic acid is oxidised by oxygen to carbon dioxide and water. The carbon dioxide is then excreted as if it had been produced by aerobic respiration. The amount of oxygen needed to oxidise the lactic acid is called an oxygen debt. The sprinters in Figure 3.16 are paying back their oxygen debt.

Figure 3.16 These sprinters have produced a lot of lactic acid in their muscles during the race. What happens to the lactic acid?

Test yourself

33 a 'Anaerobic respiration is far less efficient than aerobic respiration.' Explain what this statement means.
   b How does your body respond when you cycle up a steep hill quickly?
   c Once you finish cycling up the hill, you need a rest. Explain what happens in your muscles while you are resting. Use all the key words in the definition boxes in your answer.

Biology 2

Chapter 3 Homework questions

Homework questions

1 A student was investigating two protease enzymes labelled A and B. The table below shows a copy of his practical record.

Table 3.7

Tube number	Boiled egg white solution in cm³	Enzyme	Other solutions added (2 cm³)	Result after 10 minutes
1	5	A	Water	No change
2	5	A	Dilute HCl	Clear
3	5	A	NaCO₃	No Change
4	5	B	Water	Less cloudy
5	5	B	Dilute HCl	No change
6	5	B	NaCO₃	Clear

a Where in the human intestine would:
 (i) enzyme A be produced?
 (ii) enzyme B be produced?
 (iii) enzyme B react?
b What is the role of the hydrochloric acid?
c Explain why enzyme B did not break down the egg white when dilute hydrochloric acid was present.
d What substances in the intestine produce the optimum conditions for enzyme B to react?

2 Imagine you have just set off on a bike ride.
a Describe the changes that take place in your body as you start to exercise.
b Explain how each of the changes helps your muscles to continue pedalling your bike.
c You and your friend have a sprint finish up to the top of a steep hill. Explain why you are breathing rapidly when you stop at the top.

3 Quorn™ is produced in a continuous flow process in a 150 000 dm³ reactor vessel. The filamentous fungus *Fusarium* is cultured in a medium based on glucose, ammonium compounds and vitamin B (biotin). Cultures are maintained at 28–30°C and pH 2.0.
a Examine Figure 3.17. Is this an aerobic or anaerobic process?
b What gas is extracted from the top of the vessel?
c Why are ammonium compounds added?
d What is meant by a continuous flow process?
e Biotin is a B group vitamin. Suggest how the fungus might use this.
f Why do you think the temperature is maintained closely between 28 and 30°C?
g What do you predict would happen to the production rate if the system became more acidic?

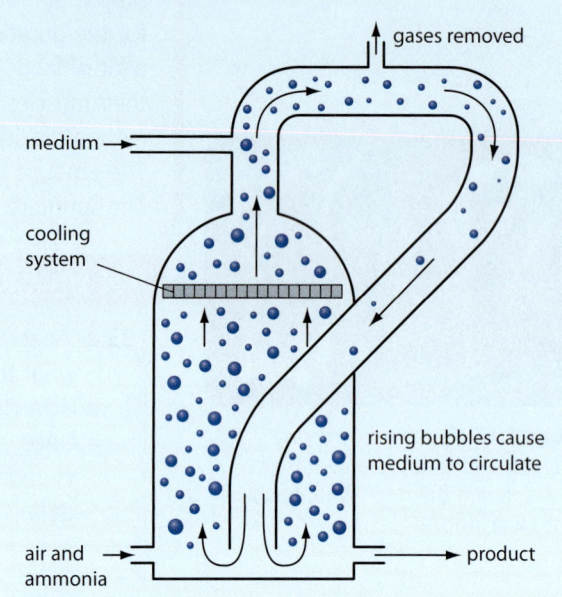

Figure 3.17

Chapter 3 Exam corner

Exam corner

Enzymes used for industrial processes have to be able to work at high temperatures to achieve a fast rate of product formation.

1 How could an enzyme which breaks down the correct substrate be modified to function at higher temperatures? *(2 marks)*

2 What are the advantages of biological washing powders to the environment? *(2 marks)*

3 In the drink industry a sugar sweetener is produced by passing starchy waste products down a continuous flow column. Give three advantages of a continuous flow column compared with a batch process in a fermenter. *(3 marks)*

Student A

1 The enzyme could be genetically modified ✓ by adding a heat-tolerant gene ✓ from another organism to the bacteria that produced the enzyme.

2 The washing can be done at a lower temperature and it gets whiter. ✗

3 When something is made in a fermenter the enzymes, products and the substrate are all mixed up and have to be separated when the batch is ready ✓ but when a continuous column is used the enzymes are fixed on the surface of beads and as the substrate trickles over the beads it is converted to the pure product ✓ which can be used as it is.

Examiner comment Student A's answer to question **1** is excellent. The answer to question **2** is incomplete — it does not explain the advantage to the environment. Make sure that you answer the question fully — always re-read and check that you have done so. Statements must be backed up.

Student A has produced a good answer to question **3** that compares the two processes. The answer includes the words enzyme, substrate and products, showing that the process is understood. The missing point is that the column can run for months without the enzyme needing replacement.

Student B

1 The enzyme would be denatured at high temperatures. ✗

2 (i) Biological detergents contain protein enzymes so they are biodegradable and are broken down in the environment doing no damage. ✓
(ii) They are used at lower temperatures so less energy is used and there is less greenhouse gas produced. ✓
(iii) Less water is used in the washing machine so this saves water, which is a scarce resource in drought times. ✓

3 The product is pure. ✗

The column can run for many weeks. ✗

It's cheaper. ✗

Examiner comment Student B has learned the word denatured carefully and was determined to use it in the answer to question **1**. Unfortunately he jumped in without reading the question properly.

In the answer to question **2** three good points are made and explained well (maximum marks are awarded for this part). This is a model answer. Take care though — it is not a good idea to introduce your own numbering system into an answer.

Examiner comment For question **3** we have a list without any comparison between the processes and two of the points do not identify which process is referred to in the statement. This answer scores no marks. The candidate needs to slow down, read the questions carefully and compose the answers much more thoughtfully.

Biology 2

Biology 2

Chapter 4
How do our bodies pass on characteristics?

ICT

In this chapter you can learn to:
- use the internet to find animations of meiosis and mitosis
- have a debate with a wide range of students using the school VLE

Setting the scene

All the information about our features and characteristics is inherited from our parents, and we start as a single fertilised cell. How can you develop from just one cell?

Practical work

In this chapter you can learn to research easy-to-recognise genetically inherited features of humans and carry out a survey.

B2 4.1 How did an Austrian monk help to shape our understanding of inheritance?

Learning outcomes

- Understand that Mendel planned his experiments.
- Understand that Mendel carried out and recorded carefully a large number of experiments
- Understand that this produced sufficient data to analyse to give reproducible results.

Figure 4.1 Gregor Mendel at work.

Test yourself

1. What does the word 'inheritance' mean when it is used in science?
2. What prompted Gregor Mendel to carry out his research into inheritance?
3. What did Mendel do to ensure that his findings were reproducible?
4. Why was it important for Mendel to obtain reproducible results?
5. Explain how Mendel's results showed that the blending theory of inheritance was unsatisfactory.

How did an Austrian monk help to shape our understanding of inheritance?

We know today how the features and characteristics that we inherited from our parents are passed on, but this was not always the case. It was only through careful experimentation that scientists gained this understanding. In this section, you will learn about the key developments that led to our current understanding of inheritance.

For many years people believed that the characteristics of parents combined in some way when they had children. For example, if a mother had black hair and a father had blonde hair, the two colours would combine to produce brown-haired children. This was also thought to be true of animals and plants. It was known as the 'blending theory' of inheritance.

An Austrian monk called Gregor Mendel, who was born in 1822, disagreed with this theory. Mendel worked in the gardens of an Austrian monastery and noticed that pea plants had either purple or white flowers. He sowed the seeds produced from these flowers. The new plants that grew only ever had purple or white flowers (the same colour as the two parents), never light mauve. He concluded that the parent plants passed on specific features to the offspring plants and not combined features.

Meticulous and mathematical are two words that can be applied to Mendel's approach to the problem. Although Mendel had not trained as a scientist, he realised that he needed to carry out controlled experiments. He could then collect reproducible evidence to analyse mathematically to prove his theory. He started by doing some preliminary experiments in which he pollinated purple and white flowered plants together. From this he confirmed that when these two types of pea plants were crossed they produced plants with white flowers and plants with purple flowers but never any plants with colours in between.

For his main experiment, Mendel took a purple flowering plant that had produced some white flowered plants, and a white flowering plant. He then pollinated these two plants together for several generations and counted the numbers of purple and white flowering plants that were produced.

Mendel analysed his results and discovered that for every three purple flowering plants there was only one white flowering plant, a ratio of 3:1. From his results, Mendel concluded that some inherited characteristics, such as purple flowers, had a stronger influence over the offspring. He called these 'stronger' influences dominant and the 'weaker' influences recessive. These findings showed clearly that the blending theory was insufficient to explain inheritance.

Mendel published his findings in 1866, expecting people to appreciate that inheritance could be much better explained using his ideas. Sadly, his ideas were largely ignored for 34 years until three other researchers published similar findings based on their own experiments. At the same time, Mendel's work was translated into English and people started to take his ideas seriously. They realised that Mendel had provided a far better explanation of inheritance than the blending theory. With the development of more powerful microscopes that enabled the chromosomes to be seen in the stages of cell division, we realise what great insight Mendel had. Mendel's theories fit and are used in the construction of genetic diagrams.

Biology 2

Chapter 4 How do our bodies pass on characteristics?

1 Mendel took a purple flower that had been produced by a pure-line cross of purple and white flowers by hand pollination.

2 He then took pollen from a white flowering plant and used it to fertilise the purple flowers.

Pollen transferred with a paint brush

4 He allowed these purple flowers to self pollinate and collected the seeds.

3 He collected the seeds and the result of this test cross was equal numbers of purple and white flowers.

5 Finally he planted the seeds that the purple flowering plant produced. When they grew into adult plants he counted how many plants had purple flowers and how many had white flowers.

705 purple flowering plants

224 white flowering plants

Figure 4.2 A flow diagram showing the procedure Mendel used to investigate inheritance in pea plants and the results he collected.

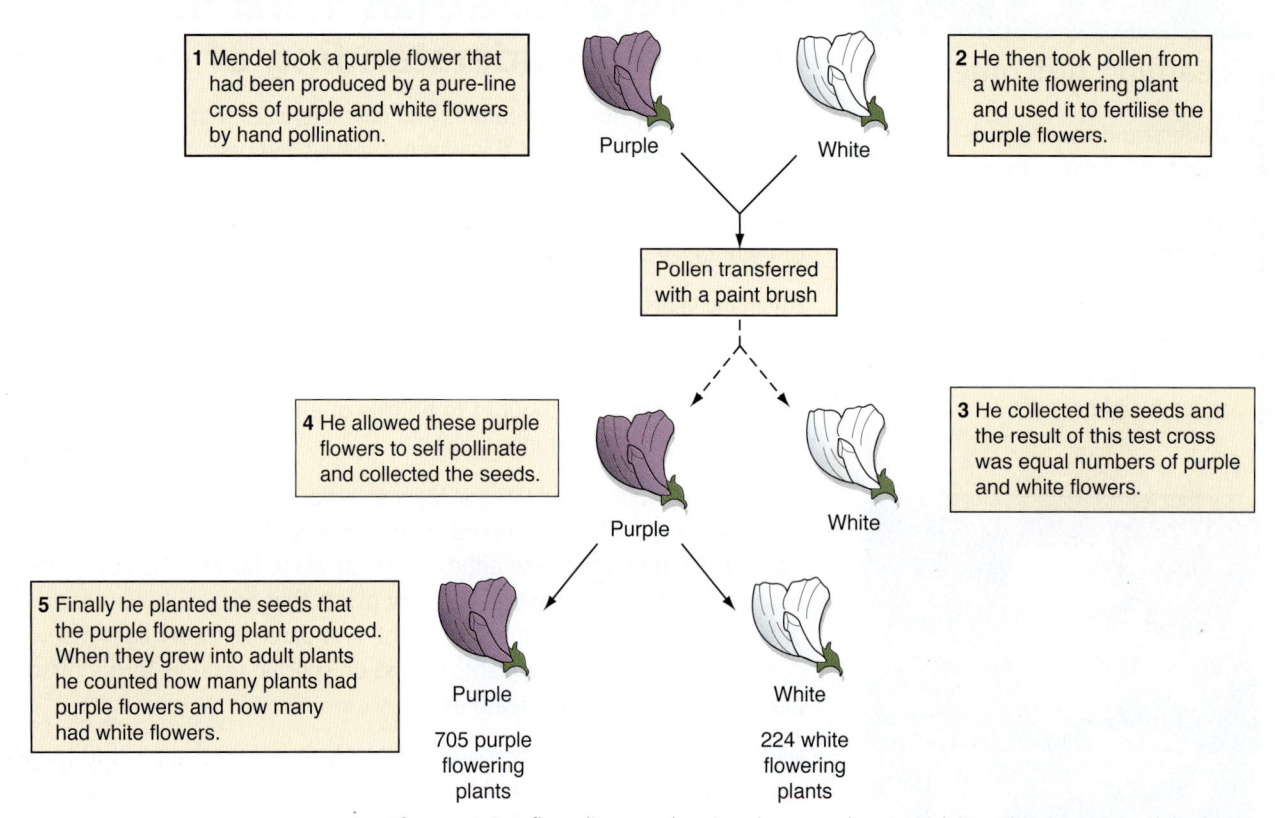

Test yourself

6 a Why do you think it took so long for Mendel's ideas to be accepted?
b Explain why the work of other people helped Mendel's explanation to replace the blending theory.

Mendel's maths

HOW SCIENCE WORKS ACTIVITY

Imagine that you are Gregor Mendel and are about to show your results to a scientific panel. Write a script to outline your findings and why you think that you have managed to explain the manner of genetic inheritance. You may also do this as a podcast or online debate.

B2 4.2 Why is cell division so important?

Learning outcomes

Understand the importance of DNA in living organisms.

Understand that mitosis is cell division that produces genetically identical cells.

Understand that meiosis is cell division that produces sex cells that are not genetically identical.

Without cell division organisms would not grow, reproduce or be able to repair damaged body tissue. The single cell formed at fertilisation repeatedly divides by normal cell division called mitosis. **Mitosis** is normally followed by differentiation as cells mature, to form their final shape for a specific function, for example muscle cells. In plants, there are certain layers of cells that can differentiate as the plant grows, such as those around the stem that produce xylem and phloem. The cells of plants produced by taking cuttings (asexual reproduction) increase by mitosis. Plants produced in this way have exactly the same genes as the single parent plant. Cell division is therefore an essential part of growing. We will examine it, starting with what happens to the genetic material.

AQA GCSE Additional Science

B2 4.2 Why is cell division so important?

What are chromosomes and genes?

In humans, there are 23 pairs of **chromosomes** inside the nucleus of a normal cell, or 46 in total. This number is different in other species. Chromosomes each contain a long molecule of **DNA** (**deoxyribonucleic acid**). A single chromosome is a molecule of DNA about 5 cm long wound around a support protein. This is amazing when you consider that chromosomes can be seen only under powerful microscopes. To achieve this each DNA molecule is thin and has to be coiled up tightly to form a chromosome.

A **gene** is a small section of a DNA molecule that carries a code. There are between 20 000 and 25 000 genes in each human nucleus. Figure 4.4 shows the double helix structure of a DNA molecule and that it contains a series of bases that are represented by the letters G, C, A and T. Bases are molecules that form part of DNA. A gene is therefore really a sequence of these bases. The sequence of bases in a gene acts as a code (instructions) for making a protein.

DNA fingerprinting

All humans have the same types of genes, so what makes us unique? Some areas of DNA, called non-coding sequences, don't contain any genes. These areas are different for each person in the world, unless you have an identical twin. This means that a person can be identified from the non-coding DNA obtained from a swab of their cells. This technique, called **DNA fingerprinting**, can be used to link people to a crime scene. If, for example, skin cells are found at a crime scene, then the DNA inside them can be removed and identified. If the non-coding DNA matches the DNA collected from a suspect then it places the suspect at the scene.

Amino acids are obtained when proteins are digested. A gene codes for the specific order that these amino acids are to be linked to make new proteins that the body needs. This order of amino acids is controlled by the order of the DNA

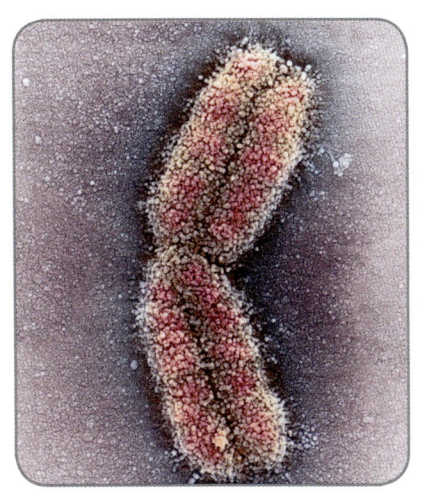

Figure 4.3 A chromosome from one human cell.

DNA fingerprinting is a technique that uses the non-coding pieces of our DNA to provide a unique identification sequence.

Test yourself

7 The DNA in one chromosome is far longer than the chromosome itself. Why is this? Think about how the shortening of DNA to a chromosome is achieved.

8 a Why do there have to be so many genes in a human cell?
   b Sometimes a gene mutates. This means that the sequence of bases on that gene is altered. What effect will this have on the protein that the gene is responsible for making? Explain your answer.

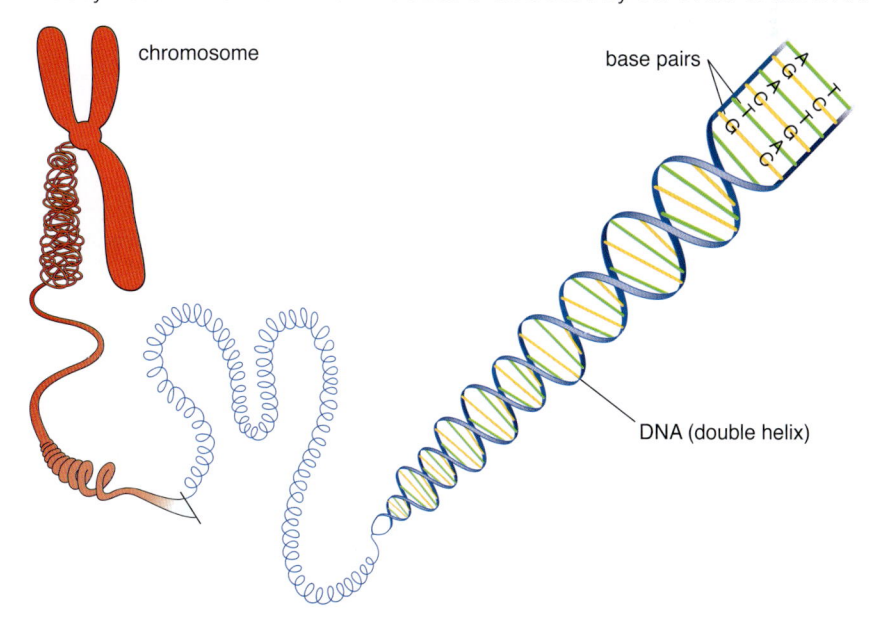

Figure 4.4 Examine the coiling of DNA to form the chromosome. The DNA is made up of a series of bases: A, T, G and C. Varying combinations of these bases make up the code for producing proteins.

Biology 2

Chapter 4 How do our bodies pass on characteristics?

> **DNA (deoxyribonucleic acid)** is a long-chain molecule, made up from a series of bases.
>
> **Chromosomes** are a protein with a molecule of DNA wound around it — the DNA molecule is a single strand. Chromosomes are found in the cell nucleus.
>
> A **gene** is a section of DNA that codes for a specific protein. Genes are in a linear sequence on the chromosome.

bases. Since each gene is a unique sequence of bases, each gene puts the amino acids in a different order and makes a different protein. This takes place in a ribosome. The proteins made can be enzymes or hormones, or they may also be used to form new body tissue, such as muscle, or even to produce the colour in your eyes.

What happens to chromosomes when cells divide?

A normal body cell has two sets of chromosomes, one set from the sperm cell and the other set from the egg cell, giving 23 pairs in humans. A cell can make a genetically identical copy of itself. This takes place when you grow and also when cells need to be replaced if body tissue becomes damaged. For example, if you were to cut yourself, the new skin that would form over the cut would be made by skin cells dividing to make new skin cells.

When a cell divides by mitosis it forms two cells that are both genetically identical to the original cell. The first stage is duplication of the chromosomes. This can happen because chromosomes have the amazing ability to make identical copies of themselves. The second step is the division of the chromosomes and cytoplasm into two cells. Figure 4.5 shows how mitosis takes place. The diagram will help you to understand what happens — you will not have to reproduce this diagram in an examination but may be required to draw

1. Nucleus disappears and chromosomes make copies of themselves. — copied chromosome / original chromosome

2. Copied chromosomes line up.

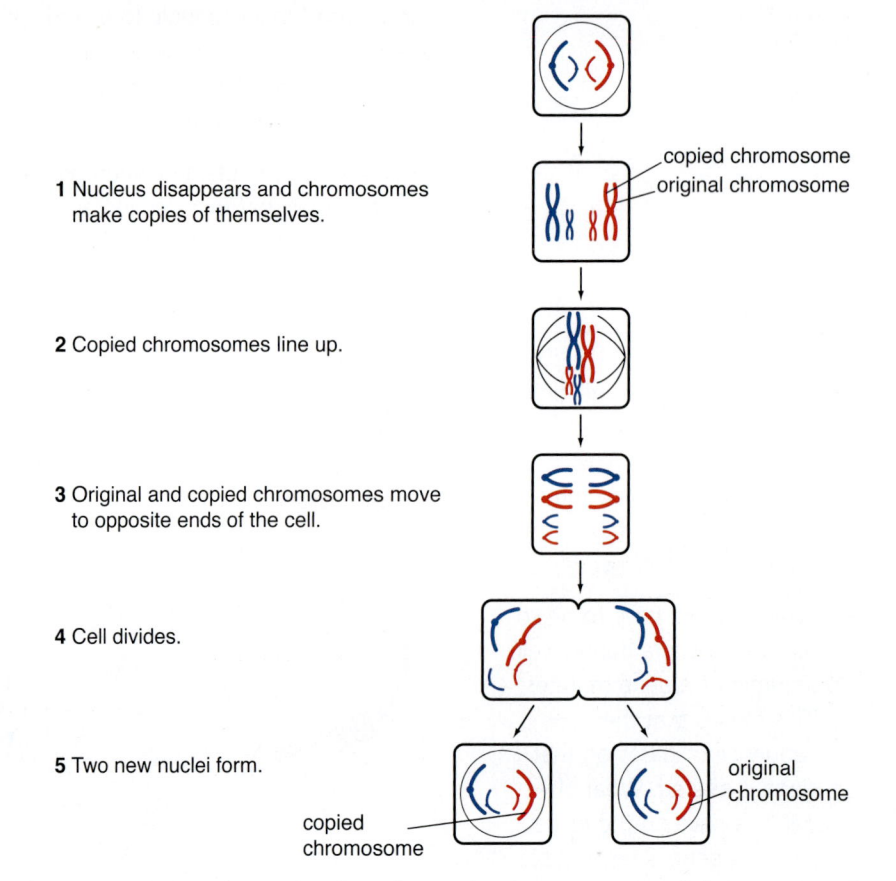

3. Original and copied chromosomes move to opposite ends of the cell.

4. Cell divides.

5. Two new nuclei form. — copied chromosome / original chromosome

Figure 4.5 The main stages in mitosis for a cell with just two pairs of chromosomes. The original and the copied chromosomes are labelled so you can see that the new cell is an exact copy of the original cell.

B2 4.2 Why is cell division so important?

> A cell divides by **mitosis** to produce two genetically identical cells with a full set of chromosomes. This is how normal body cells divide.

the products formed. If you examine a microscope slide of a root-tip showing mitosis you can see the chromosomes because they have become shorter and thicker.

Mitosis also takes place in some plants and in single-celled organisms such as bacteria and yeast during asexual reproduction. Offspring produced by this means have exactly the same genes as their parents, because they contain an exact copy of their chromosomes. They are often referred to as clones.

A second type of cell division, found in most animals and plants, has only one function — to make **gametes** (sex cells). In animals the sex cells, or gametes, are sperm cells and egg cells. In plants the male gametes are pollen cells. In humans, this type of cell division only takes place in the reproductive organs, the testes and ovaries. Cell division that forms gametes is called **meiosis**. The cells that are formed by meiosis are not identical to the original cell. They only have half the number of chromosomes. This means that in humans a gamete has 23 single chromosomes. Figure 4.9 shows a full set of human chromosomes in pairs. A gamete will only contain one chromosome from each pair.

> **Gametes** are sex cells. These include sperm cells, egg cells and pollen cells.
>
> A cell divides by **meiosis** to produce four genetically non-identical cells, each with half the original number of chromosomes. Meiosis forms cells called gametes.

Figure 4.6 will help you to understand the process — you will not need to draw this in an examination but may be required to show the products.

1 Nucleus disappears and chromosomes make copies of themselves.

2 Chromosome pairs line up, and swap pieces of information (DNA crossover).

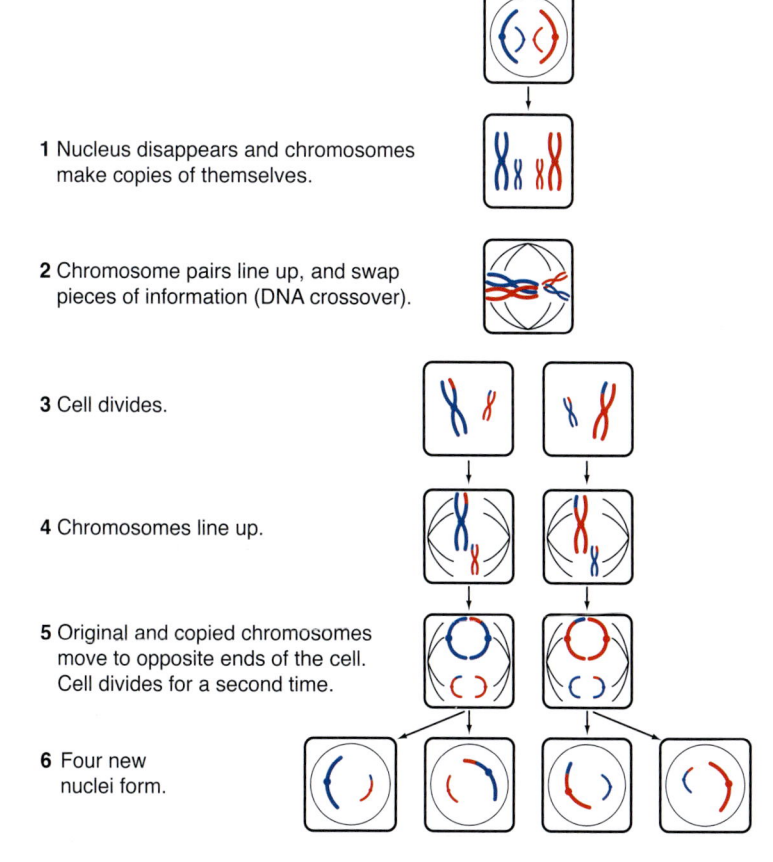

3 Cell divides.

4 Chromosomes line up.

5 Original and copied chromosomes move to opposite ends of the cell. Cell divides for a second time.

6 Four new nuclei form.

Figure 4.6 The main stages in meiosis for a cell with just two pairs of chromosomes. Once the matching chromosomes have paired up, they may swap pieces of information between them. This is called DNA crossover. The cell then divides twice and ends up with four cells (gametes), each with half the original number of chromosomes.

Test yourself

9 What are the differences between mitosis and meiosis?

10 Why do gametes have only half the number of chromosomes found in a normal body cell?

11 It is almost impossible for two human egg cells produced from one person to be identical. Explain why this is the case.

Biology 2

Chapter 4 How do our bodies pass on characteristics?

Table 4.1 Summarising cell division in animals.

	Mitosis	Meiosis
Purpose	Growth, replacement of damaged cells	Gamete production
Location	Growing points of bones, tissues. Sites of damage	Sex organs only — ovaries and testes
Products	One division only	A cell in the sex organ divides twice giving four gametes
	Two cells which have the same number of chromosomes as each other and the parent cell	Four cells which have half the number of chromosomes as the parent cell — a single set
	Two cells which are genetically identical to each other and the parent cell	Because each gamete has only one of the parent chromosomes each gamete is unique

Meiosis and mitosis   ICT

Use the internet to find animations of meiosis and mitosis. Which animation explains the ideas most clearly?

B2 4.3

Learning outcomes

- Understand and describe how a person's sex is determined.
- Understand how to complete genetic diagrams to illustrate inheritance.

How are characteristics passed on from one generation to the next?

When a sperm cell fertilises an egg cell the genetic information from the mother and father is brought together. Fertilisation produces a cell called a zygote, which contains a full set of chromosomes. Each gamete (one from the mother, one from the father) carries one chromosome from each pair found in a normal human body cell, so when the two gametes come together the full set of 23 pairs is produced. At fertilisation, all the characteristics that are controlled by your genes are determined — for example, your eye colour and hair colour. After fertilisation, the zygote divides by mitosis and develops into an embryo.

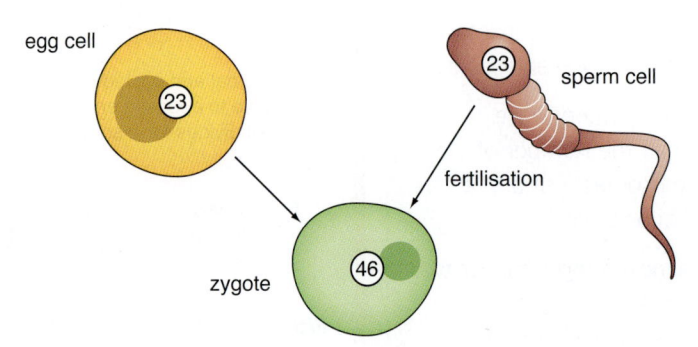

Figure 4.7 During fertilisation in humans the sperm and the egg cell bring together 23 chromosomes from each parent to form the first cell that goes on to produce a new human being. After fertilisation, this cell has the full number of chromosomes — 46.

AQA GCSE Additional Science

B2 4.3 How are characteristics passed on from one generation to the next?

Children in a family are never genetically identical unless they are identical twins. This is because each child has different chromosomes. Each sperm cell their father produces and egg cell their mother produces contains only one chromosome each from the 23 pairs in a normal body cell. If gametes contained the usual 23 pairs found in all other body cells, fertilisation would produce a cell with double the number of chromosomes each generation! This is why gametes contain only 23 chromosomes.

The one chromosome from each pair is selected randomly when gametes are formed by meiosis. If you consider that there are 23 pairs, it is incredibly unlikely that any two gametes will have exactly the same 23 chromosomes.

How is the sex of a person determined?

The first thing new parents want to know is 'is the baby a boy or a girl?', if they have not found out from a scan. One of the most obvious differences between children in a family is their sex: whether they are a boy or a girl. A person's sex is determined at fertilisation because it is controlled by the genes found on a pair of chromosomes. In Figure 4.9 find the un-numbered pair of chromosomes towards bottom right. These are labelled XY in the male and XX in the female, i.e. the last pair of 23 are the same in females, but different in males. These chromosomes are called the **sex chromosomes** and are the 23rd pair in each cell.

Figure 4.8 Which features of the children resemble those of the parents?

The 23rd pair of chromosomes in each cell are the **sex chromosomes** and they determine a person's sex — XX = female; XY = male.

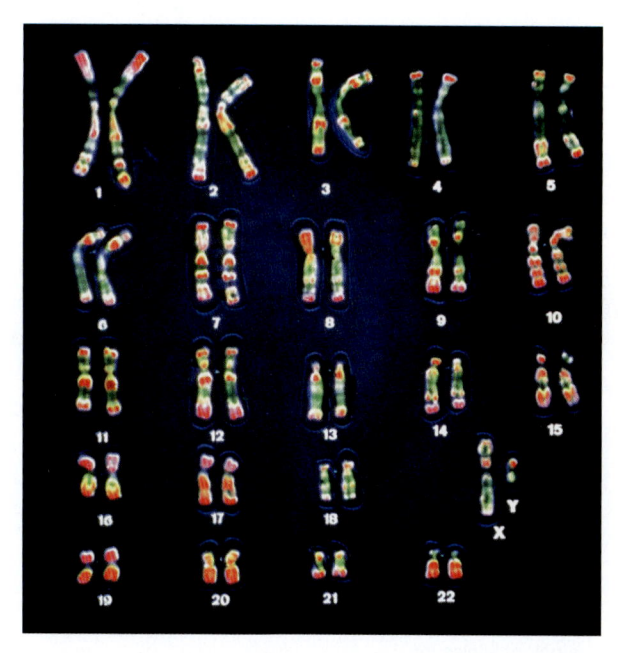

Figure 4.9 A full chromosome set from a man. Look at the last pair of chromosomes (marked X and Y). A woman's chromosomes are similar but the last pair are both X chromosomes.

In females both of the sex chromosomes are called X chromosomes. Males have one X and one Y sex chromosome. Every egg cell that a woman produces carries an X chromosome. Half the sperm cells that a man produces carry an X and half carry a Y chromosome. Remember that each gamete gets one randomly selected chromosome from each pair during meiosis. So if an X-carrying sperm cell fertilises an egg, then a baby girl will be born. If a Y-carrying sperm cell fertilises an egg, then a boy will be born.

Test yourself

12 Explain how sexual reproduction leads to variation between children in the same family.

13 Why is it essential that gametes in humans are produced by meiosis and not mitosis?

Biology 2

Chapter 4 How do our bodies pass on characteristics?

The **genetic diagram** in Figure 4.10 shows how sex chromosomes are passed on from parents to their children.

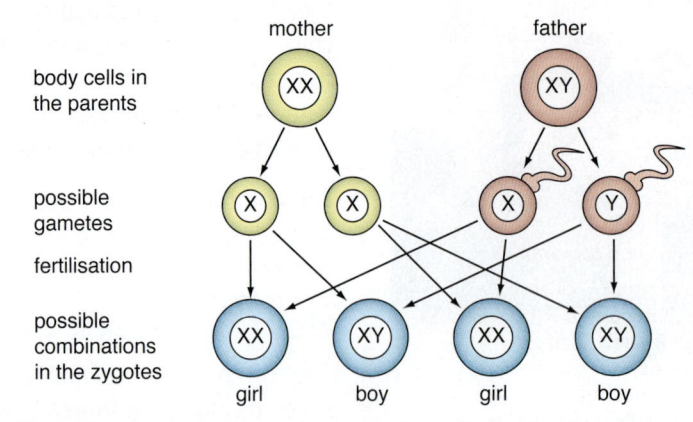

Figure 4.10 This genetic diagram shows how sex chromosomes are inherited.

Test yourself

14 'The sex of a baby is determined entirely by the father.' Do you consider this statement to be correct? Explain your answer by referring to how sex is inherited.

B2 4.4

Learning outcomes

Be able to construct genetic diagrams of monohybrid crosses to predict the outcomes.

Understand and use the terms genotype, phenotype, homozygous and heterozygous correctly.

How are some characteristics controlled by a single gene?

It is now understood that some characteristics are controlled by a pair of **alleles** — alternative forms of a gene — found on a pair of chromosomes. The flower colour in Mendel's pea plants was controlled by a single gene.

This gene for petal colour had two possible forms called alleles, known as the **dominant allele** and the **recessive allele**. Obviously flowers come in many colours and in the following example in snapdragon plants the dominant allele produces a protein that forms red flowers. The recessive allele cannot produce this protein. So if one or both of the genes in the pair is the dominant allele the plant will have red flowers. It is only when there is no dominant allele (i.e. both recessive alleles) that the flowers will be white because there is no protein produced to turn them red. You often use a capital letter (e.g. R) to represent the dominant allele and a lower case letter (e.g. r) to represent the recessive allele.

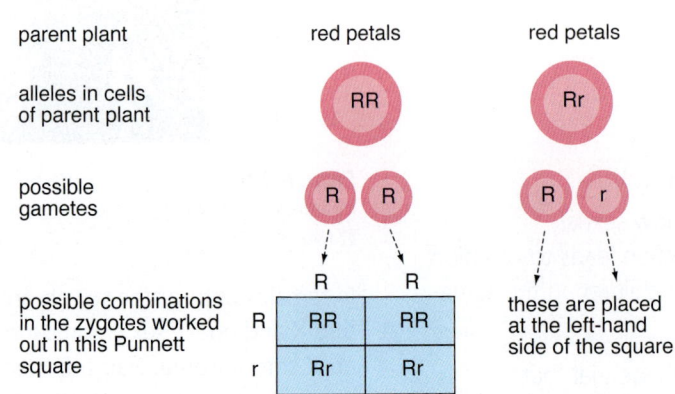

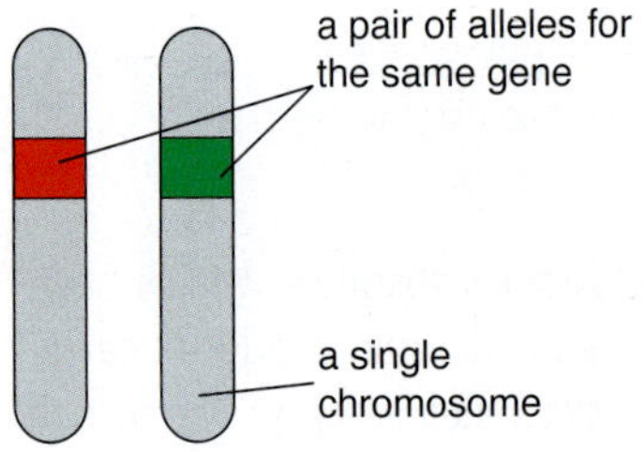

Figure 4.11 The red and the green sections on the chromosomes represent a pair of alternative alleles for the same gene.

offspring 100% red petals: each has a dominant allele R which gives red petal colour

Figure 4.12 Look at the offspring in the Punnett square. What colour flowers will they have?

B2 4.4 How are some characteristics controlled by a single gene?

> An **allele** is an alternative form of a gene found at the same position on a chromosome.
>
> A **Punnett square** is a simple method of showing a genetic cross to ensure all possible offspring are shown and the proportions can be calculated.
>
> A **dominant allele** will lead to a dominant characteristic when one or two dominant alleles are present. It is represented by a capital letter.
>
> A **recessive allele** will lead to a recessive characteristic only when no dominant allele is present. This usually means that it is present on both chromosomes in a pair. A recessive gene is represented by a small letter.

Boost your grade ✓

The only genotype that you can tell by looking at the phenotype is that of the homozygous recessive. This is an important clue to look for when working out genetics crosses or family trees.

Constructing genetic diagrams

Genetics has special words that do save complicated descriptions. It is worth looking at these and then seeing how you should use them when asked to construct genetic diagrams.

Phenotype describes the physical appearance, e.g. curly hair, red petal colour, taster or non-taster.

Genotype gives the alleles present in the parent. An individual with the same allele on both chromosomes of the pair is described as **homozygous**. An individual with alternative alleles on the two chromosomes is described as **heterozygous**.

Points to note: Homozygous is often paired with dominant or recessive — see example below — so that you know if the two alleles are dominant TT or recessive tt. Heterozygous needs no qualification or extra word because by definition it must be one dominant and one recessive Tt.

Remember we allocate a letter appropriate to the phenotype of the dominant allele so we could use R to represent the dominant allele for red petal colour with the result that r represents white petal colour in Figure 4.12 or T for taster and t for non-taster in the examples below.

Genetic diagrams are used to explain or predict outcomes of crosses. Two chemists working with a substance called PTC discovered that one found this unpleasantly bitter but the other could not taste it! Following investigations it was revealed that 70% of the population were tasters and 30% non-tasters. We will examine how to set the problem out.

Worked example

Examine the children of a non-taster wife and a man who is a taster. Taster and non-taster are their phenotypes. Let us assume that the man is a pure-bred taster since his ancestors were all tasters. This means he is homozygous with the genotype TT. His wife must be homozygous recessive as she is a non-taster and her genotype will be tt. Note that we use one letter for the gene, and use it as a capital to represent the dominant and small for the recessive allele.

Let T represent the allele for ability to taste PTC
and t represent the allele for not able to taste PTC.

	man	wife
phenotype	taster	non-taster
genotype	TT	tt
gametes	(T)	(t)
offspring		Tt
		100% tasters

NOTES TO STUDENT
1 Draw gametes in a circle
2 Only write one of each possible gamete

Figure 4.13 Setting out a genetic cross.

Boost your grade ✓

Genetic diagrams are used to explain or predict outcomes of crosses. They are drawn in a set manner that should always include all steps. Marks are allocated for showing the steps. Don't try and jump straight to the answer. It is much safer to use the Punnett square, especially when you are asked to predict possibilities.

Biology 2

Chapter 4 How do our bodies pass on characteristics?

Fortunately for Mendel, when he chose his crosses, the characteristics were controlled by one gene with a pair of alternative alleles. There are not many human characteristics controlled by one gene. Also, Mendel was able to obtain many offspring from his crosses of pea flowers. It is harder to follow the pattern of inheritance in humans because families are not big enough to show all possible combinations. However, by examining a family tree over several generations patterns can be seen.

> **Test yourself**
>
> **15 a** Set out a genetic diagram for the cross in Figure 4.12 to show the possible offspring when two heterozygous red pea flowers are cross pollinated.
> **b** From your diagram, explain in words how two red flowering pea plants can produce a seed that grows into a white flowering plant.
> **c** If you collected 400 seeds and sowed them, how many white flowering plants would you expect (assuming you are green-fingered and had 100% success)?
>
> **16** Even though Mendel did not know about the nature of genes he helped to develop our current understanding of inheritance. Explain how he did this.

> **Worked example**
>
> The cross in the previous worked example is straightforward, but what if we look at the possible children of two tasters who have already had one non-taster child? (The fact that the child is a non-taster gives us the child's genotype, which must be tt, i.e. homozygous recessive. It must receive a recessive allele from each parent — so both parents must be heterozygous.) To do this by an error-free method we use a **Punnett square**. This enables us to look at the cross and calculate the probability of inheriting a characteristic based on the alleles that the parents carry and all possible combinations. To construct a Punnett square simply draw a grid of four squares. Write the alleles of one parent above the squares at the top and the other parent at the left side. The combinations of pairs of alleles give the offspring alleles. Follow the next diagram (Figure 4.14).
>
> | | man | wife |
> |---|---|---|
> | phenotype | taster | non-taster |
> | genotype | Tt | Tt |
> | gametes | T t | T t |
>
> | | T | t |
> |---|---|---|
> | T | TT | Tt |
> | t | Tt | tt |
>
> **Figure 4.14**
>
> Use the same letters. The combinations of alleles in the zygotes (children) are shown in the Punnett square. Each square represents one-quarter of the children, so we have:
> - 75% tasters (25% TT and 50% Tt)
> - 25% non-tasters, tt
>
> So using this we could predict that there is a 3:1 chance that their next child will be a taster, but within the tasters some will be homozygous and some heterozygous. Expressed another way, there is only a 25% chance that their next child will be a non-taster.

B2 4.5 Family trees

Learning outcomes

- Be able to interpret family trees.
- Explain how simple characteristics and some diseases are inherited.

In a family tree:
- squares represent males and circles females
- individuals that are coloured in have the trait
- each horizontal row represents one generation and these are labelled I, II, III, IV, etc.
- two individuals joined by a horizontal line are parents and their children are below

When trying to work through a family tree, always start off by filling in the alleles you know — in this case all those who are shaded in have the trait and so are tt. You can also asume that:
- if an individual is tt, both of his/her parents must have a recessive gene so must be Tt if they are not tt
- if children of Q are all tasters, it would seem that Q could be TT

68 AQA GCSE Additional Science

B2 4.5 Family trees

In generation III we have two tasters, producing a non-taster daughter in IV. What alleles are both parents carrying?

Test yourself

17 Draw a genetic diagram (set out as shown in Figure 4.13) to work out the alleles in the offspring in generation II of Figure 4.15.

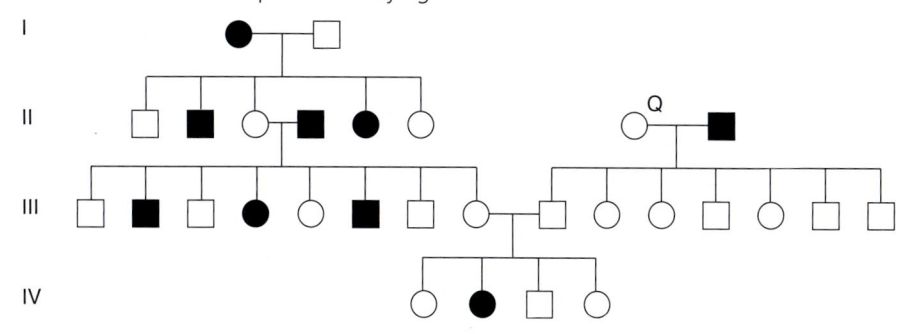

Figure 4.15 Here is the pedigree of a family showing the trait 'inability to taste PTC'.

How are some diseases inherited?

Some diseases are inherited because a defective gene from the parent is passed to their offspring. A defective gene is one that usually codes for a protein but has suffered a mutation. As a result of the mutation the order of bases has changed in the gene. This means that the code for making a specific protein is no longer correct (revise Chapter 1 if necessary). The lack of this protein causes the symptoms of the disease.

Cystic fibrosis

Cystic fibrosis is a disease that affects membranes of cells that line the lungs, gut and reproductive tract. It is caused by a recessive allele and therefore has symptomless carriers.

A **carrier** has one allele for an inherited disease caused by a recessive allele but does not have the disease because there is also a dominant gene which hides it. Two carriers can have children who inherit the disease.

Polydactyly is an inherited condition that results in an extra digit (finger or toe). It is caused by a dominant allele.

Cystic fibrosis (CF) affects more than 8500 people in the UK. CF causes problems in internal organs, especially the lungs and digestive system, by clogging them with thick sticky mucus. This makes it hard to breathe and digest food. CF makes it difficult for a sufferer to absorb enough oxygen in the lungs. It is caused by a recessive allele. A person who inherits two recessive alleles will suffer from CF.

This means that a parent can have the allele for this disease without actually suffering from the disease. In this case they would have one dominant and one recessive allele. Over two million people in the UK carry the faulty gene that causes cystic fibrosis — that is around 1 in 25 of the population. A person with this pair of alleles is called a symptomless **carrier**. Figure 4.16 shows a family tree that has the cystic fibrosis allele in it. The normal allele is shown as CF, and the recessive allele is shown by cf (NB: keep the two letters together).

Cystic fibrosis sufferers are homozygous recessive and have the genotype cfcf; carriers are heterozygous and have the genotype CFcf.

Test yourself

18 a Name three people in the family tree in Figure 4.16 who are cystic fibrosis carriers.
 b Name one person who suffers from the disease.
 c Explain why Tim is unaffected by the disease.

19 Using Figure 4.16, draw a genetic diagram to show the possible outcomes of Claire having children with somebody who is a carrier of cystic fibrosis.

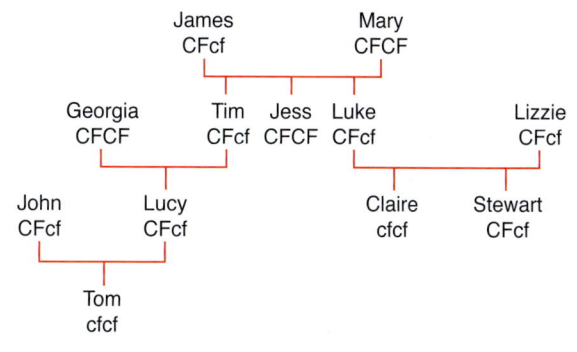

Figure 4.16 In this family the cystic fibrosis allele, cf, has passed through the generations.

Biology 2

Chapter 4 How do our bodies pass on characteristics?

Polydactyly

This is an inherited condition resulting from a dominant allele. That means that any individual inheriting a single dominant allele from the parents will have the condition, which results in six fingers or toes. This condition is rare.

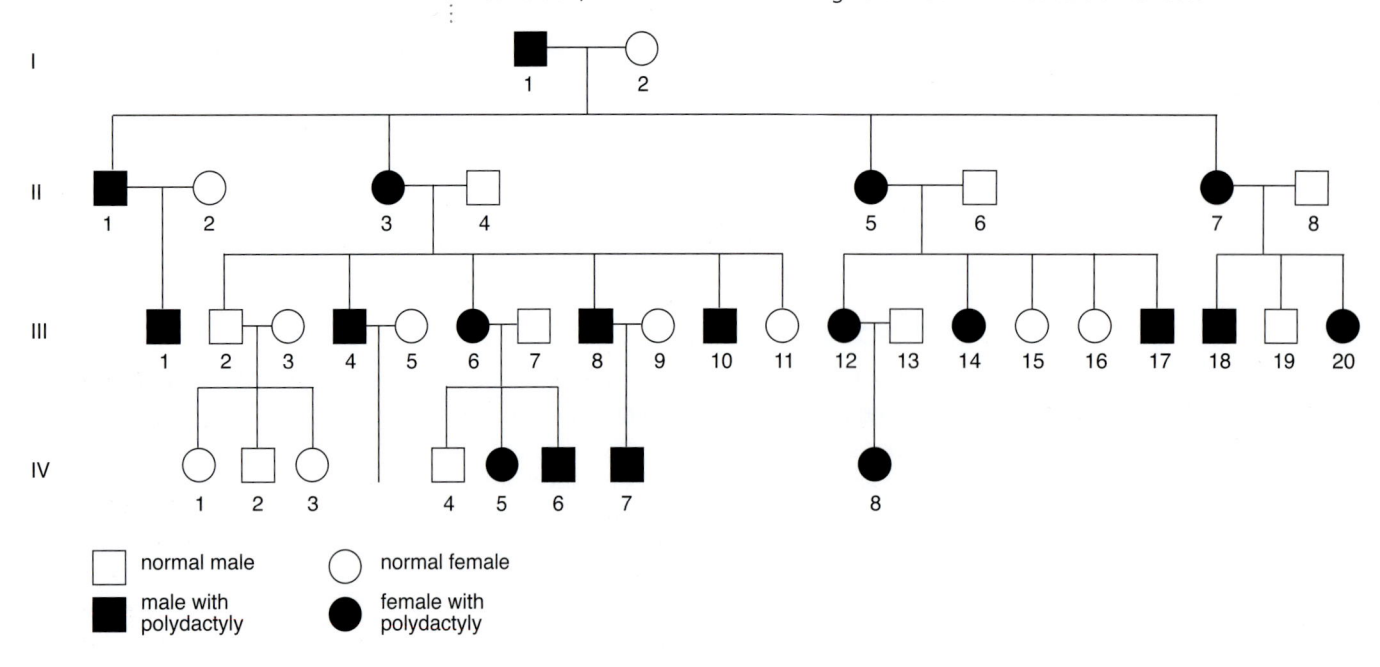

Figure 4.17 Polydactyly family tree.

Figure 4.17 shows a family tree of four generations.

Count how many males, and how many females have an extra digit.

On Figure 4.18 count the fingers and toes here.

Test yourself

20 In Figure 4.17, individuals 2, 4, 6, 8 in generation II are from the general population and are not related. (a) What alleles do they have for finger number? Let P represent the gene for polydactyly and p the normal allele.

21 If individuals 7 and 8 in generation IV (Figure 4.17) were to marry, what chance would they have of having a child with the normal number of fingers and toes? Set out a genetic cross as in Figure 4.13 to explain this. Show every step of working.

Genetic survey
HOW SCIENCE WORKS PRACTICAL SKILLS

Research 'monohybrid inheritance in humans' (i.e. a characteristic controlled by one pair of alleles). Select an easy-to-recognise genetically inherited feature that is controlled by one pair of alleles. Ask your friends about their parents, grandparents, brothers and sisters and try to determine if the gene is dominant or recessive. A good example would be tongue rolling. If you were to consider making this into a large-scale survey, how many results do you think would be necessary to form a valid conclusion?

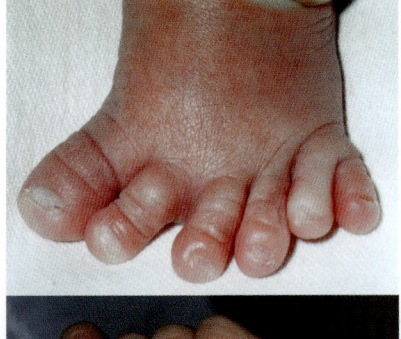

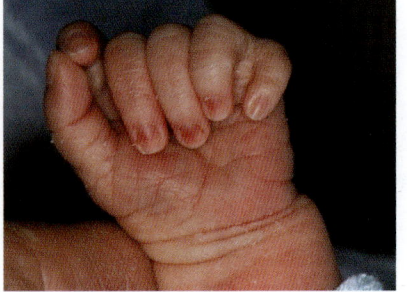

Figure 4.18 Polydactyly

70 AQA GCSE Additional Science

B2 4.6

Learning outcomes

- Know that stem cells are found in embryos and adult bone marrow.
- Evaluate the social and ethical concerns about the uses of stem cells.
- Explain why embryos might be screened.

Differentiation is the process by which an unspecialised early embryonic cell develops the shape of a specialised cell.
Undifferentiated cells have no specialised features. Stem cells are a type of undifferentiated cell.

Figure 4.19 Embryonic stem cells, like those shown here, can differentiate into completely different types of cell. Do these cells have any specialised features?

Stem cells and specialised cells

Figure 4.19 shows some human stem cells. There are two groups of stem cells: embryonic stem cells (ESC) and adult stem cells. Embryonic stem cells are found in the embryo soon after an egg cell has been fertilised. Embryonic stem cells **differentiate** into all the different specialised cells that eventually form an embryo and then a baby. They look like normal cells, so why are they so important? Stem cells can carry out one function that other cells cannot. When stem cells divide they can make specialised cells.

Embryonic stem cells are **undifferentiated**, which means that they have no specialised features to carry out a specific function. We each started as a single cell so we know that embryonic stem cells can form all types of cells. That cell divided and formed a ball of cells. From the ball of undifferentiated cells, many types of specialised cells and tissues formed such as liver, brain or heart muscle. It is still not understood exactly how they do this but, under the right conditions and with the right cell chemicals, embryonic stem cells can develop into all other types of cells.

Adult stem cells are found within organs and some body tissue, such as bone marrow. The function of adult stem cells is to repair the tissues in which they are found, so they can only make the type of specialised cells in that tissue. For example, stem cells in the pancreas produce pancreas cells that are specialised to make insulin. Nerve cells have a specialised structure that allows them to carry impulses around the body. Nerve, muscle, reproductive cells and most types of animal cell differentiate at an early stage of embryo development and then divide to produce only that type of specialised cell.

Social and ethical issues concerning the use of stem cells from embryos and embryo screening

Ethics is a thinking process that helps us to decide what is right and wrong. Your ethical thinking may be influenced by the things that were discussed by your parents, Christian or other religious teaching or your desire to help other people. The thinking process is helped by asking the right questions and the decisions are yours alone but you should be able to justify your opinion.

With regard to the ethics of embryonic stem cell (ESC) use, you might start by considering the following questions.

1. When does life begin? Is it at fertilisation? That single cell has all the genetic code it requires to form the adult but requires the correct conditions in the womb of the mother to do so. Or does life begin at birth? No one would condone killing a newborn baby because it was not 'wanted'. So if you work backwards through the lifetime of the baby/foetus/embryo, you need to decide if it is right to destroy a potential life.
2. Where have the embryos come from? Currently there are embryos 'left-over' from IVF treatment that are no longer needed or have come to the end of the period when it is considered safe to use them. These are being used for embryo research rather than being just 'flushed away'.
3. Should a woman receive medical treatment to produce many eggs which can be fertilised by IVF to make embryos for research and be paid for it? Currently this is not legal in the UK.

Biology 2

Chapter 4 How do our bodies pass on characteristics?

> **Boost your grade** ✓
>
> When asked about social or ethical issues in an examination you are expected to give a reason for your answer, which might be scientific or moral. You may even be asked to give an opinion 'for or against'. The statement 'It's not right' is not worth any marks: you must include a 'because' statement explaining the harm, damage, etc. You may need to state that other people hold the opposite opinion and justify their views. The specification words are 'make informed judgement'.

> **Therapeutic** means it is used in the treatment or cure of a disease.

4 What could be the outcomes of embryo research? The aim of the researchers is to find treatments for some serious debilitating diseases such as Parkinson's disease, motor neurone disease and also as a treatment for spinal injuries causing paralysis. No one doubts the sincerity of the researchers but does their aim warrant the destruction of embryos? Anyone with a loved one who is suffering is hopeful of a cure or control of symptoms.

5 Is there an alternative to ESCs? Could adult stem cells be used? As stated above, adult stem cells are currently limited in their use because they can only form one type of cells. However, if embryonic stem cell research leads to discovering how adult stem cells could be 'programmed' to differentiate into other cell types and how they could be activated to reproduce normally, this would then enable adult stem cells to be used instead. An advantage of adult stem cell use would be that if taken from the person needing treatment they would have the same genetic information and would not be rejected. Adult stem cells are already harvested and used for stem cell transplants for some forms of blood cancer. So the question becomes should we allow ESC research in the expectation that it will lead to the development of adult stem cell technology for use in the future?

6 Is there a risk of stem cells reproducing out of control and producing cancer? This was a real problem discovered in some research so more investigation is needed to find out about the control of cell division of stem cells.

More research has to be done before **therapeutic** stem cell transplantation becomes available. There are still some major problems to overcome. The stem cells must be able to divide and form a clone of a sufficient number of cells. The transplanted cells or tissue must be able to survive in the recipient and integrate with surrounding tissues of the organ without causing any damage or being rejected, and the transplant must be able to last as long as a normal organ. So far both embryonic and adult stem cells have been developed to form insulin-secreting cells in the laboratory, there is experimentation into multiple sclerosis and remaking heart muscle. There have been isolated experiments on humans using stem cells. However, the idea that stem cells can result in a wide range of transplantable tissues to cure a range of human diseases is still a long way off. You will see progress reported in news headlines.

Embryo screening

In the UK any embryonic screening technique has to be approved by HFEA (Human Fertilisation and Embryology Authority). Embryonic screening is used to help couples with a family history of a genetic abnormality with a high risk of transmission to their children. The technique involves a process similar to IVF (in vitro fertilisation). Multiple embryos are created and then a cell is removed from each to test for a faulty allele or a tissue match. The embryos are screened to find the ones not carrying the genetic disorders and only embryos that are free of the faulty allele are implanted into the woman's womb. One risk is that even at the 'ball of cells' stage, not every cell is identical. IVF is often used when a couple cannot have children naturally. Embryo screening is only used as part of IVF and only offered to parents with a family history of inherited disorders. It is a costly procedure.

Sometimes a parent who has a child with a serious medical problem decides to have a 'saviour sibling'. Using IVF, multiple embryos are made. If a tissue match is found between the sick child and one of the embryos, that embryo is implanted

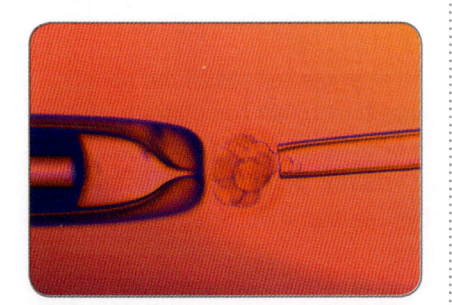

Figure 4.20 Removing a cell for embryo screening.

B2 4.6 Stem cells and specialised cells

and the other embryos are destroyed. At birth the umbilical cord blood can be used for a stem cell transplant to the sick child, or later the bone marrow.

The HFEA receives many requests to extend the embryo screening technique to other disorders. Some genes are linked to various cancers. These genes do not cause these cancers directly. They only increase the chance of developing cancer, so not all people with the gene will get the disease. Embryo screening for genes linked to an increased cancer risk is permitted. However some people are concerned that in future the technique may be extended to allow parents to screen embryos for non-medical reasons. Embryo selection is not authorised in the UK for the purposes of gender selection or 'designer babies'.

These issues raise many questions and we see that science and ethics cannot be separated. As science pushes forward the boundaries with new technologies, it becomes even more important to think about the long-term consequences and put safeguarding legislation in place. Individual choice is part of modern life, but are some of these experiments at risk of dehumanising us and using humans as commodities?

Current stem cell research involves manipulating embryonic stem cells to try to produce specialised cells. Hypothetically, these new specialised cells could replace cells that have been damaged either by disease or by an accident. Currently only small amounts of embryonic stem cells can be created using this technique, and scientists have not yet been able make them grow to create specialised cells. However, scientists believe that one day it may be possible to grow entire organs using this process.

Stem cells may one day be used to treat diseases such as Parkinson's disease. Patients with Parkinson's disease have nerve cells that are damaged. As more and more nerve cells become damaged, the patient suffers from progressive paralysis. Stem cells could be manipulated in the laboratory to produce healthy nerve cells to replace those that are damaged in the Parkinson's patient.

Opinions on stem cell research are mixed. Some people support stem cell research as they feel it could eventually give patients with previously incurable diseases a better quality of life. Other people argue against the use of stem cell research. They insist that researchers should not be allowed to experiment on embryos since these 'balls of cells' have the capacity to develop into a complete human being.

Test yourself

22 It costs about £2000 to carry out IVF with embryo screening, which is twice as much as it costs without screening. With this in mind do you think that the use of embryo screening should be extended to include all births conceived by IVF? Give reasons for your answer.

23 Can you suggest why some people object to embryo screening, apart from financial reasons?

Chapter 4 How do our bodies pass on characteristics?

Stem cell research in the news
HOW SCIENCE WORKS ACTIVITY

Stem cell treatment from cord-blood for blood disorders — acute myeloid leukaemia (2010)

Stem cell experiment lets diabetics forgo insulin

Risky transplants performed on 13 young diabetics in Brazil (2007)

Doctors have launched a trial to test whether heart disease can be treated using a patient's own stem cells (2005)

Stem-cell transplants from own bone marrow may control and even reverse multiple sclerosis symptoms if done early enough, a small study has suggested (2009)

Figure 4.21

These are just a few recent headlines from articles about experimental research that focuses on human stem cells. Of all the current biomedical research, stem cell research has probably attracted more attention than any other from the media. But why is this?

To understand the level of interest and range of views expressed, it is crucial to know exactly what happens when stem cells divide and the potential medical benefits these cells could bring.

1. Explain why stem cells are essential in human development.
2. Nerve cells, red blood cells, muscle cells and pancreatic cells are some of the specialised cells that stem cells produce. Each one has a specific structure that allows it to carry out its function. Find out and state the function of each of these specialised cells. Explain how each cell's structure enables it to carry out its function.
3. Stem cells can be collected and kept alive outside the body. With this in mind, suggest two possible uses scientists could make of stem cells. For each suggestion explain how you think scientists would carry out the procedure. Put the main points into a flow diagram.
4. In small groups, debate reasons that people may have to support or oppose embryonic stem cell research. Write these down as a list of reasons for, and a list of reasons against the research.
5. Now put the reasons in each list into order based upon which ones present the strongest arguments for and against stem cell research.
6. Finally, write a paragraph expressing your views about this issue. Support your views with reasons.

B2 4.7 How organisms change through time

Learning outcomes

- Know that new species can be formed by geographical isolation.
- Understand how fossils are evidence for evolution.
- Understand the different ways in which fossils may be formed.

Have you ever found a fossil and wondered just how old it was? This leads to thoughts of how life began on Earth. Scientists have many hypotheses. Overall, though, there is little valid and reproducible evidence. It is thought that the first life forms were bacteria and unicells, followed by soft-bodied animals. Fossils of soft-bodied animals are rarely found because they are usually eaten or decay quickly. But in May 2010, 1500 soft-bodied marine animal fossils of several sorts were found in Morocco. These well-preserved fossils date back to nearly 500 million years ago and fill in a lot of gaps in the record.

How were fossils formed?

When an organism dies, it is normally eaten by other animals or decomposed by bacteria and fungi quickly, leaving nothing to see except possibly a few bones.

B2 4.7 How organisms change through time

Normal fossil finds are from the hard parts of organisms such as bones and shells. Fossils were formed when an organism fell into a marsh or bog in which the conditions were not suitable for decay — that is, no oxygen and water with a low pH. The remains were buried under layers of vegetation and became compressed with age. (Many human remains from the Iron Age have been found in peat bogs. Although not fossils, they show how effective these conditions are for preservation.) Alternatively, fossils can also form when organisms have been covered with layers of sand, volcanic ash or silt. They become compressed and impregnated with mineral salts from water, turning them into stone. There are other interesting remains such as insects preserved in tree resin called amber, dinosaurs' bones, footprints and eggs, petrified trees, remains in ice such as mammoths and burrows in mud deposits.

> **Test yourself**
>
> 24 Explain three ways in which fossils were formed.
> 25 You are examining the layers in a chalk cliff. Where will the oldest fossils be found?

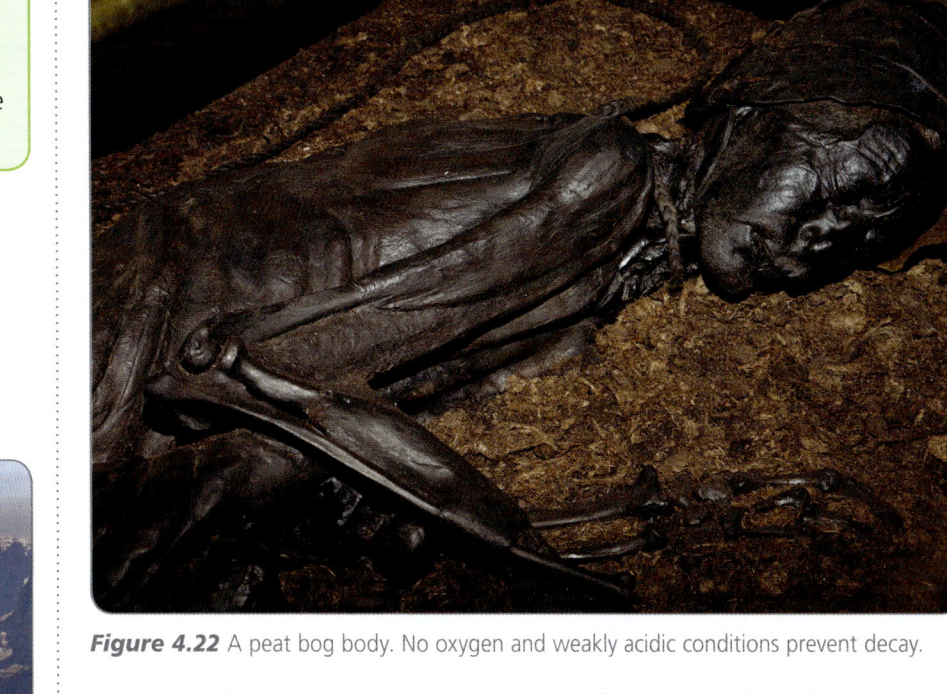

Figure 4.22 A peat bog body. No oxygen and weakly acidic conditions prevent decay.

What evidence do scientists get from the fossil record that can be related to evolutionary theory?

Scientists study fossils to find out about the animals and plants of the past. Fossils can show the time sequence of existence. In sedimentary rocks the material is laid down in layers so the deeper you go down, the older the material and the older the fossils. The Grand Canyon in Arizona has a depth of 1700 m and this represents a time span of 500 million years. Geologists can trace changes in climate and rock movements and fossil experts can follow changes in shape and structure of plants and animals that could result from climate changes.

Evidence from fossils

There is a definite order in which fossils appear.
1 The bottom layers (oldest rocks) show only a small variety of marine invertebrates and represent several million years without much change.

Figure 4.23 The Grand Canyon, Arizona. A 500 million year time sequence in the rock layers.

Biology 2

Chapter 4 How do our bodies pass on characteristics?

Figure 4.24 A trilobite fossil. The hand acts as a scale object.

Figure 4.25 This is a photograph of a coelacanth with lobed fins occasionally found by some deep sea fishermen. Species similar to this with fin-limbs were believed to have become extinct 65 million years ago.

Figure 4.26 *Archaeopteryx* fossil — how many bird and reptile features can you find?

2 Marine invertebrates, including trilobites (some as large as 40 cm), crustaceans, molluscs, primitive starfish and seaweed appear first.
3 The first land plants were dependent on water for support and were more like ferns.
4 Wingless insects spiders and crustaceans such as woodlice have been found in rocks 350 million years old.
5 The first fish were covered with bony plates; more recent fossil fish had scales. Some fish had limb-like fins that seem to be a transition towards amphibians. Sharks were around.
6 Winged insects appear next.
7 The rock layers show that there was more dry land and reptiles evolved. Their teeth show that there were both herbivores and carnivores. Fossils of turtles and primitive crocodiles were followed by dinosaurs.
8 Reptiles evolved into birds — the first had reptile-like teeth and feathers, solid bones and jointed tail bones. A gap in this part of the fossil record was filled by *Archaeopteryx* — the link form between reptiles and birds.
9 Small mammals, flowering plants and larger mammals bring us up-to-date.

The presence of a fossil in one layer and not in the one above shows that that species must have become extinct. Extinction may be caused by environmental changes and geologists can provide information about atmospheric changes and the amount of water around. These changes would be over millions of years, that is, geological time. However, some changes may have been sudden. In 2009, scientists at Leeds University found evidence of a massive volcanic eruption in China, which caused a mass extinction 260 million years ago. The eruption was in shallow sea with the lava forming a layer of rock above a layer of sedimentary rock. Further sedimentary layers were deposited from the sea on top of the volcanic lava. The layer below the lava rock has marine life and the layer above does not. The mass extinction resulted from release of sulfur dioxide and acid rain, together with heavy cloud causing low temperatures. It is suggested that collision of the Earth with an asteroid about 65 million years ago could have led to the extinction of the dinosaurs. Extinction of single species may be caused by new predators, new competitors or new diseases, as discussed in Chapter 3.

Fossils show how much or how little organisms have changed. Fossils of crocodile jaws and teeth show that they have evolved little over the last 200 million years (Figure 4.27). In comparison, the earliest fossils of our human

Figure 4.27 (a) A 110 million year old fossilised skull of the gigantic cousin of the modern crocodile compared with the 50 cm adult skull of a modern crocodile. (b) Jaw and skull of *Orrorin tugenensis* from 6 million years ago.

AQA GCSE Additional Science

B2 4.7 How organisms change through time

ancestors, such as *Orrorin tugenensis* (6 million years ago), which was the size of a chimpanzee, show that humans have evolved a great deal in just 6 million years. Evolution from *Homo habilis* to *Homo sapiens* has been about 2 million years (Figure 4.28).

Fossil records show how the modern whale could have evolved from a land-based mammal (Figure 4.29). The original land-based mammal from which whales have evolved was called Pakicetidae, about the size of a wolf. Millions of years ago Pakicetidae started to go into water to find a more abundant source of food. Within any population of the animals, some would be better swimmers.

These animals may have had wider feet or have been more streamlined. As a result, these individuals could catch more food and were more likely to survive and reproduce. The genes that controlled these features would be passed on to their offspring, who would inherit wider feet and more streamlined bodies. These individuals would be the best swimmers in the next generation. Over millions of generations of gradual change, it is possible that Pakicetidae was transformed into the modern whale. Of course, this is only a theory, but it is supported by fossil records.

Figure 4.28 *Homo habilis. Homo sapiens* has shown rapid evolution in the last 2 million years.

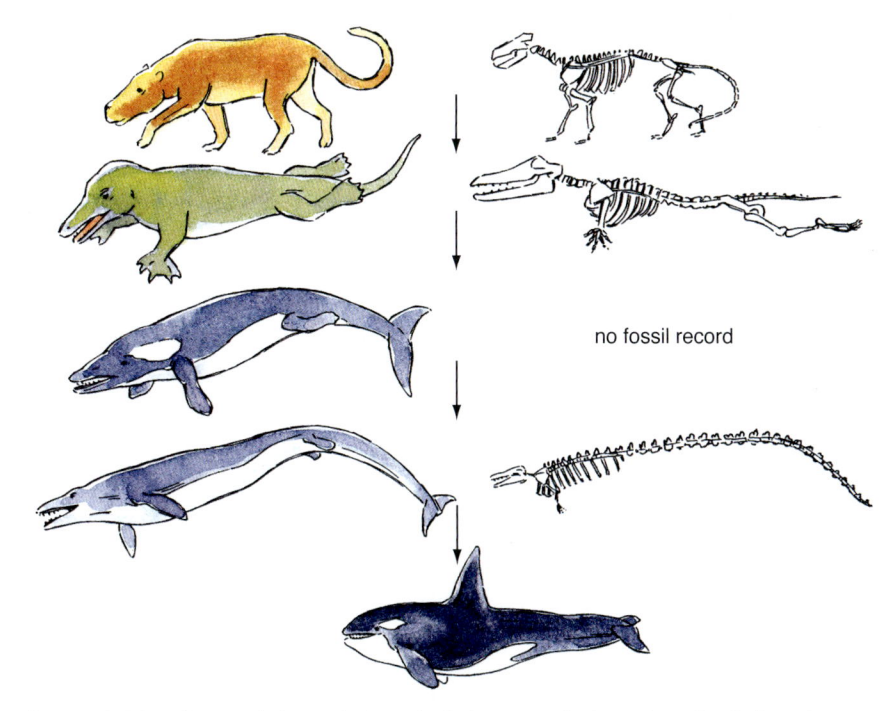

Figure 4.29 A theory of the evolution of whales as worked out from fossils found.

Test yourself

26 Fossil records for the evolution of whales provide evidence to support the theory of evolution, but they do not prove it. Explain why this is so.

27 Suggest some reasons why species have become extinct.

28 What information about life on Earth can be obtained by studying fossils?

29 Why is it important to have a series of related fossils when studying evolution?

Is the theory of evolution just a theory?

HOW SCIENCE WORKS / ICT

Open an online forum inviting contributions arguing in support of the theory and others arguing against it. Remind contributors that they must support their point of view with evidence. It is necessary to distinguish between opinion based on valid and reproducible evidence and opinion based on non-scientific ideas or prejudices.

Biology 2

Chapter 4   How do our bodies pass on characteristics?

B2 4.8

Learning outcomes

Understand how new species arise.

> The **gene pool** is all the alleles present in a breeding population.

New species can arise as a result of geographical isolation

Assume we have a population of animals such as banded snails. The colour banding is a genetic characteristic. A population is the total number of organisms of a species living in one ecosystem. This population will breed together and share one **gene pool**, which will contain all the alternative alleles in the population. If a new railway development or a new canal or motorway cut the population into two parts, what would happen? Each gene pool will now be smaller and may contain different proportions of alleles. The environment in the two parts may be different as a result of the development. We now have two separate populations.

Snail trail

HOW SCIENCE WORKS
ICT

Carry out a web search for '*Cepaea* colour and banding patterns'. This will help you work through the following section.

There is a good population of thrushes in both areas. The snails breed and the thrushes feed upon the snails that they can see most easily. Within the snail population there are many different alleles for colour. (Previously we have worked with only two alternative alleles at one position on a chromosome, here we have more. There are also genes for width of banding which will change the snail phenotype.) On the eastern side the vegetation is moist and green. More yellow and pink snails are caught and eaten but fewer green ones. On the western side the habitat is much drier and sandier with short vegetation. The green snails are eaten and the yellow and pink ones are not so easily found. After many generations, as a result of the selection by the thrushes, green snails become more common in the eastern population, so the green allele has increased as a result of natural selection. Green snails are camouflaged against the lush green vegetation. On the western side the paler yellow and pink banding alleles become more common and the darker green alleles are now less common.

Darwin observed the finches on the Galapagos Islands. They originated from Ecuador where there was a large gene pool containing many different alleles. The few that arrived at the Galapagos Islands did not have all the alleles and interbred. As a result of natural selection they became adapted to different food types and occupied different feeding areas. The different feeders became isolated, they became different sizes, may have developed different songs and displays. The result was that they did not cross-breed and became separate species and built up separate populations on the different islands.

Any natural or man-made barrier that divides a population into two parts that cannot mix, can result in a new species being formed after many generations. As a result of inbreeding in the two populations the proportions of the different alleles change. As a result of natural selection the populations begin to look different. They may become so different that if they were to be re-united the two new populations would not be able to interbreed.

Figure 4.30 Banded snails showing a wide variation in colour.

AQA GCSE Additional Science

Homework questions

1 Humans may produce two types of earwax: a wet, sticky type or a dry, hard type. The wet type is caused by a dominant allele, E, and the dry type is caused by a recessive allele, e.
   a Write down the pairs of alleles that somebody with wet, sticky earwax may have.
   b Explain why it is impossible to say what alleles a person who produces wet earwax will have.

Rob and Jo are married. Rob has wet earwax and Jo has dry earwax. They have a child who produces wet earwax.
   c What pair of alleles for earwax will their child have?
   d Rob and Jo are planning on having another child. Draw a genetic diagram to show what possible pairs of alleles for earwax their next child could inherit.

2 When a body builder trains, his muscles get bigger because more muscle cells are produced.
   a What type of cell division produces the new muscle cells?
   b What type of cell division produces his sperm cells?
   c Explain, in as much detail as you can, why the body needs two different types of cell division.

3 In a breeding experiment with peas Mendel crossed pure breeding plants that had green pods with pure breeding plants that had yellow pods. All the offspring of this cross had green pods.
   a How would he have done this cross?
   b What does this tell you about the alleles for green and yellow pods?

He allowed this generation to self-fertilise, he collected 580 seeds and grew them.
   c Draw a genetic diagram to show this cross.
   d How many of these plants would you expect to produce green pods and how many yellow pods?

4 Saviour siblings are children created using IVF with the primary aim of providing spare-part tissue for an older brother or sister.

During a debate on the HFE Bill in the House of Lords, Lady O'Cathain spoke against the creation of saviour siblings. She said: 'To manufacture a person in this way is to offend against the respect that is due to the integrity of that person, no matter how compelling the goal of trying to cure'.

Lord Tebbit said at the time: 'Because it might bring great benefits to particular people does not mean it should be done. If we accept arguments of that kind we are effectively saying that the end justifies the means'.
   a Write a paragraph explaining how you would feel if you discovered that you were a laboratory-created child for the purpose of saving the life of an older brother or sister? Compare your answer with those of other members of your group.
   b Have an online debate about this with half the class arguing for allowing 'saviour siblings' in order to save themselves from a genetic disease that they or a sibling have, and the other half arguing against it on ethical grounds.

Biology 2

Chapter 4 Exam corner

Exam corner

A young married couple, both normal, are delighted to have a baby who is normal, but their second child has cystic fibrosis.

a Explain to them how this has occurred. *(2 marks)*

They would like another child.

b Show, using a genetic diagram, the chances that the third child will have cystic fibrosis. *(3 marks)*

Student A

a The parents have the genes CF and cf and so does the first child. ✓

A second child must have the same as it has the same parents.

b Because they have one of each it would be 50/50.

Examiner comment Student A has given an ambiguous answer that must be taken that the alleles in all three are CFcf, and so is given the mark. (b) Shows a lack of understanding. This is a common form of question so it is well worth learning how to do it correctly. Student A has only achieved 1 mark for the whole question.

Student B

a Both parents are carriers ✓ of the recessive allele for cystic fibrosis. They are CFcf.

The first child inherits at least 1 dominant allele ✓ and so is OK. The second child receives two recessive alleles — one from each parent and so suffers from cystic fibrosis. ✓

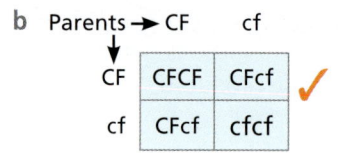

b Parents → CF   cf

	CF	cf
CF	CFCF	CFcf
cf	CFcf	cfcf

✓

Offspring: 25% CFCF normal
50% CFcf normal but carriers
25% cfcf suffer from cystic fibrosis ✓

So from this there is a 25% chance that the next child will have cystic fibrosis. ✓

Examiner comment Student B has given a good answer for (a). Use of the word 'carrier' indicates an understanding and this is backed up by giving the alleles in the parents and explanation of why the first child is normal. For part (b) 3 marks are awarded: the Punnett square is correctly set out, the results are analysed and finally there is a statement which answers the question. Too many candidates leave out this final statement.

AQA GCSE Additional Science

Biology 2

Exam questions

Chapter 1

1. a Name the labelled part of the cell in Figure 1A that releases energy from glucose. What is the name of this process? **(2 marks)**
   b Name the part of the cell that synthesises enzymes. **(1 mark)**
   c A nerve cell is a specialised cell that carries nerve impulses throughout the body.
      (i) Describe one feature of a nerve cell that allows it to carry out this function. **(1 mark)**
      (ii) Explain how this feature allows a nerve cell to carry out this function. **(1 mark)**

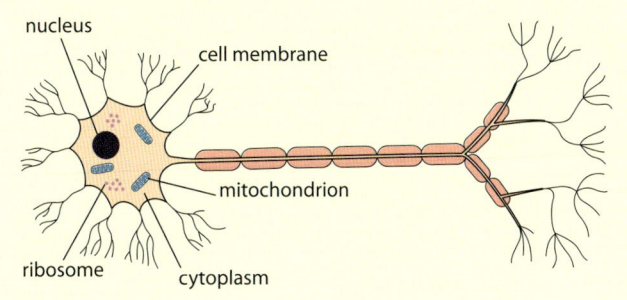

Figure 1A

2. Examine the diagram of an alveolus in the lung and a small capillary next to it (Figure 1B).
   a The blood is transporting carbon dioxide. Draw an arrow to show the direction of movement of the carbon dioxide. **(1 mark)**
   b Use the diagram to explain why the lungs are an efficient exchange surface. **(3 marks)**
   c Oxygen is used in cells for respiration. Draw a simple diagram of the part of a cell that carries out respiration. **(1 mark)**

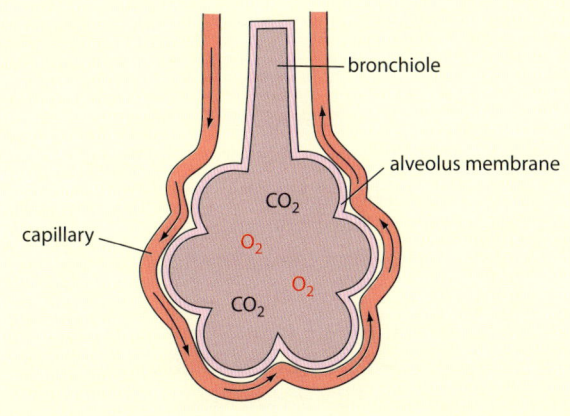

Figure 1B

   d Name a type of cell that makes protein. What part of a cell will be in large numbers in the cell you have named? **(1 mark)**
   e What role does the cell nucleus play in making proteins? **(1 mark)**

Chapter 2

1. A glasshouse owner wishes to grow many successive crops of lettuce and harvest the crops all year round.
   a Examine the computer control panel in Figure 2A and decide which factors are limiting the productivity. **(3 marks)**
   b Suggest suitable new settings. **(3 marks)**
   c The grower decides to install a paraffin heater. What problems might this cause? **(2 marks)**

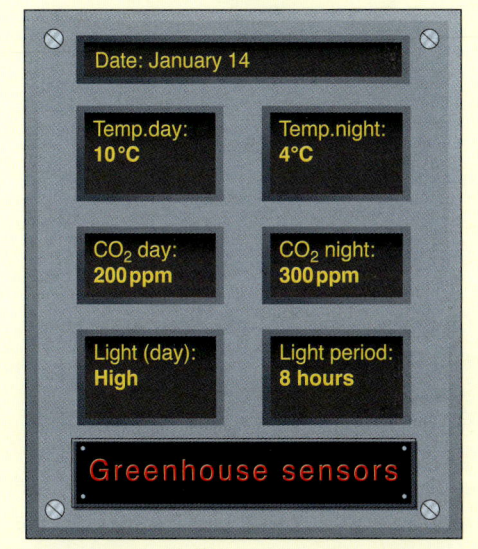

Figure 2A

2. Examine Figure 2B, which shows how the rate of photosynthesis changes with increasing light intensity at two different carbon dioxide concentrations.

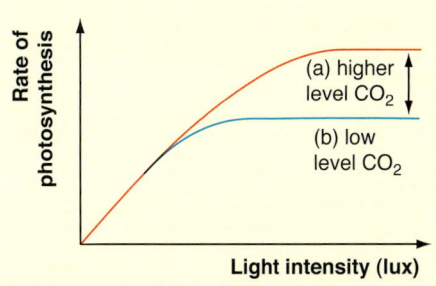

Figure 2B

Biology 2

a At low light intensity, what is the limiting factor? (1 mark)
b The rates of photosynthesis level out at a certain light intensity. Explain why? (1 mark)
c Why does the rate at low CO_2 level out at a lower light intensity than at a higher CO_2 level? (1 mark)

A variegated leaf was set up as shown in Figure 2C. It was placed in the dark for 24 hours and then into bright light for 6 hours.

Figure 2C

(labels: white, green, black stencil)

d Describe how you would test the leaf for the presence of starch. Give a reason for each step. (4 marks)
e Draw the result you would expect on the diagram and label the colours. (1 mark)

3 A student is planning a trial to find the best concentration at which to apply a liquid fertiliser. The recommended strength is 20 cm³ per 10 dm³ — which the student calls 100% strength. He decides to try the first concentration at double the recommended strength, i.e. 200%, and then a range of other concentrations. The dilutions are recorded in the table below (1 dm³ = 1000 cm³).

Volume concentrated solution in cm³	Volume water in cm³	Final concentration in %
40	9,960	200
20	9,980	100
10	9,990	
5	9,995	
2.5	9,997.5	
0	10,000	

a Complete the table to find his dilutions. (4 marks)
b Do you think the range of percentage concentrations chosen was suitable? Give a reason for your answer. (2 marks)
c The student's idea was to harvest the plants at the end of the experiment by cutting them off at soil level to find the fresh biomass. His friend suggested that he should harvest the plants and then find the dry biomass. What is the difference between fresh and dry biomass? Which do you think would give the most accurate result, and why? (3 marks)

Chapter 3

1 Pectin in fruit holds the cellulose fibres in the cell wall together. The enzyme pectinase breaks down the pectin and cell walls, releasing the juice.

A student did a laboratory experiment to extract apple juice. She grated the apple to form pulp and weighed out 50.0 g into five separate beakers. The pulp was warmed in water baths and then the pectinase added and left for 15 minutes to react. Juice was collected by straining through fine mesh lining a filter funnel into a measuring cylinder.

The results are shown in the table below.

Water bath temperature in °C	Volume of juice collected in cm³
10	12
25	25
35	50
45	78
60	45
70	12
85	12

a Plot the results on a graph. (3 marks)
b From your graph, suggest the optimum temperature for this enzyme. (1 mark)
c Describe and explain the trend shown by the results. (3 marks)
d Do you think there are enough results to make an accurate prediction of the volume of juice you would expect at 50°C? How could you improve the experiment? (3 marks)

82 AQA GCSE Additional Science

Biology 2

2 When athletes start exercising, their bodies respond.
  a State two of these responses that increase blood flow to the muscles. *(2 marks)*
  b Explain why blood flow to the muscles must increase when a person starts exercising. *(3 marks)*
  c What is the role of glycogen during exercise? *(2 marks)*

3 During vigorous exercise, aerobic respiration may be replaced by anaerobic respiration.
  a State two differences between aerobic and anaerobic respiration. *(2 marks)*
  b What happens to muscles during long periods of vigorous exercise? *(1 mark)*
  c During a period of anaerobic respiration, a runner builds up an oxygen debt. Explain what is meant by the term 'oxygen debt'. *(3 marks)*

4 A student decided to carry out investigations with a low temperature biological washing powder, which had a banner on the box stating 'suitable for delicate coloureds — no bleach added'. He produced a standard dirty piece of material with egg yolk and mayonnaise stains on white cotton. He investigated a range of temperatures between 5°C and 75°C at 10-degree intervals. He found that the washed material was whitest in the range 25°C to 35°C.
  a Name two types of enzymes you would expect to find in biological washing powder. *(2 marks)*
  b Why was the result for the higher temperatures not so good? *(1 mark)*
  c Why do you think that the result at 5°C was rather poor? *(2 marks)*
  d What are the environmental advantages of using a low temperature biological washing powder? *(2 marks)*
  e The instructions on the packet stated that gloves should be worn if clothes are hand-washed. Suggest why. *(1 mark)*

Chapter 4

1 Fur length in rabbits is controlled by a single gene that has two alleles, L and l. A pair of pure-bred rabbits bred together. The mother had long fur and the father had short fur. All their babies had long fur.
  a Give the alleles for fur length present in the body cells of:
    (i) the mother; **(ii)** the father. *(1 mark)*
  b One of the offspring bred with a rabbit with short fur. Use a genetic diagram to show how some of their babies had long fur and some had short fur. *(3 marks)*

2 Examine the section through a cliff (Figure 4A).

Figure 4A

There were many ammonite species found in layer C, which dates back to around 80 million years ago. There were none in the layer above.
  a What conclusion do you think fossil hunters will draw from this finding? *(1 mark)*
  b Explain a possible reason for the findings. *(1 mark)*

Geological evidence suggests that layer A was once under water and had fern-like vegetation growing on raised areas.

  c What conditions favoured fossil formation in bog areas? *(2 marks)*

It has been suggested that the decline of the dinosaurs was as a result of a collision between the Earth and an asteroid (giant meteorite). This collision resulted in a great cloud of dust, which blocked out the sunlight for years.

  d Explain how this might have affected the dinosaurs. *(3 marks)*

Chemistry 2

Chapter 5
Bonding in reactions

Setting the scene

We have understood what atoms are made of for only a hundred years. In this time, we have used these ideas to create millions of new materials — from memory metals that spring back to shape when we sit on our glasses to scratch-resistant coatings for our mobile-phone screens. All these new materials owe their properties to how the atoms that make them share electrons with each other.

ICT

In this chapter you can learn to:
- use computer programs to model electron transfer or sharing
- create an animation about ionic and covalent bonding
- create a resource about recycling plastic
- use the internet to research nanotechnology

Practical work

In this chapter you can learn to:
- make careful observations of crystalline substances and halogens
- look for patterns in melting and boiling point
- use models to help explain the properties of silicon dioxide
- compare the properties of plastics and metals

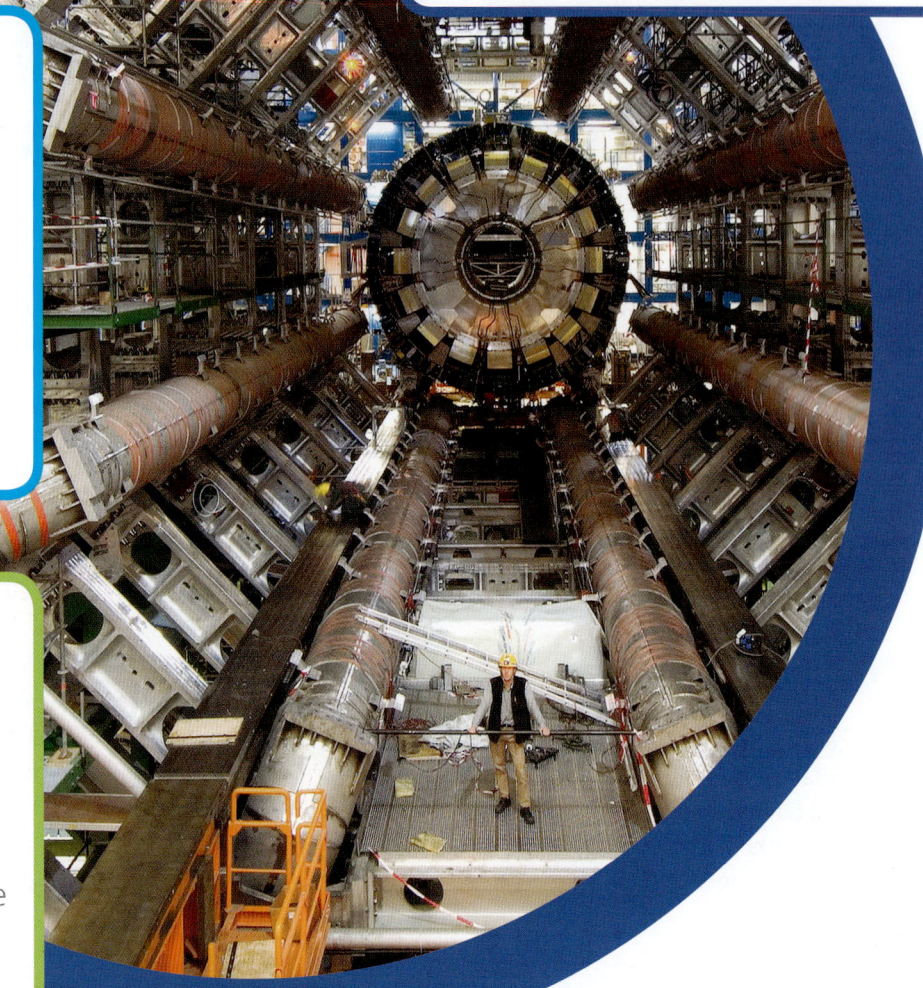

C2 5.1 What happens to electrons when elements react?

C2 5.1
Learning outcomes

- Know that all materials are made from a small number of types of atoms.
- Be able to explain that chemical bonding involves either transferring or sharing electrons in the outer shells in atoms.

Making substances from atoms

Atoms are the basic building blocks of all materials. Of the 118 elements listed in the latest version of the periodic table, only 94 occur naturally. The other 24 (with high atomic numbers) have been created by scientists in tiny quantities in particle accelerators.

Of the 94 elements found naturally, 14 are found in only tiny quantities as a result of the radioactive decay of elements with bigger atoms. This leaves a total of 80 elements that are stable and present in the Earth's crust — and they make up everything we come into contact with. These are the elements with atomic numbers 1–82 with the exception of technetium (43) and promethium (61). These two elements have no stable isotopes. Of these 80 elements, 50 are so scarce that they make up only one-millionth of the mass of the Earth's crust.

So we are only likely to come across 30 elements in our lives. These 30 elements make up virtually all pure substances, and the materials made from those substances. To make these substances, the atoms in their elements combine according to strict rules. The way the atoms combine dictates the properties of the substances and materials made from them. This chapter is about understanding those rules.

Electrons and chemical bonds

Compounds are pure substances made in chemical changes. The particles of a compound are made up of two or more types of atoms joined in a simple ratio. The sharing or transferring of electrons between different types of atoms is what combines the atoms chemically. This **chemical bond** involves only electrons in the outermost electron shells (highest energy levels).

The electrons are shared or transferred to give the atoms the electronic configuration of a noble gas (group 0), with all the shells/energy levels filled up.

> A **chemical bond** holds two atoms together after a chemical reaction.

Test yourself

1. Look at the periodic table on page 283.
   a. Make a list of all the elements for which you can name a use in everyday life.
   b. How many of these are metals?

2. a. Make a list of the electronic structures of the group 0 elements.
   b. What is special about this group of elements?

Boost your grade ✓

Remember that in all bonding the aim is to give each particle a full outer shell of electrons, like the noble gases at the end of each row of the periodic table.

Chemistry 2   85

Chapter 5 Bonding in reactions

C2 5.2 Ions and ionic bonding — transferred electrons

Learning outcomes

- Know that metal atoms only react with non-metals when forming new substances.
- Know that electron transfer results in the formation of positive and negative ions held together by ionic bonds.
- Be able to draw diagrams to show how ions are formed.

Metal elements, like sodium and magnesium, only form compounds with non-metal elements like chlorine and sulfur. Two or more metal elements may combine to make a mixture called an alloy — but this is not a compound or a pure substance (Section 5.7).

Metals and non-metals join chemically to make *ionic* compounds. The metal atoms always lose electrons and form positive ions; the non-metal atoms gain electrons and form negative ions.

The ions are charged particles. They are similar to atoms, but losing or gaining electrons has given them the electronic configuration of a noble gas (with filled outer shells). The basic particles in ionic compounds are ions, not atoms or molecules.

In ionic compounds, the positive and negative ions are separate particles. It is the attraction of the oppositely charged ions that holds these particles together in a giant solid lattice. A suitable solvent, often water, will separate the ions and hold them in solution as separate particles. In the solution, the positively charged ions balance the negatively charged ions.

The ionic bond — what does it look like?

A good example of a simple ionic compound is sodium chloride — table salt.

Figure 5.1 shows what happens when sodium chloride (Na^+Cl^-) is formed. Full electron structures are shown for the electrons in all the shells of the sodium and chlorine atoms and their respective ions.

sodium atom (2,8,1) + chlorine atom (2,8,7) → sodium ion (2,8)$^+$ chloride ion (2,8,8)$^-$

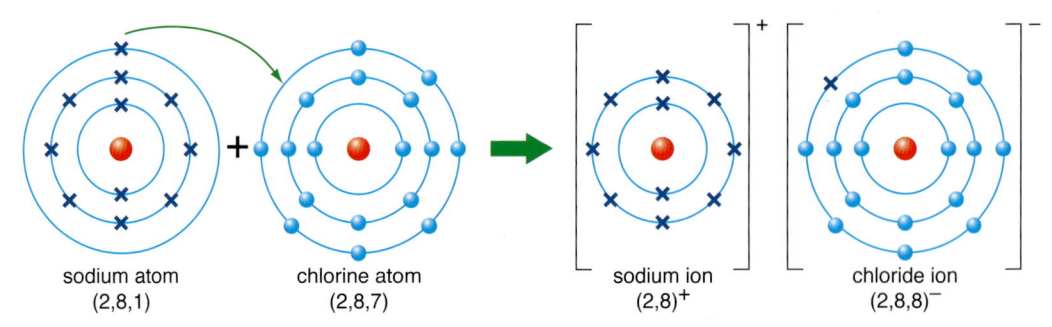

Figure 5.1 Electron transfer during the formation of sodium chloride.

The sodium atom has one electron in its outer shell and the chlorine atom has seven. During the reaction, a sodium atom transfers its one electron into the outer shell of the chlorine atom. This produces:
- a sodium ion (Na^+) with the same stable electron structure as the noble gas neon
- a chloride ion (Cl^-) with the same stable electron structure as the noble gas argon

So, the formation of sodium chloride involves the complete transfer of an electron from a sodium atom to a chlorine atom forming Na^+ and Cl^- ions.

The formation of sodium chloride (Figure 5.1) can be summarised by showing only the outer electrons, as in Figure 5.2.

AQA GCSE Additional Science

C2 5.2 Ions and ionic bonding — transferred electrons

Na× + °Cl° → [Na]⁺ [°Cl°×]⁻
(2,8,1)   (2,8,7)   (2,8)⁺   (2,8,8)⁻

Figure 5.2 A dot–cross diagram for the formation of sodium chloride.

> **Ionic bonds** are formed when metals react with non-metals. Metals lose electrons and form positive ions. The electrons are transferred to non-metals, which gain the electrons and form negative ions. The attraction of the opposite charges on the ions forms the ionic bond. These forces of attraction are strong at the short distances between the ions in the lattice. So the lattice of ions is held together strongly. An ionic bond results from the electrical attraction between oppositely charged ions.

This is called a 'dot–cross' diagram because the electrons of the different atoms are shown as either dots or crosses. The electrons are not really different — they have been shown as dots or crosses to help in understanding what happens to electrons when the elements react.

Solid ionic compounds have a giant, repeating structure of ions held together by strong forces of attraction between the oppositely charged ions. These forces of attraction act in all directions and form a lattice of ions held together — this is called **ionic bonding**.

Figure 5.3 shows two more examples of electron transfer between metal elements and non-metal elements forming ionic compounds.

Magnesium oxide

magnesium atom (2,8,2) + oxygen atom (2,6) → magnesium ion (2,8)²⁺ + oxide ion (2,8)²⁻

Calcium chloride

chlorine atom (2,8,7) + calcium atom (2,8,8,2) + chlorine atom (2,8,7) → chloride ion (2,8,8)⁻ + calcium ion (2,8,8)²⁺ + chloride ion (2,8,8)⁻

Figure 5.3 Electron transfers in the formation of magnesium oxide and calcium chloride.

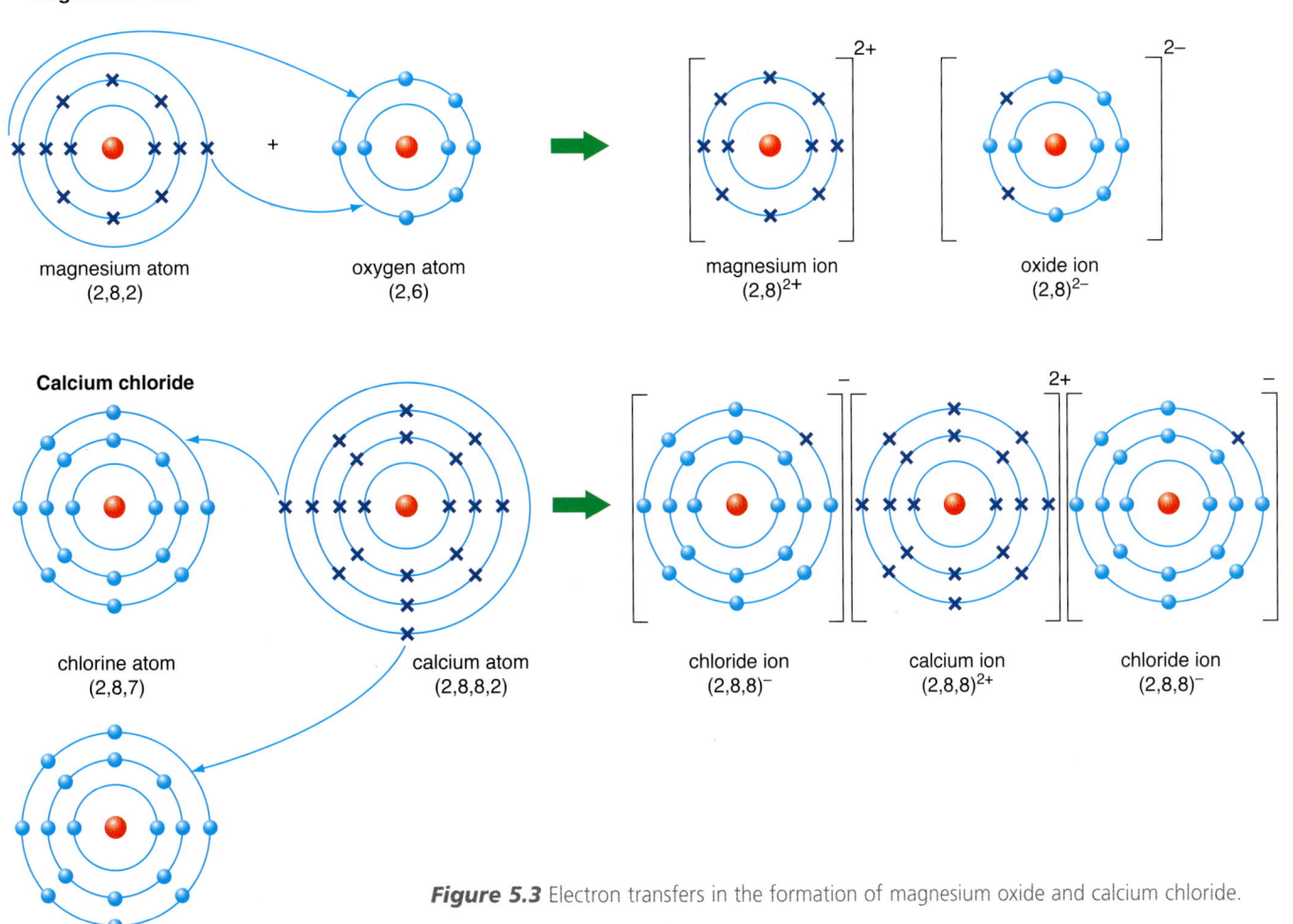

The giant structure, bonding and properties of ionic compounds are dealt with in more detail in Section 5.4.

Chemistry 2

Chapter 5 Bonding in reactions

Boost your grade

In ionic bonding **metals** lose *all* the electrons in their outer shell to form positive ions. **Non-metals** use enough of these lost electrons to fill up their outer shell to form negative ions.

Looking at ionic crystals

HOW SCIENCE WORKS / PRACTICAL SKILLS

Look at a selection of ionic substances in crystalline form and compare their appearance. Use a hand lens to look at them closely.

Test yourself

3 Look at the formation of magnesium oxide in Figure 5.3.
  a How many electrons do magnesium and oxygen atoms gain or lose in forming magnesium oxide?
  b What is the charge on:
    (i) a magnesium ion?
    (ii) an oxide ion?
  c What is the formula of magnesium oxide?
  d Which noble gas has an electron structure like the ions in magnesium oxide?

4 Look at the formation of calcium chloride in Figure 5.3.
  a How many electrons do calcium and chlorine atoms gain or lose in forming calcium chloride?
  b What is the charge on:
    (i) a calcium ion?
    (ii) a chloride ion?
  c What is the formula of calcium chloride?
  d Why does one calcium atom react with two chlorine atoms in forming calcium chloride?

5 Draw a dot–cross diagram for the formation of potassium oxide.

Modelling ions

ACTIVITY

Using Plasticine, counters or people, model the ionic bonding of sodium and chlorine. Note: you may just want to show the outer shells.

Computer modelling of ions

ICT

Use a computer program to model electron transfer and the formation of ions in a compound.

AQA GCSE Additional Science

C2 5.3 Covalent bonding — sharing electrons

Learning outcomes

- Know that electron sharing results in the formation of uncharged molecules, or giant covalent structures, in which atoms are held together by covalent bonds.
- Be able to draw diagrams to show the formation of covalent bonds.
- Know that covalent compounds can exist as molecules, or giant covalent structures that extend for millions of atoms in each direction.

> Covalent bonds are formed when non-metal elements react with each other. A **covalent bond** is formed by the sharing of a pair of electrons between two atoms. Each atom contributes one electron to the bond. Covalent bonds are strong and hold molecules together well. Covalent bonds in a giant lattice hold the whole material together strongly.

Two or more different non-metals atoms can react chemically to make a compound. The non-metal atoms have to share electrons so they all can have an outer shell/highest energy level full of electrons.

The non-metal atoms share (one or more) pairs of electrons — each pair has one electron from each atom and each pair makes up one **covalent bond**. The positive nucleus of each non-metal atom attracts the shared negative electrons. This holds the two nuclei together and forms the covalent bond.

Covalent bonds can hold a small group of atoms together in a small unit called a molecule. In some covalent compounds the bonds can make a big network holding many atoms together — this is called a giant covalent structure.

The covalent bond — what does it look like?

A chlorine atom is unstable — its outer shell contains just seven electrons. At normal temperatures, chlorine atoms join up in pairs to form Cl_2 molecules by sharing electrons (Figure 5.4).

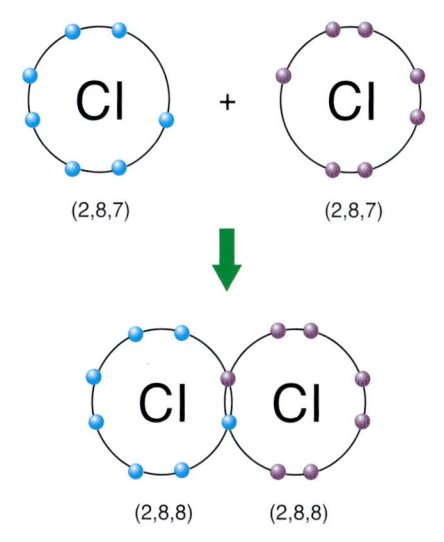

Figure 5.4 Electron sharing in the covalent bond in a chlorine molecule.

If the two chlorine atoms come close together, the electrons in their outer shells can overlap. Each chlorine atom can share one of its electrons with the other atom (Figure 5.4). The two chlorine atoms share a pair of electrons and get a stable electron structure — like argon atoms in this case.

The positive nuclei of both atoms attract the shared electrons and this holds the atoms together. This attraction forms a covalent bond.

Notice in Figure 5.4 that:
- The shared pair of electrons form part of the outer shell of both chlorine atoms.
- Circles are used to join up the electrons in the outer shell of each chlorine atom.
- The chlorine atoms are bonded together forming an uncharged molecule.

Chapter 5 Bonding in reactions

Covalent bonds between atoms are strong bonds. Some covalently bonded substances — such as chlorine (Cl_2), hydrogen (H_2), oxygen (O_2), hydrogen chloride (HCl), water (H_2O), ammonia (NH_3) and methane (CH_4) — consist of simple molecules containing a small number of atoms. Other covalently bonded substances — such as diamond, silicon dioxide and many polymers — have giant structures (giant molecules) containing thousands, sometimes millions, of atoms. The structure and properties of these giant structures are discussed further in Section 5.6. The electron structures and covalent bonding in some simple molecules are shown in Figure 5.5.

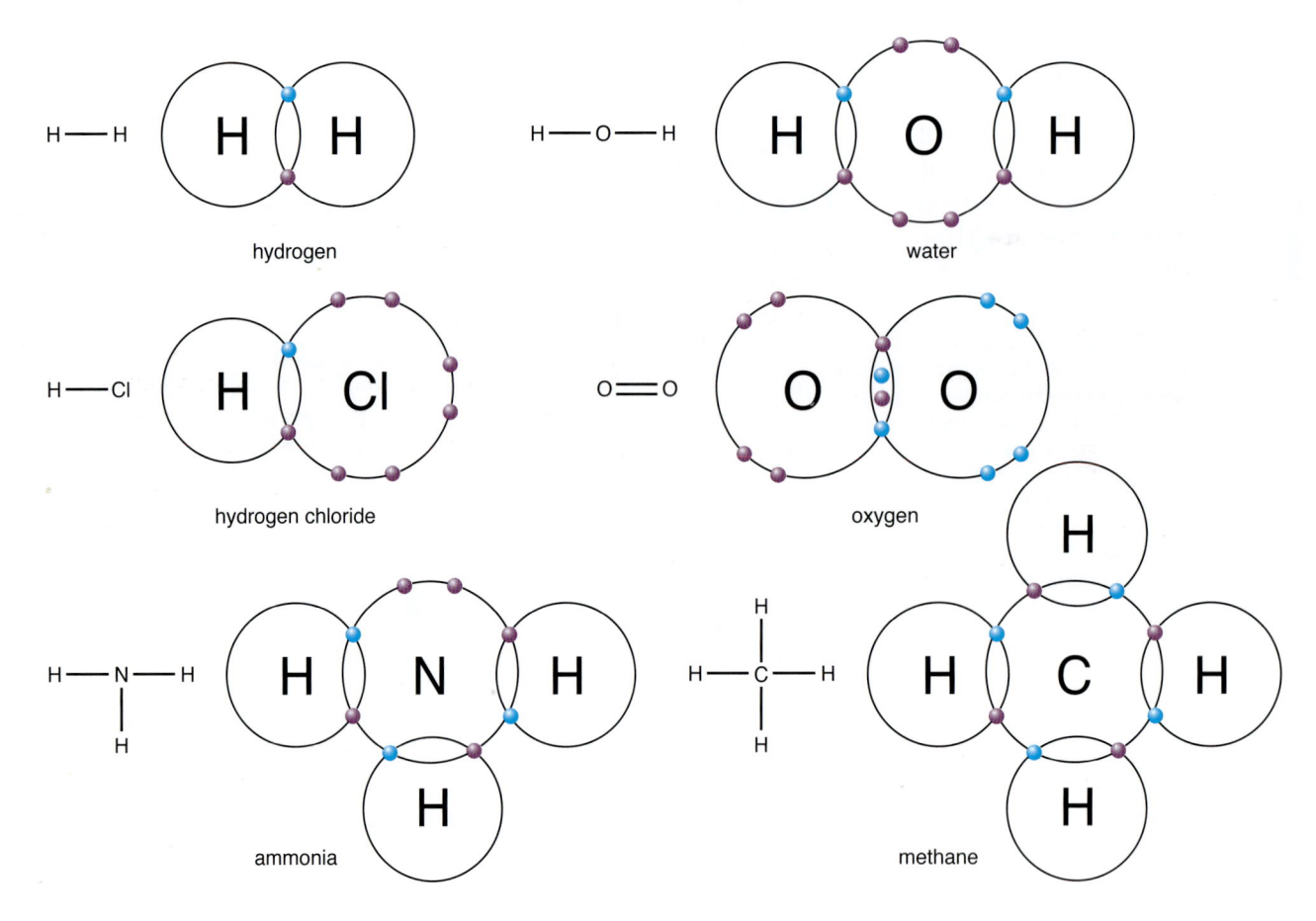

Figure 5.5 Covalent bonding and electron structures in some simple molecules.

In Figure 5.5 notice that:
- The electron structures of the simple molecules can be related to their **structural formulae**, which is shown on the left-hand side of each dot–cross diagram. In the structural formulae, each covalent bond is shown as a line between atoms. For example, H—O—H for water.
- After they have formed bonds: hydrogen atoms (H) always have one bond; chlorine atoms (Cl) always have one bond; oxygen atoms (O) always have two bonds, nitrogen atoms (N) always have three bonds; and carbon atoms (C) always have four bonds.

> The **structural formula** of a compound shows which atoms are bonded to each other in a molecule of the compound.

AQA GCSE Additional Science

C2 5.3 Covalent bonding — sharing electrons

Giant covalent structures

Some covalent compounds consist of neat small groups of molecules. But in some covalent compounds the covalent bonds link together a network of atoms into a giant covalent structure. For example, carbon and silicon are both group 4 elements and they each react with oxygen, a group 6 element. When carbon reacts with oxygen, the carbon atom shares two outer electrons with each of two oxygen atoms, and gets a share of two electrons from each of the two oxygen atoms in return. This makes a small, neat molecule (Figure 5.6).

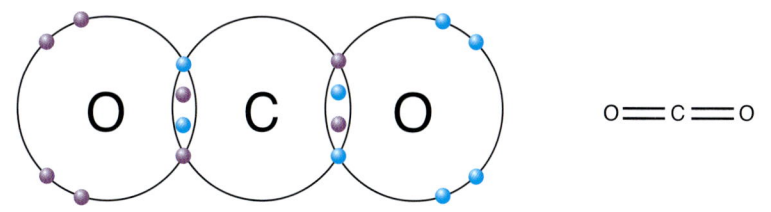

Figure 5.6 A carbon dioxide (CO_2) molecule. Each carbon atom has two covalent bonds with each of two oxygen atoms.

When silicon reacts with oxygen, the silicon atom is too big to share two electrons with each oxygen atom. It must share one of its outer electrons with each of four oxygen atoms and receive a share of one electron from each oxygen in return. The oxygen atom needs a share of two electrons to have a noble gas electronic configuration. So, an oxygen atom reacts with two silicon atoms. This effectively links all the atoms together strongly in a network called a giant structure. These giant covalent structures are sometimes called macromolecules (Figures 5.7 and 5.8).

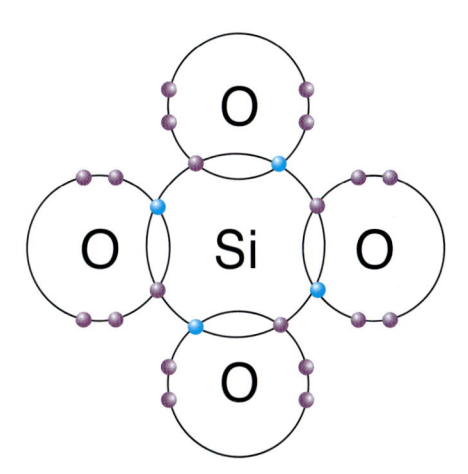

Figure 5.7 Silicon dioxide's electronic structure. Each silicon atom has four covalent bonds with each of four oxygen atoms. Each oxygen atoms has two covalent bonds with two silicon atoms.

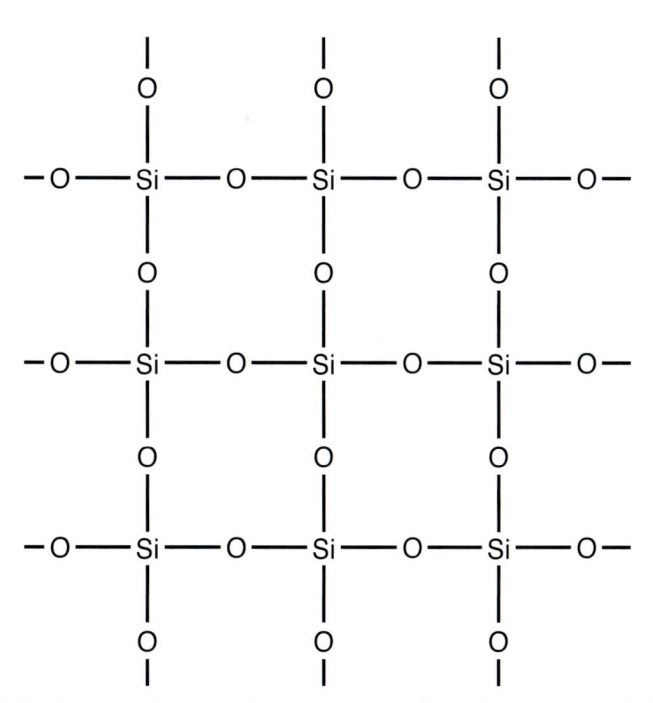

Figure 5.8 This is a two-dimensional representation of the giant structure of silicon dioxide SiO_2.

Chemistry 2

Chapter 5 Bonding in reactions

Modelling covalent bonding — ACTIVITY

Using Plasticine, counters or people, model the covalent bonding of hydrogen atoms to form a hydrogen molecule. Repeat this for one other example of an element (or elements) that bond with covalent bonds. Note: you may just want to show the outer shells.

Computer modelling of molecules — ICT

Use a computer program to model electron sharing and the formation of molecules in a compound.

Test yourself

6 Copy and complete the following statement:

A covalent bond is formed by the s_____ of a p_____ of electrons between t_____ atoms. Each atom contributes one e_____ to the b_____.

7 Look at Figure 5.5.
  a Which noble gas has an electron structure like the hydrogen atoms in H_2?
  b In oxygen, O_2, there is a double covalent bond between the two oxygen atoms. How many electrons are shared in a double covalent bond?
  c The structural formula for carbon dioxide is O=C=O. Draw the dot–cross diagram for carbon dioxide.

8 Draw a dot–cross diagram for the formation of:
  a an ethane molecule, C_2H_6
  b a nitrogen molecule, N_2

Chemical bonding theory

These ideas about ionic and covalent bonding form the basis of the **electronic theory** of chemical bonding. When elements react to produce compounds, chemical bonds are formed. These chemical bonds involve either the transfer or the sharing of electrons in the outer shell (highest energy level) of atoms. When atoms react, they lose, gain or share electrons in order to get a more stable electron structure. These more stable electron structures are like those of the noble gas atoms.

> The **electronic theory** of chemical bonding says that when atoms react, they lose, gain or share electrons in order to get a more stable electron structure.

Simple covalent compounds — ACTIVITY

Make a display of simple covalent compounds and note the difference in appearance between the molecular compounds and giant covalent structures.

C2 5.4 Ionic compounds — giant structures

Learning outcomes

- Know that the alkali metals show typical metal reactions when they react with halogen atoms to make simple ionic salts.
- Know that ionic compounds are made up of a three-dimensional lattice of individual ions held together by strong electrostatic forces of attraction.
- Know that the ionic lattice extends for a huge number of ions in each direction and is made up of symmetrical repeating units.

Figure 5.9 Sodium metal burning in chlorine gas.

Ionic compounds are formed when metal elements react with non-metal elements. During these reactions, electrons are transferred from the metal atoms to the non-metal atoms, forming positive metal ions and negative non-metal ions (Section 5.2). This gives ions that are more stable than the metal and non-metal atoms. These ions have a noble gas electronic structure.

For example, calcium reacts with oxygen to form calcium oxide. In this case, two electrons are transferred from each calcium atom to each oxygen atom.

$$\text{calcium} + \text{oxygen} \rightarrow \text{calcium oxide}$$
$$\text{Ca} + \text{O} \rightarrow \text{Ca}^{2+} + \text{O}^{2-}$$
$$\text{calcium atom} \quad \text{oxygen atom} \quad \text{calcium ion} \quad \text{oxide ion}$$

$Ca \rightarrow Ca^{2+} + 2e^-$ each Ca atom loses two electrons and becomes like argon (2,8,8)

$O + 2e^- \rightarrow O^{2-}$ each O atom gains two electrons and becomes like neon (2,8)

Alkali metal/halogen compounds

Typical reactions to make ionic compounds are seen when alkali metal atoms (group 1) react with halogen atoms (group 7) to make compounds.

A typical alkali metal atom is sodium. A sodium atom always reacts to become a sodium ion because the easiest way to get a full outer shell of electrons (like the noble gas neon) is to lose the one electron it has in its third shell. This behaviour is the same for all alkali metals — they lose their one electron in the outer shell.

Alkali metals lose their outer electron easily. But the alkali metal atom has to find other atoms to give its outer electron to — it cannot just lose an electron into space.

sodium atom (2,8,1)   sodium ion (2,8)$^+$

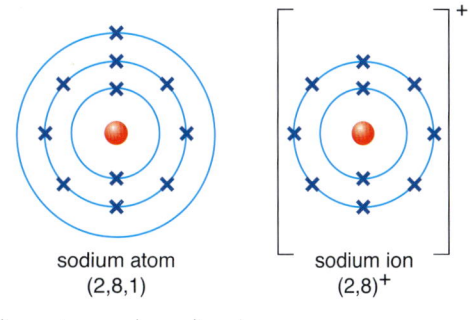

Figure 5.10 A sodium atom and a sodium ion.

A typical halogen atom is chlorine. A chlorine atom has an outer shell of electrons that has one space to fill. Chorine has a big appetite for reacting with other elements to fill this space in its outer shell — that is why it is such a reactive element.

Chlorine will grab electrons from any atom it can to fill that hole in its outer shell. The extra electron turns the chlorine atom into a chloride ion (note the change of name from -ine to -ide), as shown in Figure 5.11.

Chemistry 2

Chapter 5 Bonding in reactions

chlorine atom
(2,8,7)

chloride ion
(2,8,8)⁻

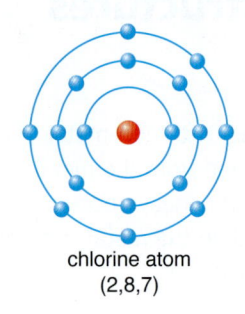

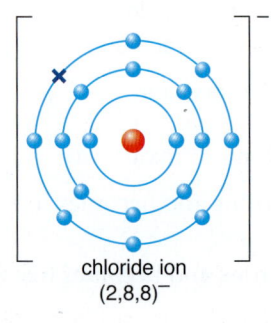

Figure 5.11 A chlorine atom and a chloride ion.

Now it is easy to see why sodium, with one electron to lose, and chlorine, with one electron to gain, make perfect partners. All the group 1 alkali metals react strongly with all the group 7 halogens to make stable ionic compounds.

All the alkali metal halides are hard, white crystalline solids with high melting and boiling points. These compounds do not decompose on melting but make a stable molten compound. Their melting points give an indication of their similarity and stability.

Table 5.1 Melting points (in °C) of alkali halides.

	Li⁺	Na⁺	K⁺
F⁻	845	993	858
Cl⁻	605	801	770
Br⁻	552	747	734
I⁻	469	661	681

Alkali metals with other non-metals

The alkali metals owe their high reactivity to their atoms having to lose only one electron to make an ion with a stable electronic structure — like a noble gas.

Alkali metals react easily with all non-metals — the alkali metal atom loses its outer electron to form an ion with a single positive charge. The non-metal atoms pick up as many of the electrons as they need to make stable negative ions with the electronic configuration of a noble gas.

For example, sodium and chlorine combine in a 1 : 1 reaction. The sodium ion (Na⁺) achieves the electronic structure of neon (2,8) and chloride ion (Cl⁻) has the electronic structure of argon (2,8,8) (see Figure 5.1 for details).

Figure 5.12 shows that sodium and oxygen combine in a 2 : 1 reaction — the oxide has a formula of Na_2O. Sodium reacts as usual and the sodium ion achieves the electronic structure of neon (2,8). The oxygen atom gains two electrons from two sodium atoms to also become like neon (2,8).

oxygen atom
(2,6)

oxide ion
(2,8)²⁻

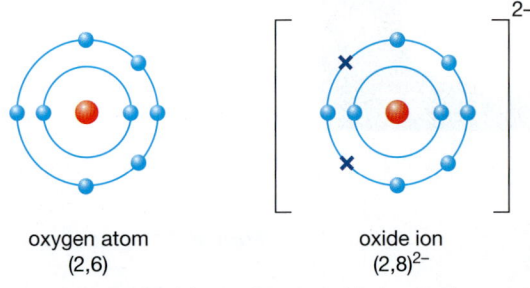

Figure 5.12 An oxygen atom (2,6) becoming an oxide ion (2,8).

$$4Na(s) + O_2(g) \rightarrow 2Na_2O(s)$$
sodium metal + oxygen gas → sodium oxide solid

Sodium and nitrogen combine in a 3 : 1 reaction (Figure 5.13) — the ionic nitride has the formula Na_3N. Sodium reacts as before. Nitrogen atoms gain three electrons each from three sodium atoms to become like neon (2,8).

C2 5.4 Ionic compounds — giant structures

Test yourself

9 Look at Table 5.1 on page 94, showing the melting points of the alkali metal halides.
  a What trends can you see in the data in the table?
  b Explain why a high melting point is an indication of a stable compound.

10 Draw a simple dot–cross diagram of the formation of magnesium chloride.

$$6Na(s) + N_2(g) \rightarrow 2Na_3N(s)$$
sodium + nitrogen → sodium nitride

nitrogen atom (2,5)

nitride ion $(2,8)^{3-}$

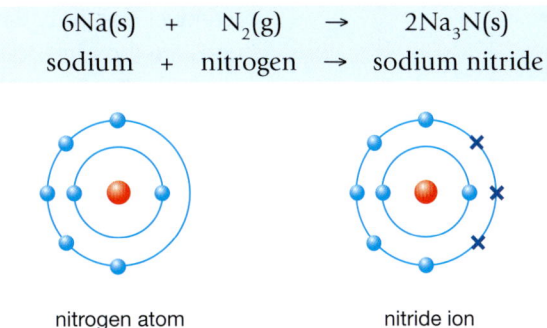

Figure 5.13 A nitrogen atom (2,5) becoming an nitride ion (2,8).

All alkali metals show this kind of electron transfer reaction with non-metals. All that changes is the number of electrons transferred. This number is sometimes called the combining power of an element.

Structure and properties of ionic compounds

In ionic compounds, large numbers of positive ions and negative ions are packed together in a regular pattern. These giant ionic lattices contain billions of ions and are examples of giant structures.

Giant ionic lattices form a regular structure of oppositely charged ions.

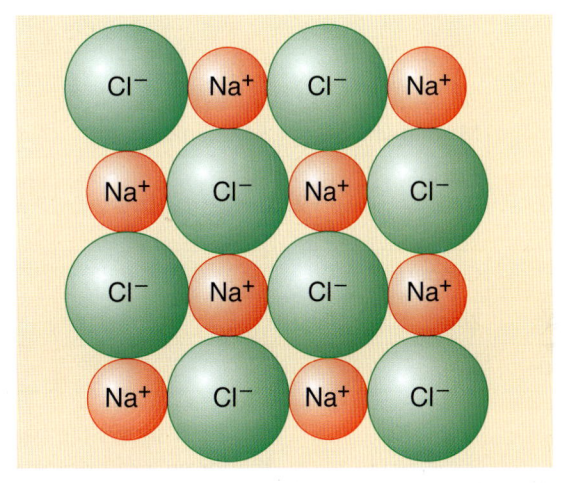

Figure 5.14 Arrangement of ions in one layer of a sodium chloride (salt) crystal.

Figure 5.14 shows how the ions are arranged in one layer of sodium chloride (NaCl) and Figure 5.15 is a three-dimensional model of its structure.

Notice that in three dimensions each Na⁺ ion is surrounded by six Cl⁻ ions, and that each Cl⁻ ion is surrounded by six Na⁺ ions. This means that there are strong electrostatic forces in all directions in the lattice between oppositely charged ions. These forces of attraction between the oppositely charged ions are called ionic bonds (see Section 5.2). The ratio of sodium and chloride ions is 1 : 1.

The ions are bonded together strongly in ionic compounds. This explains why ionic compounds:
- are hard substances
- have high melting points and high boiling points
- do not conduct electricity when solid — their ions have charges but cannot move away from fixed positions in the giant structure

Figure 5.15 A three-dimensional model of the structure of sodium chloride. The larger green balls represent Cl⁻ ions. The smaller red balls represent Na⁺ ions.

Chemistry 2

Chapter 5 Bonding in reactions

- conduct electricity when they are melted or dissolved in water — the charged ions are then free to move. Positive ions move towards the negative terminal and negative ions move towards the positive terminal, and they carry the current through the liquid. The conduction of electricity by ionic compounds is discussed in more detail in Chapter 8

Formulae of ionic compounds

The formulae of ionic compounds can be worked out simply by balancing the charges on positive and negative ions. For example, the formula of calcium chloride is $Ca^{2+}(Cl^-)_2$, or just $CaCl_2$. The two positive charges on one calcium ion (Ca^{2+}) are balanced by the single negative charges on two chloride ions (Cl^-).

The number of charges on an ion is a measure of its combining power or **valency**. Na^+ has a combining power of 1, whereas Ca^{2+} has a combining power of 2. Na^+ can combine with only one Cl^- to form Na^+Cl^-, whereas Ca^{2+} can combine with two Cl^- ions to form $Ca^{2+}(Cl^-)_2$.

Elements such as iron, which has two different ions (Fe^{2+} and Fe^{3+}), have two valencies. So, iron can form two different compounds with chlorine — iron(II) chloride, $FeCl_2$, and iron(III) chloride, $FeCl_3$.

> **Valency** is the number of chemical bonds (ionic or covalent) an atom makes.

Table 5.2 Formulae and combining power of some ions.

Charges on ion											
+1		+2		+3		−3		−2		−1	
Lithium	Li^+	Magnesium	Mg^{2+}	Aluminium	Al^{3+}	Nitride	N^{3-}	Oxide	O^{2-}	Fluoride	F^-
Sodium	Na^+	Calcium	Ca^{2+}	Iron(III)	Fe^{3+}	Phosphate	PO_4^{3-}	Sulfide	S^{2-}	Chloride	Cl^-
Potassium	K^+	Iron(II)	Fe^{2+}					Sulfate	SO_4^{2-}	Bromide	Br^-
Hydrogen	H^+	Copper(II)	Cu^{2+}					Carbonate	CO_3^{2-}	Iodide	I^-
Silver	Ag^+	Zinc	Zn^{2+}							Hydroxide	OH^-
Ammonium	NH_4^+									Nitrate	NO_3^-

Table 5.3 shows the names and formulae of some ionic compounds. Notice that the formula of calcium nitrate is $Ca(NO_3)_2$. The brackets around NO_3^- show that it is a single unit containing one nitrogen and three oxygen atoms with one negative charge. Thus, two NO_3^- ions balance one Ca^{2+} ion. Other ions, such as SO_4^{2-}, CO_3^{2-} and OH^-, should also be put in brackets when there are two or more of them in a formula.

Table 5.3 The names and formulae of some ionic compounds.

Name of compound	Formula		
Calcium nitrate	$Ca^{2+}(NO_3^-)_2$	or	$Ca(NO_3)_2$
Zinc sulfate	$Zn^{2+}SO_4^{2-}$	or	$ZnSO_4$
Magnesium carbonate	$Mg^{2+}CO_3^{2-}$	or	$MgCO_3$
Potassium iodide	K^+I^-	or	KI
Iron(III) chloride	$Fe^{3+}(Cl^-)_3$	or	$FeCl_3$
Copper(II) bromide	$Cu^{2+}(Br^-)_2$	or	$CuBr_2$

> **Test yourself**
>
> 11 Use Table 5.2 to write the formulae for:
> a copper chloride
> b potassium sulfate
> c ammonium bromide
> d aluminium fluoride
> e sodium carbonate

C2 5.4 Ionic compounds — giant structures

Comparing ionic substances
HOW SCIENCE WORKS — PRACTICAL SKILLS

1. Take a selection of ionic substances in crystalline form and compare their properties. Use sodium chloride, potassium iodide, copper chloride, zinc chloride and magnesium bromide.
   - Examine the crystals carefully with a hand lens.
   - Heat them strongly in a test tube to see if they melt.
   - Test them to see if they are soluble in water.
   - Test them to see if they conduct electricity — either as a solid or in solution. Use a low-voltage supply set to about 5 volts and carbon electrodes or mild steel nails.

   Make a table of the results.

2. Burn sodium metal in a gas jar of chlorine gas. Only do this in a fume cupboard — chlorine is poisonous. Set fire to the sodium metal on a combustion spoon and plunge this into a gas jar of chlorine. Make careful observations of what you see at each stage.

3. Compare the reactions of the halogens in gaseous form with hot iron wool. Only do this in a fume cupboard. Prepare a gas jar full of the halogen vapour. (Not fluorine; for bromine and iodine, preheat the gas gar with boiling water, or stand the jar on a electric hotplate set to a low heat.)

   Put about 2 cm of water in the bottom of the gas gar to protect it from hot combustion products. Hold some fresh iron wool (about 20 g) in tongs. Heat the iron wool in a Bunsen flame for 2 seconds. Plunge the iron wool into the gas jar.

   Make careful observations of what you see at each stage.

Test yourself

12. Which of the following substances conduct electricity:
    a when liquid?
    b when solid?

    A diamond   B potassium chloride   C copper   D sulfur

13. Look carefully at Figures 5.14 and 5.15.
    a How many Cl^- ions surround one Na^+ ion in one layer of the NaCl crystal?
    b How many Cl^- ions surround one Na^+ ion in the three-dimensional crystal?
    c How many Na^+ ions surround one Cl^- ion in the three-dimensional crystal?

14. Sodium fluoride (NaF) and magnesium oxide (MgO) have the same crystal structure and there are similar distances between ions. The melting point of NaF is 992°C but that of MgO is 2640°C.
    a Write the formula for sodium fluoride showing charges on the ions.
    b Write the formula for magnesium oxide showing charges on the ions.
    c Why is there such a big difference in the melting points of NaF and MgO?

Chemistry 2

Chapter 5 Bonding in reactions

C2 5.5 Simple covalent molecular substances

Learning outcomes

- Know that most common molecular substances are gases or liquids, or soft solids with a low melting point.
- Know that molecules are held together by strong internal forces, but that forces between molecules are weak.
- Be able to represent covalent bonds in dot–cross and ball-and-stick diagrams.

Most non-metallic elements, and the compounds formed when different non-metal elements react with each other, are simple molecular substances. Oxygen, hydrogen, chlorine, water, hydrogen chloride and methane are all simple molecular substances. They have simple molecules containing a few atoms. Their formulae and structures are shown in Figure 5.16.

Name and formula	Structural formula	Model of structure
Hydrogen, H_2	H—H	
Oxygen, O_2	O=O	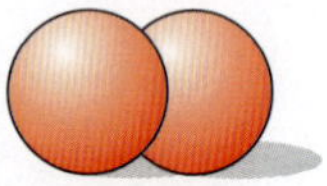
Water, H_2O	H—O—H	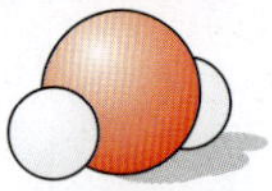
Methane, CH_4	H—C—H (with H above and H below)	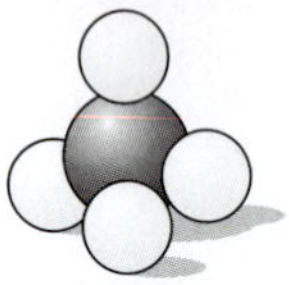
Hydrogen chloride, HCl	H—Cl	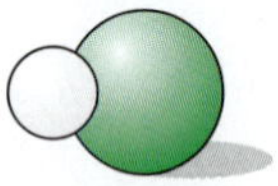
Chlorine, Cl_2	Cl—Cl	
Carbon dioxide, CO_2	O=C=O	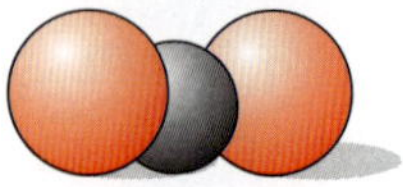
Iodine, I_2	I—I	

Figure 5.16 Formulae and structures of some simple molecular substances.

Sugar (sucrose, $C_{12}H_{22}O_{11}$) has much larger molecules than the substances in Figure 5.16, but it still counts as a simple molecule. In these **simple molecular substances**, the atoms are held together in each molecule by strong covalent bonds (Figure 5.17).

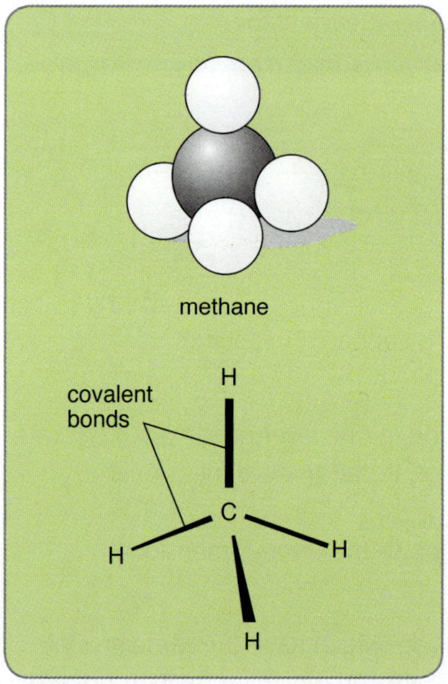

But there are only weak forces between the separate molecules (Figure 5.18). The weak forces between the separate molecules in simple molecular substances are called **intermolecular forces** ('inter' means 'between'). In the solid state, it does not require much energy to pull the molecules away from each other.

Figure 5.17 Methane (CH_4) is a simple molecular substance. In methane, the carbon atom and four hydrogen atoms are held together by strong covalent bonds.

AQA GCSE Additional Science

C2 5.5 Simple covalent molecular substances

Properties of simple molecular substances

The properties of simple molecular substances can be explained in terms of their structures and the weak forces between their molecules.

In molecular substances, there are no ions. These substances consist of simple molecules with no overall charge. So, there are no obvious electrical forces holding the molecules together. However, some simple molecular substances — such as sugar, water and iodine — do exist as solids and liquids, so there must be some forces holding their molecules together.

Low melting points and low boiling points

There are only weak forces between the molecules in simple molecular substances. These weak intermolecular forces are overcome when simple molecular substances melt or boil on heating. It takes much less energy to separate the simple molecules than to separate the oppositely charged ions in ionic compounds or the atoms in most metals.

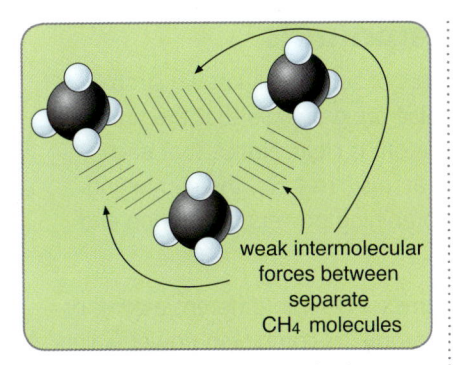

Figure 5.18 The intermolecular forces between molecules of methane are weak.

Simple molecular substances are small molecules with just a few atoms compared with polymers or giant molecules.
Intermolecular forces are weak forces between one molecule and another.

Figure 5.19 This butcher is using 'dry ice' (solid carbon dioxide) to keep meat cool during mincing. 'Dry ice' is a simple molecular substance. After mincing, the 'dry ice' sublimes (changing directly from solid to gas) without spoiling the meat.

Soft solids

The separate molecules in simple molecular substances are usually further apart than atoms in metals and ions in ionic structures. The forces between the molecules are weak and the molecules are easy to separate. Because of this, simple molecular solids, such as iodine and wax, are softer than other materials.

Non-conductors of electricity

Simple molecules have no overall electric charge. Unlike metals, they do not have electrons that move easily from atom to atom. And there are no ions like in ionic compounds. All this means that they do not conduct electricity.

Figure 5.20 Candle wax (paraffin wax) is a typical molecular solid — it is soft with a low melting point.

Chemistry 2

Chapter 5 Bonding in reactions

> The combining power or **valency** of an atom is the number of covalent bonds that it forms with other atoms.

Formulae of molecular compounds

Figure 5.16 shows the formulae and structures of some well-known simple molecular compounds. The structural formulae are drawn so that the number of covalent bonds (drawn as a line) to each atom is clear. Notice that each hydrogen atom forms one bond with other atoms (H—). So, the combining power, or **valency**, of hydrogen is 1. The combining powers of chlorine and iodine are also 1.

Oxygen atoms form two bonds to other atoms (—O—) to different atoms, or O= if both bonds are formed to the same atom). Its combining power (valency) is therefore 2. Carbon atoms form four bonds to other atoms, so the combining power of carbon is 4. Using these ideas of combining power (the number of bonds that an atom forms), you can predict the formulae of most (but not all) molecular compounds.

Table 5.4 Combining power of some atoms.

Valency			
1	2	3	4
Hydrogen	Oxygen	Nitrogen	Carbon
Fluorine	Sulfur	Phosphorus	Silicon
Chlorine			
Bromine			
Iodine			

Making molecular models — ACTIVITY

Make models of the simple molecules H_2O, CO_2, NH_3, H_2, Cl_2, O_2, CH_4 and C_2H_6.

Use Molymod plastic modelling kits, or the SEP chemistry jigsaw kits. You could make your own modelling kit with polystyrene or Plasticine balls and cocktail sticks.

Investigating melting points — HOW SCIENCE WORKS / PRACTICAL SKILLS

1. Investigate the melting points and boiling points of simple molecular materials. Some of them will need a data search to find the values. Some, such as those below, can be measured experimentally. Make careful observations of what you see at each stage.
   - Melting point of ice and paraffin wax — heat in a test tube in hot water.
   - Boiling point of water, ethanol and cyclohexane — heat the ethanol and hexane in tubes in a water bath with an electrical hot plate. Do this in a fume cupboard, away from naked flames.

C2 5.5 Simple covalent molecular substances

- Obtain some solid carbon dioxide (dry ice) — use an electric temperature probe (thermocouple) to find the temperature at which it sublimes.
- Compare these properties with the properties of a giant covalent structure — for example silicon dioxide (sand).

2 Make a model of an S_8 ring molecule of elemental sulfur using a Molymod molecular modelling kit.

a Investigate the heating of powdered sulfur — how it first becomes a runny liquid consisting of liquid S_8 ring molecules. On further heating, the rings begin to break up into S_8 chains, and the liquid becomes thick and viscous as the chains tangle with each other.

b Investigate pouring the hot molten sulfur into cold water — the rapid cooling prevents the reformation of the S_8 ring molecules and the solid is a plastic mass of amorphous sulfur atoms.

Make details observations at each stage and produce a report.

Test yourself

15 Look at Figure 5.21. What properties does butter have that show it contains simple molecular substances?

Figure 5.21 Butter contains a mixture of simple molecular substances.

16 Many simple molecular substances have a smell, but no metals have a smell. Why is this?

17 Solid X is a poor conductor of electricity. It melts at 150°C and boils at 350°C. Which of the following could X be?
A a metal element
B a non-metal element
C a compound of a metal and a non-metal
D a compound of non-metal elements
E an ionic compound
F a simple molecular compound

18 Using the usual combining power (valency) of the elements, draw structural formulae for the following compounds (show each bond as a line —):
a dichlorine oxide (Cl_2O)
b tetrachloromethane (CCl_4)
c hydrogen peroxide (H_2O_2)
d ethane (C_2H_6)

Chemistry 2

Chapter 5 Bonding in reactions

C2 5.6 Giant covalent structures

Learning outcomes

- Know that covalent bonds can link millions of atoms together in a network.
- Know that the atoms share outer electrons and are linked together strongly.
- Know that covalent giant structures are typically hard solids with high melting points.

Figure 5.22 Diamond, the hardest material known, owes its properties to a network of covalent bonds holding carbon atoms together strongly.

Atoms that share electrons in covalent bonds can also form **giant covalent structures**, or macromolecules. One important group of compounds with giant covalent structures are polymers, such as polyethene and polyvinylchloride (PVC). Polymers such as these consist of long, chain molecules made by addition reactions between smaller monomer molecules. The polymers consist of hundreds, often thousands, of atoms joined by strong covalent bonds. The properties of polymers depend on their structure.

Diamond, graphite and silicon dioxide (Section 5.3) are other examples of giant covalent structures — these form three-dimensional lattices with billions of atoms in them.

Diamond and graphite are both made of carbon atoms and nothing else — they are different crystalline forms of pure carbon. The two solids have different properties and uses — diamond is hard and clear, whereas graphite is soft and black. Diamonds are used to cut stone and engrave glass, whereas graphite is used by artists to achieve a soft, shaded effect for their sketches.

Diamond and graphite have different properties and different uses because they have different structures. The carbon atoms are packed together in different ways in these substances. The arrangements of carbon atoms in diamond and graphite have been studied by X-ray crystallography. This is a way of finding out about the arrangement of atoms in substances.

Diamond

In diamond, carbon atoms are joined to each other by strong covalent bonds. Inside the diamond structure, each carbon atom forms covalent bonds with four other carbon atoms. Check this for yourself in Figure 5.24. The strong covalent bonds extend through the whole diamond, forming a giant, strong, three-dimensional structure. A perfect diamond, without flaws or cracks, is a single

> **Giant covalent structures** contain large numbers of atoms joined to each other by strong covalent bonds.

Figure 5.23 This artist is using a soft graphite pencil. Graphite is a giant covalent structure.

AQA GCSE Additional Science

C2 5.6 Giant covalent structures

giant molecule (macromolecule) with covalent bonds linking one carbon atom to the next.

Only a small number of atoms are shown in Figure 5.24. In a real diamond, this arrangement of carbon atoms is extended billions and billions of times.

Properties and uses of diamond

- **Hard** — it is difficult to break the strong covalent bonds between carbon atoms in diamond. Another reason for diamond's hardness is that the atoms are not arranged in layers — so they cannot slide over one another like the atoms in metals. Most industrial uses of diamond depend on its hardness. Some diamonds are not perfect enough to be cut and polished to make jewellery. These lower-quality diamonds are still just as hard so they are used in glass cutters, in diamond-studded saws and as abrasives for smoothing hard materials.
- **High melting point** — carbon atoms in diamond are linked together in the giant structure by strong covalent bonds. This means that the atoms cannot vibrate fast enough to break away from their neighbours until the temperature is about 3800°C. At this temperature, carbon will burn in air.
- **Non-conductor of electricity** — unlike metals, diamond has no free electrons. All the electrons in the outer shell of each carbon atom are held strongly in one place in covalent bonds — there are no free electrons to carry an electric current.

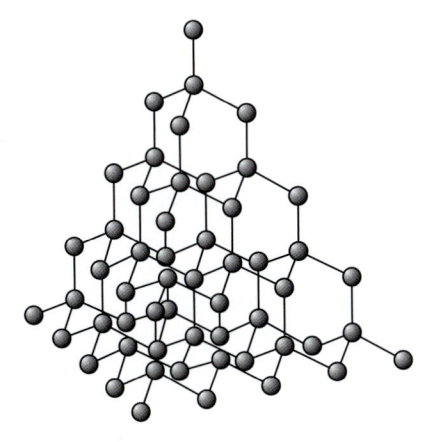

Figure 5.24 An 'open' model of the giant structure in diamond. Each grey ball represents a carbon atom and each line is a covalent bond.

Graphite

Figure 5.25 shows a model of part of the structure of graphite. The carbon atoms are arranged in layers. Each layer contains billions and billions of carbon atoms arranged in hexagons. Each carbon atom is held in its layer by strong covalent bonds to three other carbon atoms. So, every layer is a giant covalent structure. The distance between carbon atoms in the same layer is only 0.14 nm, but the distance between the layers is 0.34 nm — more than twice as far.

The hexagonal layers of carbon atoms in graphite only account for three covalent bonds. The fourth electron is held loosely between the layers. This fourth electron is not held by any carbon atom in particular but is 'delocalised'. This means this electron is able to move easily from atom to atom.

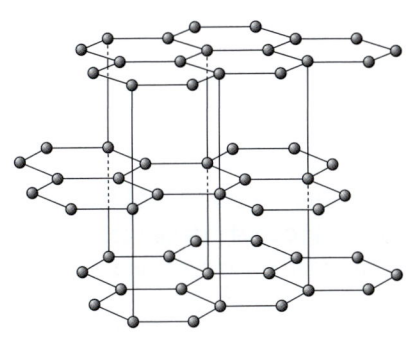

Figure 5.25 A model of the structure of graphite. Notice how graphite has layers of carbon atoms — within the layers, the carbon atoms are arranged in hexagons.

Properties and uses of graphite

- **Soft and slippery** — in graphite, each carbon atom is linked by strong covalent bonds to three other atoms in its own layer. But the layers are more than twice as far apart as the carbon atoms in a layer. This means that the forces between the layers are weak. If you rub graphite, it feels soft and slippery as the layers slide over each other and onto your fingers. This property has led to the use of graphite as the 'lead' in pencils and as a lubricant in engine oils.
- **High melting point** — although the layers of graphite move over each other easily, it is difficult to break the strong covalent bonds between carbon atoms within one layer. Because of this, graphite does not melt until 3730°C and it is used to make crucibles for hot molten metals.
- **Conducts electricity and heat well** — carbon atoms have four outer-shell electrons. In graphite, three of these electrons form covalent bonds with electrons from three other carbon atoms in the same layer. This leaves one

Chemistry 2 103

Chapter 5 Bonding in reactions

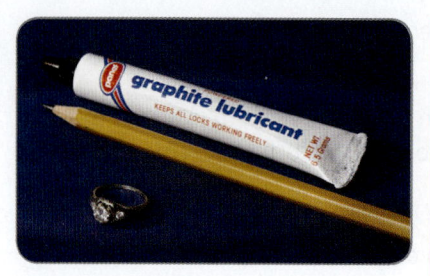

Figure 5.26 A tube of lubricating oil containing graphite. The layers in graphite slide over one another easily — this makes graphite an excellent material to improve the lubricating action of oils.

Figure 5.28 The buckyball has a similar structural shape to a football.

> **Allotropes** are different crystalline forms of the same material.
>
> A **fullerene** is an unusual allotrope of carbon that exists as balls, sheets or tubes.
>
> A **buckyball** is a fullerene with a formula of C_{60} and is shaped like a football.

unshared electron on each carbon atom. These unshared electrons are delocalised (Section 5.7) over the atoms in their layer. So, they allow graphite to conduct electricity and heat in a similar way to metals. Because of this property, graphite is used for electrodes in industry and as the positive terminal in dry cells (batteries). Because of these delocalised electrons, graphite has a structure that has some similar properties to a metal.

Buckyballs and fullerenes

Carbon has a tendency to form a hexagonal pattern of atoms, like in graphite. This makes a neat, low-energy arrangement. This is the property behind some carbon substances called **fullerenes**, which are all giant macromolecules. They are forms of pure carbon, but with different structures and atomic arrangements from diamond or graphite (Figure 5.27). These are referred to as **allotropes** of carbon.

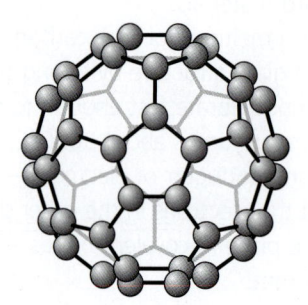

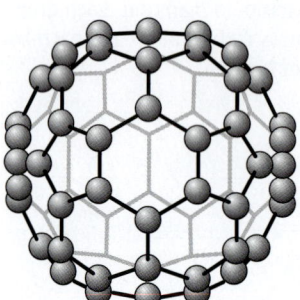

Figure 5.27 The original C_{60} buckyball and the C_{70} variation of this — these are both examples of fullerenes. They are one nanometre in diameter.

The first fullerene was discovered in 1985. It consisted of ball-like macromolecules with a formula of C_{60}. Their structure is like that of a standard soccer ball (Figure 5.28) containing twenty hexagons and twelve pentagons.

This compound was named buckminsterfullerene, after the inventor of the geodesic dome (Richard Buckminster Fuller). This was soon shortened to **buckyball**. A whole family of similar materials has been discovered under the general name of fullerenes, including C_{70} (Figure 5.27) and nanotubes.

Nanotubes

Many variations of this type of pure carbon structure have been discovered in the last 30 years. One of the most useful has been the carbon nanotube. This looks like a sheet of graphite that has been curved into a long tube.

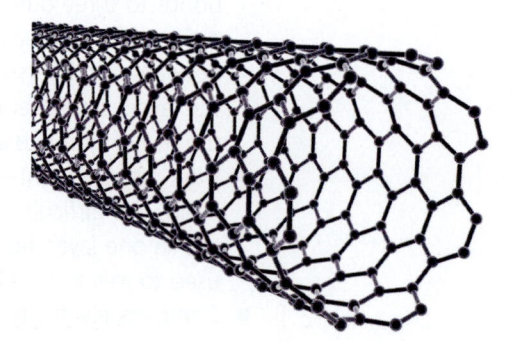

Figure 5.29 A carbon nanotube.

AQA GCSE Additional Science

C2 5.6 Giant covalent structures

Uses of fullerenes

It is said that as soon as you write down a list of uses of fullerenes, it goes out of date because more uses have been found. The balls can be used as a lubricant because they act as tiny spheres between two sliding surfaces. They are often used to trap materials in microscopic quantities and take them to where they are needed — for example, pharmaceuticals in the body. They are also used to trap and hold catalysts to speed up chemical reactions.

Nanotubes are used for many purposes from tiny electronic components to structural materials with a high strength to weight ratio in armour. There are many materials that are now produced with reinforcement from carbon nanotubes (Figure 5.30).

Using models

Models are important in science. They help us to think about the way substances behave and how things work. They also help us to test new ideas and hypotheses and to make predictions. The kinetic theory is a model that helps us to understand how the particles in solids, liquids and gases move and interact. Sometimes the models that scientists use are complex computer programs, like those that help weather forecasters. In this chapter, we have been using models of the structures of different substances to help us understand their properties and uses.

Figure 5.30 Carbon nanotubes are used to reinforce the shafts of golf clubs, badminton rackets and even broken bones.

Making models ACTIVITY

Silicon dioxide (silica, SiO_2) has a giant covalent structure (Figure 5.31). It can occur as large, beautiful crystals of quartz (Figure 5.32) or more often as impure gritty bits of sand.

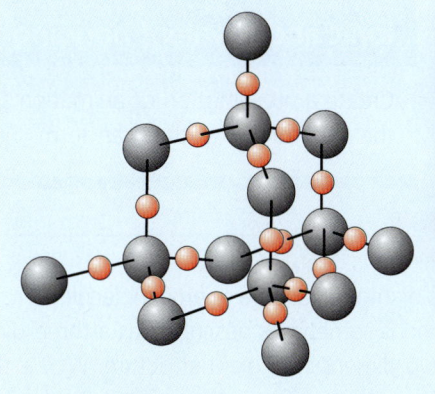

Figure 5.31 A ball-and-stick model of silicon dioxide.

Figure 5.32 Crystals of quartz. Quartz is silicon dioxide (SiO_2).

1. Use a ball-and-stick model kit to build a structure of silicon dioxide like that in Figure 5.31. Use black balls (normally carbon atoms) from the kit to represent silicon atoms, and red balls to represent oxygen atoms.
    a How many covalent bonds are formed by the silicon atoms in the middle of your structure?
    b How many covalent bonds are formed by the oxygen atoms in your structure?

Chemistry 2

Chapter 5 Bonding in reactions

2 Explain why silicon dioxide:
  a is hard
  b has a high melting point
  c does not conduct electricity

Silicon is the second most common element in the Earth's crust. Most rocks including sandstones and clays contain silicon, and the structures of these silicate rocks are based on that of silicon dioxide.

3 Although models are usually helpful, it is important to remember that they are not real representations. For example, metal atoms are not really like marbles or polystyrene spheres and silicon dioxide doesn't have tiny balls and sticks.

Why are models important in science?

4 Investigate the physical and chemical properties of graphite and silica (silicon dioxide/sand).
  a Heat the substances as much as possible in a Bunsen flame, in a crucible with a lid (to restrict oxygen) to see if they melt.
  b Heat the substances on an open tin lid to see if they burn.
  c Try to dissolve the substances in water and ethanol to see if they are soluble. Compare these properties with the properties of an giant ionic structure (salt/sodium chloride) and a molecular covalent compound (sugar/sucrose).

5 As a class project, make a model of a buckyball using Molymod molecular modelling kits. This will take several kits. Make a series of 12 pentagons and join these by using extra 'bonds' to make hexagons. Compare this with a football structure.

Ionic and covalent bonding   ICT

Create a presentation or animation showing ionic and covalent bonding and the structures that they can form.

Test yourself

19 a Why is diamond called a 'giant' molecule?
   b Diamond is sometimes described as a three-dimensional giant structure, and graphite as a two-dimensional giant structure. Why is this?

20 a Make a table to show at least three similarities and three differences between diamond and graphite.
   b Why do diamond and graphite have some differences?

21 a Why does a zip-fastener move more freely after being rubbed with a soft pencil?
   b Why is it better to use a pencil rather than oil to make a zip-fastener move more freely?

22 Why is graphite better than oil for lubricating the moving parts of hot machinery?

23 Draw a small section of the surface of a buckyball showing how each hexagon of carbon atoms is surrounded by six pentagons of carbon atoms.

C2 5.7 Metals — giant metallic structures

Learning outcomes

- Know that metals contain a regular arrangement of atomic centres packed close together.
- Know that some of the outer electrons in metals are delocalised and move easily through the structure.
- Know that delocalised electrons enable metals to conduct electrical current and thermal energy.
- Be able to explain how layers of atoms in metals can slide past each other without breaking the metal.
- Know that smart metal materials can remember their original shape.

Metals are important and useful materials. Just look around you and notice the uses of different metals — vehicles, girders, bridges, pipes, taps, radiators, cutlery, pans, jewellery and ornaments. Metals have always been important materials in the development of human society — look up the Bronze Age and the Iron Age in historical sources.

X-ray studies show that the atoms in most metals are packed together to make a **giant metallic structure**. This arrangement is called close packing. Figure 5.33 shows a model of a few atoms in one layer of a metal crystal.

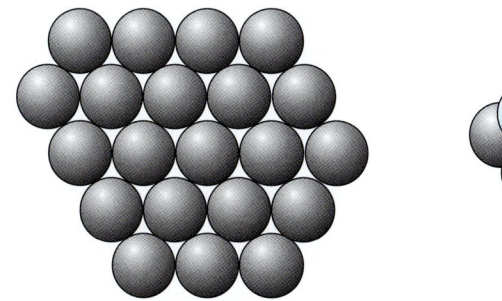

Figure 5.33 Close packing of atoms in one layer of a metal.

Figure 5.34 Atoms in two layers of a metal crystal.

Notice that each atom in the middle of the layer 'touches' six other atoms in the same layer. When a second layer is placed on top of the first, atoms in the second layer sink into the dips between atoms in the first layer (Figure 5.34). This close packing allows the atoms in one layer to get as close as possible to those in the next layer.

In this giant structure of atoms, electrons in the outer shell of each metal atom are free to move through the whole structure — they are not held by one or two atoms, but are free to move everywhere. These electrons are described as being 'delocalised'. The metal can be thought of as a network of positive metal ions with negative electrons moving around and between them (Figure 5.35). The electrostatic attractions between the 'sea' of delocalised electrons and the positive metal ions result in strong forces that hold all the metal atoms together.

> In **giant metallic structures**, metal atoms are close-packed in a giant lattice through which the outer shell electrons can move freely.

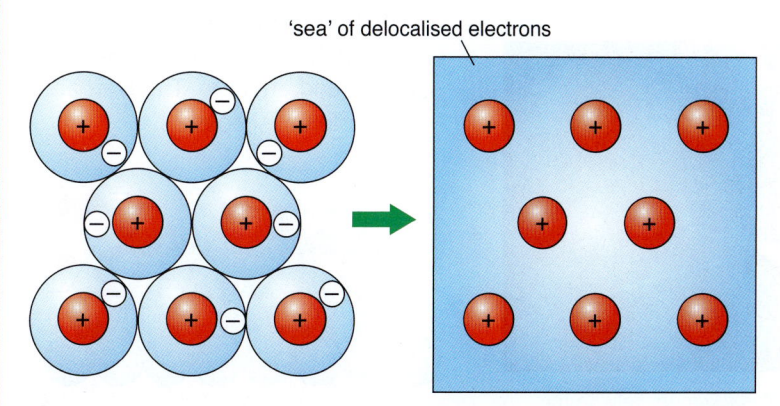

Figure 5.35 Attractions between negative delocalised electrons and positive ions result in strong forces between metal atoms.

Chemistry 2

Chapter 5 Bonding in reactions

Properties of metals

In general, metals:
- have high densities
- have high melting points and high boiling points
- are good conductors of heat and electricity
- are **malleable** — can be bent or hammered into different shapes without breaking

> A **malleable** material is one that can be hammered into a different shape without breaking. This is a property of metals.

All the properties of metals can be explained by the close-packed structure of metal ions in a sea of electrons.
- High density — the close packing means that all the atoms get as close as possible to atoms in their own layer and to atoms in the layers above and below them, resulting in a high density.
- High melting points and high boiling points — the atoms are closely packed with strong forces between them. So, in most metals it takes a lot of energy to separate the particles and change state.
- Good conductors of heat — when a metal is heated, energy is transferred by the electrons. The delocalised electrons move around faster and conduct the thermal (energy) rapidly to other parts of the metal.
- Good conductors of electricity — when a metal is connected in a circuit, the delocalised electrons move towards the positive terminal (Figure 5.36). This flow of electrons through the metal forms an electric current.

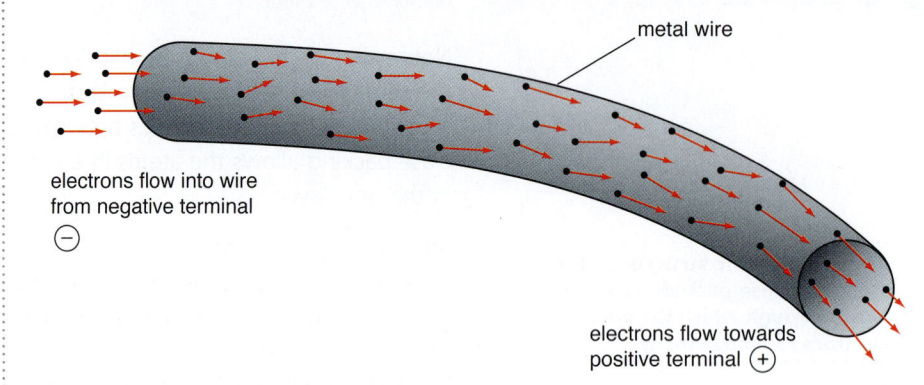

Figure 5.36 Electrons flowing along a metal wire form an electric current.

- Malleable — the bonds between metal atoms are strong, but they are not fixed and rigid. When a force is applied to a metal, the layers of atoms can slide over each other. This is known as 'slip'. After slipping, the atoms settle into close-packed positions again. Figure 5.38 shows the positions of atoms before and after slip. This is what happens when a metal is bent or hammered into different shapes.

Figure 5.37 Blacksmiths rely on the malleability of metals to hammer and bend them into useful shapes.

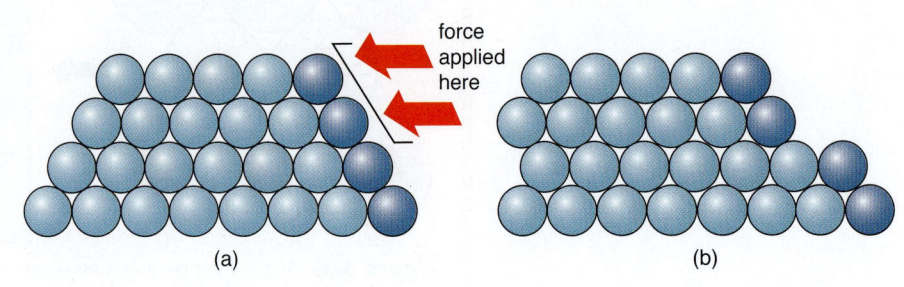

Figure 5.38 The positions of atoms in a metal: (a) before and (b) after slip has occurred.

C2 5.7 Metals — giant metallic structures

Explaining the properties of alloys

X-ray crystallography shows that the atoms in most metals are packed as close together as possible (Figures 5.33 and 5.34). An alloy is a mixture of a metal with other elements — usually other metals but it can be carbon as well. The introduction of other atoms of different sizes has a big effect on the properties of the metal, particularly the 'slip' property.

An good example of this is in a common alloy of iron called steel. In steel, the regular arrangement of iron atoms is disrupted by adding smaller carbon atoms. The different-sized carbon atoms distort the layers of iron atoms in the structure of the pure metal. This makes it more difficult for the layers to slide over each other, so the steel alloy is harder and much less malleable (Figure 5.39).

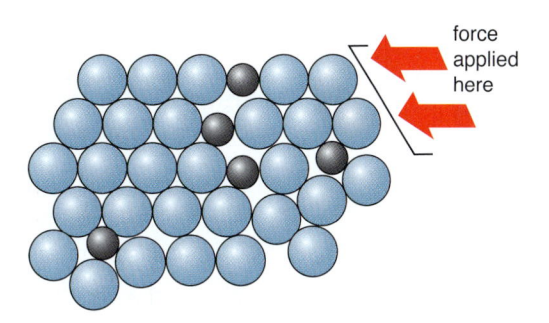

Figure 5.39 Slip cannot occur so easily in a steel alloy because the atoms of different sizes cannot slide over each other.

Similar properties are seen in other alloys:
- Brass is an alloy of zinc and copper.
- Bronze is an alloy of tin and copper.
- Solder is an alloy of tin and lead.
- Aluminium alloys contain aluminium and magnesium or copper.

Memory metals

In recent years, smart alloys have been developed — they are sometimes called **memory metals**. These can return to their original shape after being deformed. Smart alloys are excellent for use in spectacle frames and in the tooth braces fitted by dentists. The smart alloy braces are made so that, after fitting, they will return to their original shape and pull the teeth into better alignment.

> A **memory metal** is a metal alloy that when heated returns to its original shape after deformation.

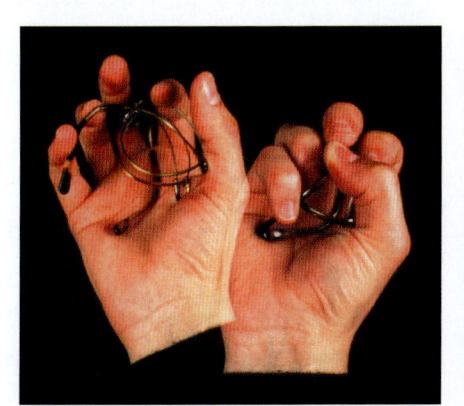

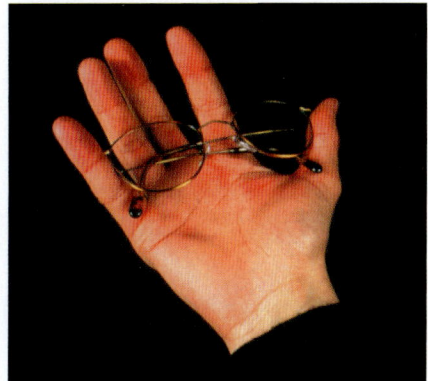

Figure 5.40 Spectacle frames made of smart alloys revert to their original shape in a second, after being bent and twisted.

Chemistry 2

Chapter 5 Bonding in reactions

These smart materials are alloys of nickel and titanium, usually called nitinol. The alloys have two different states — a state in which they can be bent at a low temperature, and an elastic state at a higher temperature. When the metal alloy is heated past this transition temperature, the metal snaps from one state to another. With spectacles and dental braces, in normal use the metal is above its transition temperature and always returns to its original shape.

Stents

A stent is a coiled spring that is put inside a blood vessel to hold it open. Stents are now made of smart memory metals such as nitinol. The stent is cooled and completely straightened out. It is then placed in the blood vessel to be treated. As the stent warms up to body temperature it regains its coiled-spring shape and holds the blood vessel open to the correct extent.

Boost your grade ✓

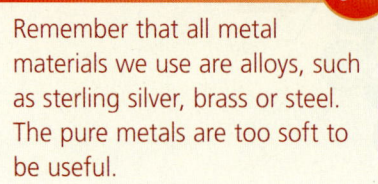

Remember that all metal materials we use are alloys, such as sterling silver, brass or steel. The pure metals are too soft to be useful.

Investigate the properties of metal and alloys
HOW SCIENCE WORKS — PRACTICAL SKILLS

1. Compare a bar-breaker bar (made of cast iron) and a similar length cut from a 10 cm mild steel nail. Examine their appearance and compare their densities, masses and colours.

2. Use the bar-breaker apparatus to break the bar in two. Compare the effect of a similar procedure on the mild steel sample. Examine the broken end of the cast iron sample with a hand lens.

3. Use wires made from steel, copper, nichrome alloy, solder and any others easy to obtain. Compare samples of metal and alloy wires for:
   a flexibility
   b hardness — use a blow from a sharp edge and examine how large the v-shaped dent is
   c conductivity — use a multimeter on a resistance scale on wire samples with similar dimensions)
   d melting point — use a crucible and pipe clay triangle
   e tensile strength — hang masses from similar samples till they break.

4. Make metal crystals in displacement reactions. For example, try:
   - copper wire in silver nitrate solution
   - magnesium ribbon in zinc nitrate solution
   - a strip of zinc foil in lead nitrate or lead ethanoate solution

5. Get a piece of memory metal wire. Straighten out the metal wire, then warm it gently in hot water. Observe what happens.

AQA GCSE Additional Science

C2 5.7 Metals — giant metallic structures

The memory metal spring in Figure 5.41 shows how memory properties work. The smart memory spring can be stretched by applying a load. When the spring is heated (for example, by passing a current through it), it returns to its original shape with considerable force lifting the mass upwards.

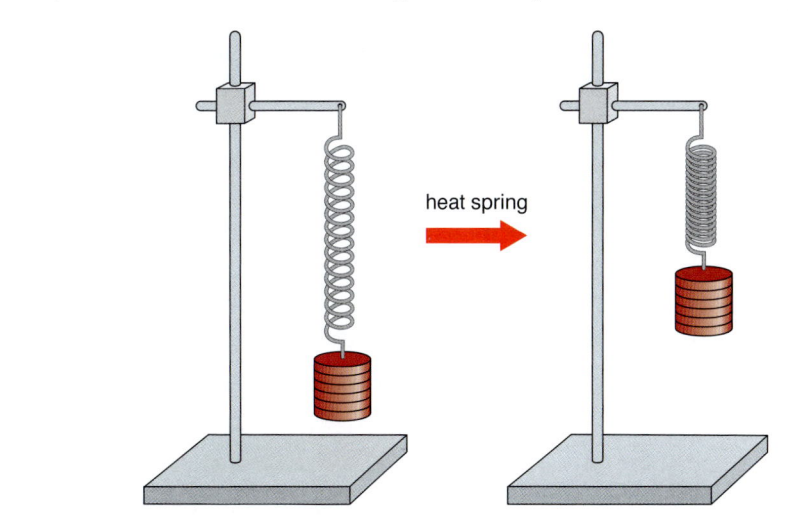

Figure 5.41 Shape-memory metal spring.

Choosing metals for different uses

HOW SCIENCE WORKS ACTIVITY

Various properties of five metals are shown in Table 5.5.

Table 5.5 Properties of five commonly used metals.

Metal	Density in g/cm³	Relative strength compared with iron	Melting point in °C	Relative electrical conductivity compared with iron	Relative thermal conductivity compared with iron	Rate of corrosion	Cost per tonne in £
Aluminium	2.7	0.33	660	3.7	3.0	Very slow	1325
Copper	8.9	0.62	1083	5.8	4.8	Very slow	4400
Iron	7.9	1.00	1535	1.0	1.0	Quick	230
Silver	10.5	0.39	962	6.1	5.2	Very slight	189000
Zinc	7.1	0.51	420	1.6	1.4	Slow	1800

1 Use the information in Table 5.5 to explain the following statements.
  a The frames and poles of tents are made from aluminium.
  b Electrical cables in the National Grid are made of aluminium, even though copper is a better conductor.
  c Bridges are built from steel, which is mainly iron — even though it corrodes (rusts) faster than the other metals.
  d Metal gates and dustbins are made from iron coated (galvanised) with zinc.
  e Silver is no longer used to make coins.
  f High-quality saucepans have copper bottoms rather than steel (iron).

2 a What is the range in values for (i) density, (ii) melting points in Table 5.5?
  b Do you think any of the values given for density or melting point are anomalous (unusual or irregular)? Explain your answer.

Chemistry 2

Chapter 5 Bonding in reactions

③ Luke and Kathy were working together on a metals project. Luke said, 'If the atoms in a metal pack closer, the density should be higher, the bonds between atoms should be stronger and so the melting point should be higher.' Kathy agreed with Luke and predicted that there could be a pattern (relationship) between the densities and melting points of metals.
  a Use the data in Table 5.5 to plot a graph of density against melting point for the five metals.
  b Is there a pattern between the densities and melting points of the metals? Say 'Yes' or 'No' and explain your answer.

④ The explanation of both electrical conductivity and thermal conductivity in metals relies on the idea of delocalised electrons. This suggests that there should be a pattern (relationship) between electrical and thermal conductivities for metals. Use the relative electrical and thermal conductivities in Table 5.5 to check whether there is a pattern or not.
  a Describe briefly what you did to investigate whether there is a pattern.
  b Is there a pattern? Say 'Yes' or 'No' and explain your answer.

> **Test yourself**
>
> 24 Explain, in terms of atoms, why blacksmiths usually get metals red hot before hammering them into the shapes they want.
>
> 25 Which of the following statements provide evidence for classifying copper as a metal? Copper:
>   A burns to form an oxide
>   B reacts with non-metals
>   C reacts only with non-metals
>   D is non-magnetic
>   E conducts electricity
>   F has two common isotopes
>
> 26 Look at the suit of armour in Figure 5.42.
>   a What particles make up the structure of steel in armour?
>   b How are these particles arranged in steel?
>   c What properties of steel are important for its use in armour?
>   d Explain, in terms of particles, why steel has one of the properties you gave in answer to c.
>
> **Figure 5.42** Steel was used to make suits of armour.
>
> 27 Why are the electrons in the outermost shell of metal atoms described as 'delocalised'?
>
> 28 Write a paragraph that could be given to hospital patients to explain how a memory metal stent will keep their arteries open.
>
> 29 Invent a practical household application that would use a memory metal spring to detect overheating or a fire in the home.

C2 5.8 Polymer materials

> ### Atomic models for giant metallic structures — ACTIVITY
>
> You need about 25 marbles or 25 polystyrene spheres for this activity. The marbles or polystyrene spheres represent atoms.
>
> 1. Put three books flat on a bench so that they form a triangle around 10 atoms. Move the books carefully so that they enclose a layer of close-packed atoms like those in Figure 5.33.
> 2. Now add another layer of atoms on top of the first one.
> 3. Put further layers on top of this.
> 4. Look carefully at your structure, which represents a good model for most metals. Push one layer of atoms all at the same time (use a ruler) and observe what happens when the layers 'slip'
>   a. How many atoms does an atom in the middle of the structure touch?
>   b. Why is this called a 'close-packed' structure?
>   c. Why is this a giant structure?
>   d. What happens to the position of the atoms when 'slip' occurs?
>   e. How does this model help us to understand metals?
>
> The structure that you have built for metals is sometimes described as a 'space-filling' model because it shows all the space occupied by metal atoms.
>
> Earlier in this chapter, we used space-filling models to represent the structures of simple molecular substances in Figure 5.16.
>
> These simple molecular substances can also be studied by using ball-and-stick models like that in Figure 5.31. Ball-and-stick models are useful because the 'sticks' show the bonds that hold the atoms together.

C2 5.8 Polymer materials

Learning outcomes

- Know that polymers are long-chain covalent macromolecules made from smaller molecules.
- Know that some polymers soften on heating (thermoplastic polymers).
- Know that some polymers are rigid three-dimensional networks (thermosetting polymers).
- Know that different manufacturing conditions can lead to polymers with different properties.

LD and HD polyethene

Polyethene (polythene) is probably the most common plastic. Joining ethene **monomer** units together makes polyethene. LD (low-density) polyethene is used for making plastic bags. HD (high-density) polythene is used for making garden chairs. These two different uses demand different properties from the **polymer** that is used. This can be achieved by altering the conditions when the ethene molecules are polymerised to make the plastic. The processes that make the two different types of polythene use the same raw material (ethene gas), but the conditions are different.

LD polyethene is made in a high-pressure chamber with a little oxygen to get the polymerisation started. In the chamber, the high pressure squashes the ethene molecules together so much that they join up. This is a fairly random process and so produces a plastic with lots of little branches and side chains along its length. The polymer molecules do not pack closely together and the structure of the final product is random and not too closely packed. This means that it has a low density and is easy to bend.

Chemistry 2

Chapter 5 Bonding in reactions

> A **polymer** is a molecule made up of many repeating units.
>
> A **monomer** is a simple molecule that joins with similar molecules to make a polymer.

HD polyethene is made in a much more controlled way using a titanium-based catalyst to align the ethene molecules as they join. Hence the polymer molecules are much more ordered and contain fewer side-chain branches. These polymer molecules can pack more closely together giving a product that is denser, stiffer and stronger than LD polythene.

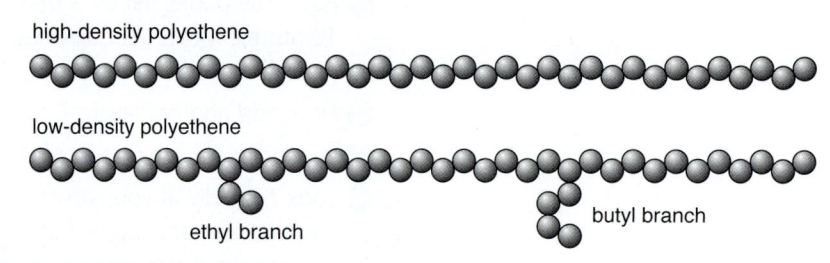

Figure 5.43 LD and HD polyethene molecules.

Table 5.6 Properties of LD and HD polyethene.

Property	LD polyethene	HD polyethene
Recycling symbol	4	2
Density	0.92 g/cm^3	0.95 g/cm^3
Max. temperature for use	80°C	110°C
Flexibility	Very flexible	Quite stiff
Amount of side chain branching	Considerable	Hardly any
Tensile strength	Lower	Higher
Resilience	Higher	Lower
Uses	Plastic bags, food trays, food and packaging wrap, six-pack rings, juice cartons	Pipes and ducts for water, sewerage, cables; tables chairs, sheds, electrical insulation, bottles

Investigating the strength of plastics *HOW SCIENCE WORKS ACTIVITY*

Plan and carry out an investigation to find which supermarket plastic bag is the strongest.

Thermosetting and thermosoftening polymers

Some polymers, like polyethene and polypropene, are only useful at lower temperatures. These are called **thermosoftening** polymers. Other plastics are rigid and stiff no matter what temperature they are at. These are called **thermosetting** polymers.

> A **thermosetting** material goes rigid when it is formed and its shape remains fixed.
>
> A **thermosoftening** material softens on heating and can be reshaped.

C2 5.8 Polymer materials

Thermosoftening polymers

A plastic bucket made of one of these and filled with hot water becomes flexible. Some thermosoftening plastics soften and become useless as the temperature rises. Try heating polyethene bags or polystyrene packaging gently.

Thermosoftening polymers are a mass of individual tangled polymer chains. As they get hotter, the polymer chains slide past each other more and more easily, and this is what makes the plastic behave as a soft material. The only forces holding the thermoplastic molecules together are weak intermolecular forces.

Thermosetting polymers

Some polymer production requires the plastic to be cured, or set, in the final stage of the process. Examples of this are epoxy resins and the vulcanisation of latex rubber. The first part of the production process makes flexible long-chain polymers, like polyethene or polystyrene.

The final stage involves another chemical reaction that makes links between the polymer chains, which are then locked together into a rigid three-dimensional lattice. The chemical bonds formed lock the lattice together. If the polymer molecules cannot slide past each other, the structure becomes rigid.

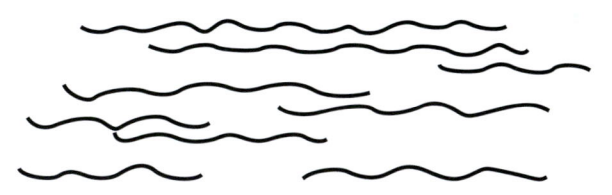

flexible polymer molecules

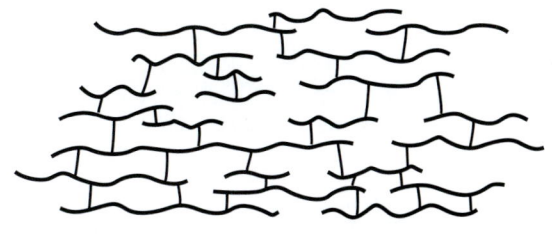

crosslinks make the structure more rigid

Figure 5.44 Thermosoftening polymers are made of long flexible molecules. In rigid thermosetting polymers, chemical bonds have formed between neighbouring molecules. These bonds join different parts of the polymer together in a three-dimensional network. The cross-links lock the molecules together and stop them from being flexible.

Recycling plastics

HOW SCIENCE WORKS / ICT

Create a resource (paper-based or electronic) to be used to enable members of the public to recognise the different types of plastics for recycling.

Slime

Slime is a cross-linked polymer. Polyethenol consists of long-chain molecules that are free to move around in solution. When sodium tetraborate (borax) solution is

Figure 5.45 A car tyre would not last 20 kilometres if it was made out of untreated latex. Sulfur is added to the latex to cross-link and harden the rubber.

Chemistry 2   115

Chapter 5   Bonding in reactions

added, links are made between the chains. As more cross-linking borax solution is added, the 'slime' produced behaves more like a solid (Figure 5.46).

Figure 5.46 Polymers — two runny liquids make a semi-solid.

> ### Comparing plastics
> **HOW SCIENCE WORKS — PRACTICAL SKILLS**
>
> 1. Show the difference between LD and HD polyethene by using a 50% ethanol/water mixture.
> 2. Make slime using polyethenol and investigate the effects of different amounts of the borax cross-linking agent on the stiffness of the resulting material.
> 3. Compare the effect of gentle heating (use hot water) on the flexibility of samples of different polymers. Compare:
>   - LD polyethene from plastic bags
>   - HD polyethene from domestic objects — look for '2' on the recycling label
>   - fibreglass objects
>   - polystyrene objects — from model kits
>   - PVC objects — use off-cuts from PVC window frames
>   - Perspex
>   - polycarbonate — used fizzy drinks bottles

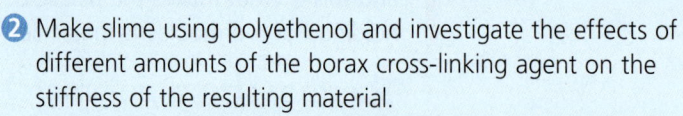

Test yourself

30. Copy and complete Table 5.7. Put ticks in the correct column for the type of polymer to be used, and write a brief explanation.

    Table 5.7

Application	LD polyethene	HD polyethene	Thermosetting polymer	Why this plastic is used
Electric plug socket				
Book covering plastic				
Small portable step to reach a sink				
Control knob for a cooker				

31. Draw a diagram and annotate it to explain why solid LD polyethene with branched polymer chains has a lower density than the straight-chain HD polythene.

32. Invent a hazard label for a thermosoftening polymer bucket warning users not to carry hot liquids in it.

33. a Explain what happens when a polymer forms cross-links.
    b Fibreglass is glass-reinforced plastic (GRP), When the resin used to make it is set, a hardener is mixed with the semi-runny plastic liquid. What does the hardener do?
    c Draw a simple diagram to show the long-chain polymers in fibreglass resin and what they would look like after the resin has set solid.

C2 5.9 Nanomaterials

Learning outcomes

- Know that nanomaterials have different properties from normal-sized chunks of the same material because of their small size.
- Be able to evaluate the advantages and disadvantages of using nanomaterials.

Nanotechnology involves working with and controlling atoms and molecules with sizes between 1 nm and 100 nm.

Nanoparticles are manufactured particles with sizes measured in nanometres (nm).

Nanomaterials and **nanotechnology** belong to the twenty-first century. Few people had heard of these terms before the year 2000. Since then, a wide range of materials has been developed and used in the form of **nanoparticles**, which are between 1 and 100 nanometres (nm) in size — between one millionth and one ten-thousandth of a millimetre.

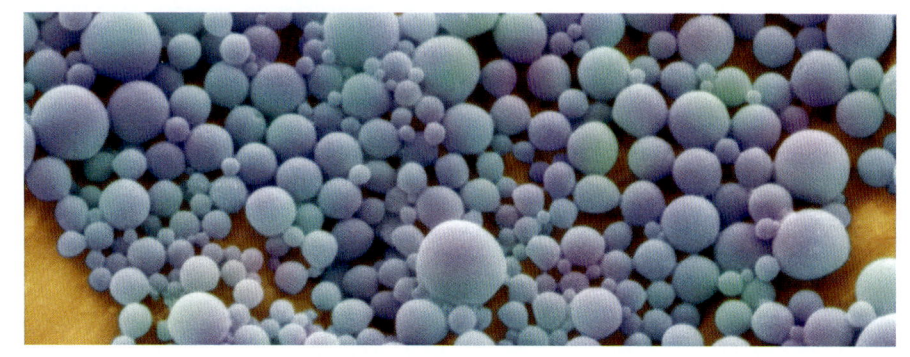

Figure 5.47 These polymer nanoparticles have been designed to carry drug molecules into the body. Just 100 nm to 500 nm across, they can pass through vein walls, and so the drug can be released into tissues in exactly the part of the body that needs treating.

Nanoparticles are groups of just a few hundred atoms. They include metals, carbon, plastics and polymers. Nanoparticles show different properties from the same material in bulk (large pieces) because of the precise way in which their atoms are arranged and because of their high surface area to volume ratio.

Nanotechnology is already being used to manufacture products that take advantage of the ability to make layers of material just a few atoms thick, or particles with a large surface area to mass ratio. Here are some of the current applications.

- Cosmetics is a boom area for nanotechnology. Anti-ageing creams can use nanotechnology to deliver anti-wrinkle preparations through the skin. Sunscreen creams with nanoscale pieces of zinc oxide or titanium dioxide look transparent so are more acceptable. Nanoparticle deodorants can deliver tiny particles of silver that kill bacteria on the skin to prevent smells developing.
- A huge development for nanotechnology is in stain-resistant clothing. Fabrics can be coated with nano-sized polymer particles. Any liquid spilt on the fabric turns into 'beads' and rolls off without staining the fabric. The same type of coating is put on some windows and they rarely need to be cleaned.
- Ultrathin, yet extremely strong, coatings of scratch-resistant polymers are being used to protect lenses for spectacles, cameras and contact lenses. The coatings are so thin that the polymer material is transparent.
- New photovoltaic cells have been developed with nano-sized metal crystals on their surface. These have a vast surface area on which to collect the Sun's energy more efficiently.
- Ordinary tennis rackets are weakened by tiny gaps between the carbon fibres in their structure. These tiny gaps can be plugged and reinforced with nano-sized crystals of silicon dioxide.
- Nanoparticles have also led to important developments in catalysts because of their large surface area compared with ordinary particles of materials. These

Chemistry 2

Chapter 5 Bonding in reactions

developments include the use of gold particles to oxidise poisonous carbon monoxide to carbon dioxide. The gold particles are totally inactive until they are smaller than about 8 nm in diameter. Their high surface area means you need only a tiny amount of gold — which is much cheaper.

In development are nanotechnology applications for:
- smaller and more sensitive sensors in electronic devices, and potentially smaller and faster computers and disk drives with layers of magnetic nanoparticles to increase data storage capacity
- lighter yet stronger construction materials made from composite materials containing nanotubes — carbon nanotubes are 10 times lighter than steel but 250 times stronger
- 'smart' clothing with nano-sized sensors to measure pulse rate and temperature

Are nanomaterials safe?

Although nanomaterials open up exciting possibilities for new materials, there are concerns about the effects of nanomaterials when they get into the human body. Nanoparticles could be more reactive, and so more toxic than larger masses of the same material. Although most areas of nanotechnology pose no risks, scientists agree that more research into the health risks of nanomaterials is needed. They also suggest that:
- nanoparticles should be treated as new chemicals
- an independent scientific safety committee should approve new substances at the nano-scale before they are used in consumer products
- nanoparticles that have direct contact with our bodies, such as those in cosmetic and sunscreen creams, are the most likely to cause long-term damage
- there needs to be workplace regulations to protect workers in industry who are exposed to nanoparticles

---

Nanomaterials and the future *HOW SCIENCE WORKS — ACTIVITY*

1. Explain what is meant by nanoscience in no more than 25 words.
2. Scientific and technological developments offer different opportunities to different groups of people. Which groups of people stand to benefit most from the development of nanomaterials?
3. Describe three benefits to society from the development of nanomaterials.
4. Describe two potential risks from their development.
5. Should scientists or society make decisions about the development of nanomaterials?

Researching nanotechnology *HOW SCIENCE WORKS — ICT*

Nanotechnology is developing on a daily basis. Research and present the latest uses of nanotechnology.

C2 5.10 From structure to properties and uses

Learning outcomes

- Know that the structure and bonding of a substance determine its properties, and that these properties determine its uses.
- Know that the particles in substances may be atoms, ions or molecules.
- Know that these three different particles give rise to four different structures (giant metallic, giant covalent, giant ionic and simple molecular) with different properties.
- Be able to relate the properties of substances to their uses
- Be able to identify the structure of a substance from its properties.

If you know how the particles in a substance are arranged (its structure) and how they are held together (its bonding), you can explain the properties of the substance. For example, copper is composed of close-packed atoms with freely moving outer electrons. These electrons move through the structure when copper is connected to a battery, so it is a good conductor of electricity. Atoms in the close-packed structure of copper can slide over each other and because of this copper can be drawn into wires. These properties of copper make it useful in electrical wires and cables.

It is important to realise that:
- the structure and bonding of a substance determine its properties
- the properties of a substance determine its uses

The links between structure/bonding and properties help us to explain the uses of substances and materials. They explain why metals are used as conductors, why graphite is used in pencils and why clay is used to make pots.

Figure 5.48 Wet clay is soft and easily moulded by the potter. After he has moulded the clay, it is heated (fired) in a furnace. This makes it hard and rigid so it can be used for making pots and crockery.

All substances are made up from only three kinds of particle — atoms, ions and molecules.

These three particles give rise to four different solid structures:
- giant metallic
- simple molecular
- giant covalent (macromolecular)
- giant ionic

Table 5.8 shows the particles in these four structures, the types of substances they form and examples of these substances.

Chemistry 2

Chapter 5 Bonding in reactions

Table 5.8 The four types of solid structures and the particles they contain.

Type of structure	Particles in the structure	Type of substance	Structure	Bonding	Properties
Giant metallic	Atoms close-packed	Metals, e.g. Na, Fe, Cu and alloys such as steel		Atoms are held in a close-packed giant structure by the attraction of positive ions for delocalised electrons.	• High melting points and boiling points • Conduct electricity • High density • Hard but malleable
Giant covalent (macro-molecular)	Very large molecules containing thousands or billions of atoms (giant molecules)	A few non-metals (e.g. diamond, graphite) and some non-metal compounds (e.g. polyethene, PVC, silicon dioxide (sand))		Large numbers of atoms are joined together by strong covalent bonds to give a giant 3D lattice or a very long, thin molecule.	• High melting points and boiling points • Do not conduct electricity (except graphite) • Hard but brittle (3D structures) or flexible polymers
Simple molecular	Small molecules containing a few atoms	Most non-metals and non-metal compounds, e.g. O_2, I_2, H_2O, CO_2, sugar		Atoms are held together in small molecules by strong covalent bonds. The bonds between molecules are weak.	• Low melting points and boiling points • Do not conduct electricity • Soft when solid
Giant ionic	Ions	Metal/non-metal compounds, e.g. Na^+Cl^-, $Ca^{2+}O^{2-}$, $Mg^{2+}(Cl^-)_2$		Positive and negative ions are held together by the attraction between their opposite charges — ionic bonds.	• High melting points and boiling points • Conduct electricity when melted or dissolved in water • Hard but brittle

Test yourself

34 Match each of the following substances to one of the solid structures in Table 5.8:
  a chlorine (Cl_2)
  b limestone (calcium carbonate, $CaCO_3$)
  c gold (Au)
  d PVC (polyvinyl chloride)
  e copper sulfate ($CuSO_4$)
  f methane in natural gas (CH_4)

35 Look at the photo of a lead sulfide crystal in Figure 5.49.

  a What elements does lead sulfide contain?
  b Is each of these elements a metal or a non-metal?
  c What particles does lead sulfide contain: atoms, ions or molecules?
  d How do you think the particles are arranged in lead sulfide? Explain your answer.

36 Copy and complete these sentences using the words given below.

bond   hard   molecules   moulded   three   two   water

Wet clay is soft and easily _____ because _____ molecules can get between its flat _____-dimensional particles. When the clay is fired, all the water _____ are driven out. Atoms in one layer _____ to those in the layers above and below. This gives the clay a _____-dimensional structure making it _____ and rigid.

Figure 5.49 A natural crystal of lead sulfide.

Homework questions

1. Use the information in Table 5.2 and Table 5.4 to answer this question.

   The pairs of elements in Table 5.9 react to make compounds. For each compound made:
   a. Decide if it would have ionic or covalent bonding.
   b. If it is ionic draw a simple representation of the ions combined in the correct ratio to give a neutral product.
   c. If it is covalent draw an electron-sharing diagram to show how a molecule of the compound is bonded.

   Table 5.9

A	phosphorus	with	bromine
B	zinc	with	chlorine
C	carbon	with	chlorine
D	sodium	with	phosphate
E	silicon	with	hydrogen
F	iron(II)	with	sulfate
G	calcium	with	chlorine
H	nitrogen	with	iodine
I	aluminium	with	fluorine
J	silver	with	bromine
K	hydrogen	with	fluorine
L	magnesium	with	nitrogen

2. Butane (C_4H_{10}) is a colourless gas that boils at 0.5°C and burns easily in air.
   a. Explain how the information above gives evidence that butane is a covalent compound made of molecules.
   b. Draw an electron-sharing picture of a molecule of butane.
   c. What unique property of carbon allows it to form long molecules like butane.

3. Carbon dioxide: gas at room temperature, sublimes at −77°C

   Silicon dioxide: solid at room temperature, melts at 1650°C

   a. How many covalent bonds does carbon form? What is the formula of carbon dioxide?
   b. How many covalent bonds does silicon form? What is the formula of silicon dioxide?
   c. What does the information above tell you about the similarity of carbon dioxide and silicon dioxide?
   d. Explain the differences between carbon dioxide and silicon dioxide.
   e. Explain how the bonding between atoms is arranged.
   f. Draw simple diagrams to show the difference in the structure and bonding in carbon dioxide and silicon dioxide.

4. Explain the meaning of the following terms and give an example of a use of each type of material. Explain how the properties of the material make it especially useful. If possible illustrate your answer with diagrams.
   a. Buckyballs
   b. Carbon nanotubes
   c. Alloys
   d. Shape-memory metals
   e. HD polyethene and LD polyethene
   f. Thermosetting polymers

5. a. Make a list of six common metallic elements. For each one explain what it is used for as a material.
   b. Draw the structure of a metal such as magnesium. Show the positions of the atomic nuclei and the electrons orbiting them.
   c. Make a list of the properties you would expect of metals.
   d. What is different about the positioning of electrons in metals compared to that in non-metallic materials?
   e. Explain how the properties of metals are related to their structure.

6. Make a list of all the objects you use at home or school that are made from plastics. For each object describe the natural materials you would need to make it if there were no plastics. Which property would each material need?

Chapter 5 Exam corner

Exam corner

a (i) Describe how sodium sulfide (Na₂S) is formed and how the atoms have joined together. *(4 marks)*

Student A

(i) Sulfur and sodium ions. ✓

Examiner comment The mark scheme looks for: ionic compound (1); sodium atom transfers 1 electron to sulfur atom (1); sulfur atom takes two electrons from two sodium atoms (1); ions stay together by electrostatic attraction (1). Student A gets 1 mark for 'ions'.

Student B

(i) [Na]⁺ → [S]²⁻ ← [Na]⁺ ✓✓✓✓
electrostatic force ← → electrostatic force

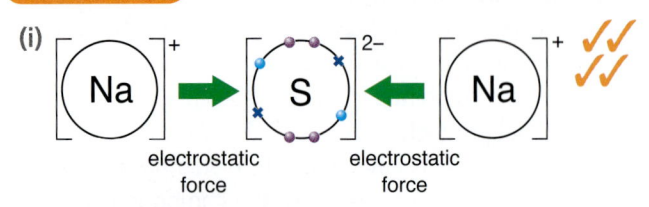

Examiner comment This is a full answer. Even though drawn as a diagram, all the mark points have been met.

(ii) The physical properties and particles in sodium sulfide are different to those of hydrogen sulfide. Explain why. *(4 marks)*

Student A

(ii) One is ionic, one is covalent ✗✗

Examiner comment The mark scheme is looking for: low melting point (1); molecules (1) in hydrogen sulfide. High melting point (1) and giant ionic lattice (1) in sodium sulfide. Student A does not mention properties or particles.

Student B

(ii) Sodium compound is a giant ionic lattice ✓ with a high melting point. ✓

Hydrogen compound is made of small molecules ✓ so has a low melting point. ✓

Examiner comment This is almost a full answer — 'sodium' and 'hydrogen' are not the subjects of the question. But this is not penalised.

---

b What other type of covalent structure is there and how is it different from, for example, hydrogen sulfide? *(2 marks)*

Student A

b Giant lattice ✓

Examiner comment Correct structure type but no mention of the difference.

Student B

b The covalent bonds are in a three dimensional lattice ✓ throughout the whole bulk of the compound. This means it is a hard solid with a high melting point ✓

Examiner comment Full answer with an acceptable difference.

122 AQA GCSE Additional Science

Chemistry 2

Chapter 6
Mass, moles and yield

Setting the scene

American pancakes — makes six large pancakes

Ingredients:
- 300 g of flour
- Half a bottle of milk
- 2 eggs — beaten
- 25 cm^3 of oil
- 1 small packet of baking soda

Pour the milk into the flour and stir well. Add the oil and eggs. Stir again. The batter mix must be fluid. If not, add a little water. Oil a frying pan. Pour in some batter mix and spread it over the bottom. Heat!

If you needed to make enough for 30 pancakes what do you do? Not all the ingredients are in the same units. Could this be a problem to a cook? What would you do the make the units more 'scientific'? How would you find the mass of ingredients in one pancake?

ICT

In this chapter you can learn to create a multimedia presentation on instrumental methods of analysis.

Practical work

In this chapter you can learn to:
- identify potential risks when carrying out an experiment
- identify anomalous results in a set of data
- understand why the resolution of a balance is important

Chapter 6 Mass, moles and yield

C2 6.1 Atomic number and mass number

Learning outcomes

- Know that atoms are too small to be weighed and that their mass is given relative to the fixed mass of a carbon-12 atom.
- Know that the atomic number is the number of protons in the nucleus of an atom.
- Know that the mass number is the number of protons and neutrons in the nucleus of an atom.
- Know the relative masses and relative charges of protons, neutrons and electrons.

We now know that all atoms are made up of protons, neutrons and electrons. The relative masses, relative charges and positions within atoms of these subatomic particles are summarised in Table 6.1.

Table 6.1 Relative masses, charges and positions within atoms of protons, neutrons and electrons.

Particle	Relative mass (atomic mass units)	Relative charge	Position within atoms
Proton	1	+1	Nucleus
Neutron	1	0	Nucleus
Electron	$\frac{1}{2000}$	−1	Orbiting the nucleus

Atoms of different elements have different numbers of protons. Hydrogen atoms are the only atoms with one proton; helium atoms are the only atoms with two protons; lithium atoms are the only atoms with three protons; and so on. This means that the number of protons in an atom determines which element it is. Scientists call this the **atomic number** or **proton number**. So, hydrogen has an atomic number of 1, helium has an atomic number of 2, lithium has an atomic number of 3, and so on.

Hydrogen, the first element in the periodic table, has atoms with one proton. Helium, the second element in the periodic table, has atoms with two protons. Lithium, the third element in the periodic table, has atoms with three protons, and so on. So, the position of an element in the periodic table tells us how many protons it has. Aluminium is the thirteenth element in the periodic table. This means it has 13 protons and its atomic number (proton number) is therefore 13.

> The **atomic number** is the number of protons in an atom.

The number of protons in an atom can tell you which element it is, but it cannot tell you its mass. The mass of an atom depends on the number of protons plus the number of neutrons. This number is called the **mass number** of the atom.

Hydrogen atoms (with 1 proton and 0 neutrons) have a mass number of 1. Helium atoms (2 protons and 2 neutrons) have a mass number of 4, and sodium atoms (11 protons and 12 neutrons) have a mass number of 23.

We can write the symbol $^{23}_{11}$Na (Figure 6.1) to show the mass number and the atomic number of a sodium atom. The mass number is written at the top left and the atomic number at the bottom left of the symbol.

> The **mass number** is the total number of protons plus neutrons in an atom.

mass number = 23

Na

atomic number = 11

Figure 6.1 Mass number and atomic number shown with the symbol for sodium.

C2 6.2 Comparing the masses of atoms

> **Test yourself**
>
> 1 Look at Figure 6.1.
>   a How many protons, neutrons and electrons are there in:
>   (i) one Na atom?
>   (ii) one Na⁺ ion?
>   b What do 27, 13, 3+ and Al mean with reference to the symbol $^{27}_{13}Al^{3+}$?
>
> 2 Write the symbols for the following in the same style as $^{23}_{11}Na$, using the periodic table on page 283:
>   a magnesium
>   b hydrogen
>   c a potassium ion
>   d an oxide ion
>
> 3 Work out, as a decimal fraction or as a percentage, the amount of extra mass the electrons contribute to a sodium atom with 11 protons, 12 neutrons and 11 electrons.

C2 6.2 Comparing the masses of atoms

Learning outcomes

- Understand how atomic number and mass number relate to isotopes and relative atomic masses.
- Know that isotopes of an element have different numbers of neutrons in their atoms.

A **mass spectrometer** is a device that sorts particles and molecule fragments and determines their relative mass.

Relative atomic masses

A single atom is so small that it cannot be weighed on a balance. However, the mass of one atom can be *compared* with that of another atom using an instrument called a **mass spectrometer** (Figure 6.2).

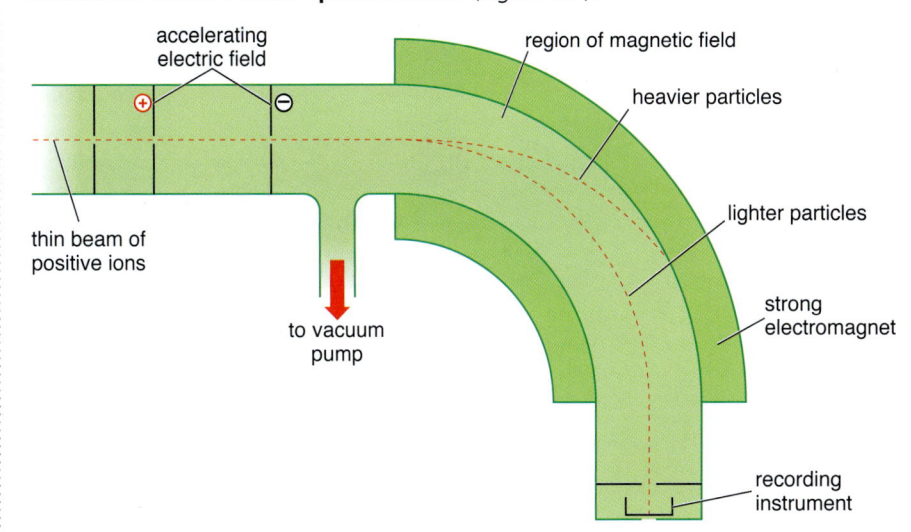

Figure 6.2 A mass spectrometer produces positive ions of the elements being analysed. A thin beam of these ions passes through an electric field, which accelerates the ions. The ions then pass through a magnetic field, which deflects them. How much they are deflected depends on the mass of the particles — lighter particles are deflected more than heavier particles. From the amount of deflection, we can compare the masses of different atoms and make a list of their relative masses.

Chemistry 2

Chapter 6 Mass, moles and yield

> **Relative atomic masses** tell us the relative masses of the atoms of different elements, relative to a atom of $^{12}_{6}C$ having a mass of 12.000.

> One **mole** of a substance contains 6×10^{23} particles. One mole weighs the relative atomic or molecular mass in grams.

> **Boost your grade** ✓
>
> 1 mole is the number we use to convert the tiny world of atoms into quantities that we can weigh out.

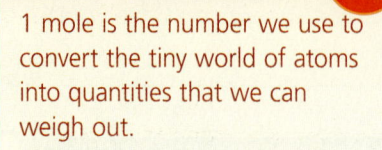

A mass spectrometer must be calibrated before it is used. Atoms of known mass are analysed in the instrument to find by how much they are deflected. Scientists can then find the masses of other atoms from their extent of deflection. This is an example of calibration, which involves marking a scale on a measuring instrument.

The relative masses of different atoms are called **relative atomic masses** (A_r). The relative atomic mass of hydrogen is 1, the relative atomic mass of helium is 4 and the relative atomic mass of carbon is 12.

The relative atomic mass of an element in grams is sometimes called one **mole**.

So,   1 g of hydrogen = 1 mole

12 g of carbon = 1 mole

and   24 g of carbon = 2 moles

The symbol for relative atomic mass is A_r. So, we can write $A_rC = 12.0$, $A_rH = 1.0$ and so on — or simply C = 12.0 and H = 1.0 for short.

$$\text{number of moles} = \frac{\text{mass}}{\text{relative atomic mass}}$$

Relative atomic masses show that one atom of carbon is 12 times as heavy as one atom of hydrogen. This means that 12 g of carbon will contain the same number of atoms as 1 g of hydrogen. An atom of oxygen is 16 times as heavy as an atom of hydrogen, so 16 g of oxygen will also contain the same number of atoms as 1 g of hydrogen.

The relative atomic mass in grams (1 mole) of every element (1 g of hydrogen, 12 g carbon, 16 g oxygen etc.) contains the same number of atoms. This number is called the Avogadro constant. The term 'Avogadro constant' was chosen in honour of the Italian scientist, Amedeo Avogadro. Experiments show that the Avogadro constant is 6×10^{23}. Written out in full, this is 600 000 000 000 000 000 000 000 or six hundred thousand billion billion. So, 1 mole of an element always contains 6×10^{23} atoms.

> **Test yourself**
>
> 4 a Why is the relative atomic mass of helium 4 and that of hydrogen 1?
>   b Carbon has 6 protons, so why is the relative atomic mass of carbon 12?
>   c Prove that a sulfur atom is twice as heavy as an oxygen atom.
>   d What is the ratio of the mass of a helium atom to the mass of a calcium atom?
>
> 5 a What is the mass of 1 mole of nitrogen atoms?
>   b What is the mass of 1 mole of nitrogen molecules, N_2?
>
> 6 a How many particles are there in half a mole of a substance?
>   b How many particles are there in one-hundredth of a mole of substance?
>   c How many particles are there in 100 moles of a substance?
>   d Why is it easier to measure numbers of particles in moles rather than actual numbers of particles?

C2 6.2 Comparing the masses of atoms

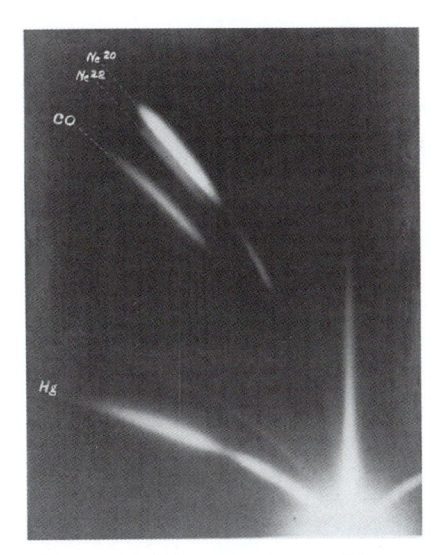

Figure 6.3 This photo shows evidence for the two isotopes of neon, neon-20 and neon-22.

Different **isotopes** of the same element have different mass numbers because they have different numbers of neutrons in the nucleus.

Boost your grade ✓

The word 'isotope' is often associated with radioactive isotopes. Remember, however, that most isotopes of elements are stable and not radioactive.

Isotopes and atomic mass

The mass of an atom depends on the number of protons and neutrons in its nucleus. Because protons and neutrons have a relative mass of 1.00, the relative atomic masses of many elements are close to whole numbers.

But, some elements have relative atomic masses that are nowhere near whole numbers. For example, the relative atomic mass of chlorine is 35.5 and that of copper is 63.5. At one time, scientists could not understand why the relative atomic masses of these elements were not close to whole numbers. The answer soon became clear when the first mass spectrometer was built in 1919. When atoms of elements such as chlorine or copper were ionised and passed through a mass spectrometer, the beam of ions separated into two or more paths. This suggested that one element could have atoms with different masses. These atoms of the same element with different masses are called **isotopes**.

All the isotopes of one element have the same number of protons. So, they have the same atomic number and the same number of electrons. This gives them the same chemical properties because chemical properties depend on the number of electrons in an atom.

Isotopes have different numbers of neutrons, and this gives them different masses. This means that isotopes have the same atomic number but different mass numbers (Table 6.2).

Table 6.2 Isotopes: similarities and differences.

Isotopes have the same:	Isotopes have different:
• number of protons • number of electrons • atomic number • chemical properties	• numbers of neutrons • mass numbers • physical properties

For example, neon has two isotopes (Figure 6.4). Each isotope has 10 protons and 10 electrons — and therefore an atomic number of 10. But one of these isotopes has 10 neutrons and the other has 12 neutrons — their mass numbers are therefore 20 and 22. They are sometimes called neon-20 and neon-22.

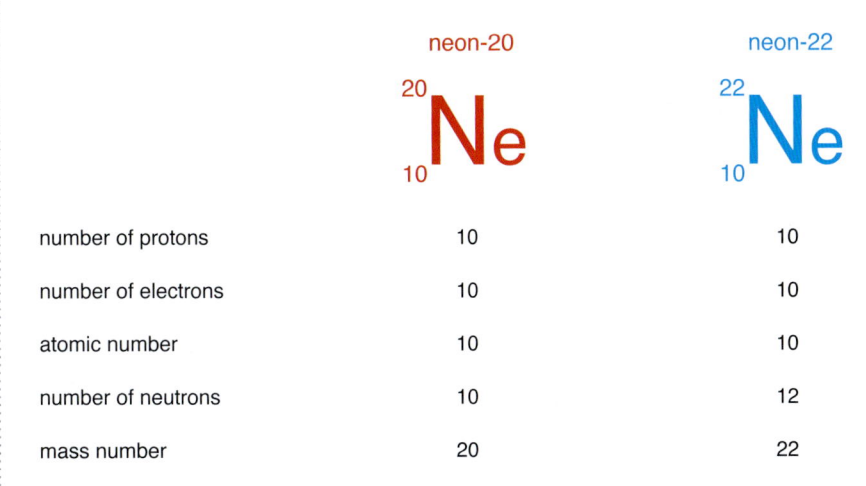

	neon-20 $^{20}_{10}Ne$	neon-22 $^{22}_{10}Ne$
number of protons	10	10
number of electrons	10	10
atomic number	10	10
number of neutrons	10	12
mass number	20	22

Figure 6.4 The two isotopes of neon.

Chemistry 2   127

Chapter 6 Mass, moles and yield

These two isotopes of neon have the same chemical properties because they have the same number of electrons. But they have different physical properties because they have different masses. So, $^{20}_{10}$Ne and $^{22}_{10}$Ne have different densities, different melting points and different boiling points.

> **Test yourself**
>
> 7 Look closely at Figure 6.3.
>   a The trace from neon-20 is much stronger than that from neon-22. What does this tell you about the two isotopes?
>   b The beam of CO has been deflected less than both neon isotopes.
>   (i) What substance is CO?
>   (ii) Why is CO deflected less than the neon isotopes?

Modelling a mass spectrometer — ACTIVITY

Model a mass spectrometer in groups of at least 6 students. Your model needs to be able to show what happens when:
- an atom of an element passes through it and is detected
- an atom of the same element with a higher velocity passes through the spectrometer
- a heavier atom passes through it at the same velocity

At one time, the standard for relative atomic masses was hydrogen. Because of the existence of isotopes, it became necessary to choose one particular isotope as the standard. Today, the isotope carbon-12 ($^{12}_{6}$C) is chosen as the standard and given a relative mass of exactly 12.

The relative masses of other atoms are then obtained by comparison with carbon-12. On this scale, the relative atomic mass of hydrogen is still 1.0 and that of helium is 4.0. Some relative atomic masses are listed in Table 6.3.

Table 6.3 Relative atomic masses of common elements.

Element	Symbol	Relative atomic mass
Carbon	C	12.0
Hydrogen	H	1.0
Helium	He	4.0
Oxygen	O	16.0
Magnesium	Mg	24.0
Sulfur	S	32.1
Calcium	Ca	40.1
Iron	Fe	55.8
Copper	Cu	63.5
Gold	Au	197.0

C2 6.2 Comparing the masses of atoms

Test yourself

8 Look at Table 6.3 on page 128.
   a How many times heavier are carbon atoms than hydrogen atoms?
   b How many times heavier are magnesium atoms than carbon atoms?
   c Which element has atoms almost four times as heavy as oxygen atoms?
   d Write the following elements in order of increasing deflection in a mass spectrometer: helium, iron, gold.

Measuring relative atomic masses

Most elements contain a mixture of isotopes. This explains why their relative atomic masses are not whole numbers. Using modern mass spectrometers, it is possible to obtain printouts showing the relative amounts of the different isotopes in a sample of an element and the relative atomic masses of each isotope. One of these printouts is shown in Figure 6.5. This shows that chlorine consists of two isotopes, $^{35}_{17}Cl$ and $^{37}_{17}Cl$, with relative atomic masses of 35 and 37 respectively.

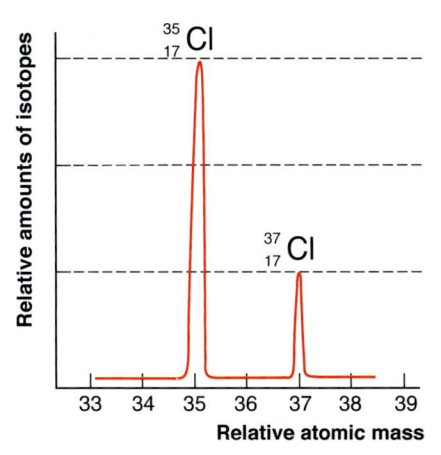

Figure 6.5 A mass spectrometer printout for chlorine.

The relative atomic mass of an element is the average mass of one atom of the element. This average must take into account the relative amounts of the different isotopes. For example, if chlorine contained 100% $^{35}_{17}Cl$, its relative atomic mass would be 35.0.

If it contained 100% $^{37}_{17}Cl$, then its relative atomic mass would be 37.0. A 50:50 mixture of the isotopes $^{35}_{17}Cl$ and $^{37}_{17}Cl$ would have a relative atomic mass of 36.0

Figure 6.5 shows that naturally occurring chlorine contains three times as much $^{35}_{17}Cl$ as $^{37}_{17}Cl$ — i.e. $\frac{3}{4}$ or 75% of chlorine is $^{35}_{17}Cl$, and $\frac{1}{4}$ or 25% is $^{37}_{17}Cl$. So, the relative atomic mass can be calculated by:

$$\text{(75\% chlorine-35)} + \text{(25\% chlorine-37)}$$
$$= \left(\frac{75}{100} \times 35.0\right) + \left(\frac{25}{100} \times 37.0\right)$$
$$= 26.25 + 9.25$$
$$= 35.5$$

Test yourself

9 Silver consists of just over 50% $^{107}_{47}Ag$ with the rest being $^{109}_{47}Ag$.
What is the value for the relative atomic mass of silver?

Chapter 6 Mass, moles and yield

C2 6.3 Using relative atomic masses to find out about compounds

Learning outcomes

- Know that the sum of atomic masses in a substance's chemical formula is called its formula mass.
- Know that 1 mole of a substance is its formula mass in grams.
- Be able to calculate relative formula masses and the percentages of elements in compounds.
- Be able to calculate the percentages by mass of elements in a compound, given the formula of the compound.
- Be able to calculate empirical formulae from mass data.

Relative formula masses can be used to compare the masses of different compounds.

Figure 6.6 The formula for table sugar is $C_{12}H_{22}O_{11}$.

Boost your grade ✓

When you are doing calculations involving mass and moles, you should always convert masses or percentages by mass into numbers of moles as the first step.

Comparing the masses of molecules

Relative atomic masses are used to compare the masses of different atoms. Relative atomic masses can also be used to compare the masses of molecules in different compounds. The relative masses of compounds are called **relative formula masses** (symbol M_r). The relative formula mass of a compound is the sum of the relative atomic masses of all the atoms in its formula.

For example, the relative formula mass of water, $M_r\,H_2O$:

= (2 × relative atomic mass of hydrogen) + relative atomic mass of oxygen
= (2 × 1.0) + 16.0
= 18.0

The relative formula mass of iron(III) oxide, $M_r\,Fe_2O_3$:

= (2 × relative atomic mass of iron) + (3 × relative atomic mass of oxygen)
= (2 × 55.8) + (3 × 16.0) = 159.6

The relative formula mass of a compound in grams is known as 1 mole of that substance. So, 1 mole of water is 18.0 g and so it follows that:

0.1 mole of water = 0.1 × 18.0 g
= 1.8 g

1 mole of iron(III) oxide is 159.6 g, and so:

5 moles of iron(III) oxide = 5 × 159.6 g
= 798.0 g

Percentage composition of compounds

Relative atomic masses can be used to calculate the percentage of the different elements in a compound. First, you work out the relative amount of each element in the relative formula mass of the compound. For example, carbon dioxide (CO_2) contains 12.0 g of carbon and 32.0 g (2 × 16.0 g) of oxygen in 44.0 g of carbon dioxide.

So, it contains $\frac{12}{44}$ parts carbon and $\frac{32}{44}$ parts oxygen. Changing to percentages, this is:

$\frac{12}{44} \times 100 = 27\%$ carbon

and

$\frac{32}{44} \times 100 = 73\%$ oxygen

Worked example

Calculate the relative formula mass of sugar, $C_{12}H_{22}O_{11}$.
The relative formula mass of sugar, $M_r\,C_{12}H_{22}O_{11}$:
= (12 × A_r C) + (22 × A_r H) + (11 × A_r O)
= (12 × 12.0) + (22 × 1.0) + (11 × 16)
= 144 + 22 + 176
= 342

AQA GCSE Additional Science

C2 6.3 Using relative atomic masses to find out about compounds

Finding formulae using relative atomic masses

All formulae of compounds are obtained by doing experiments to find the relative number of moles of the different elements that react to form a compound. These formulae are called **empirical formulae**. When water is decomposed into hydrogen and oxygen, results show that:

> 18.0 g of water gives 2.0 g of hydrogen and 16.0 g of oxygen
> = 2 moles of hydrogen + 1 mole of oxygen
> = (2 × 6 × 10^{23} atoms of hydrogen) + (1 × 6 × 10^{23} atoms of oxygen)

This shows that 12 × 10^{23} hydrogen atoms combine with 6 × 10^{23} oxygen atoms, so two hydrogen atoms combine with one oxygen atom. Therefore, the formula of water is H$_2$O. These results are set out in Table 6.4.

> An **empirical formula** shows the simplest whole number ratio for the atoms of different elements in a compound. It can be calculated from the masses of elements or the percentages of elements by mass in a compound.

Table 6.4 Finding the formula of water.

	H	O
Masses reacting	2.0 g	16.0 g
Mass of 1 mole	1 g	16 g
Moles reacting	2	1
Ratio of atoms	2	1
Formula	H$_2$O	

Test yourself

10 What is the relative formula mass of:
  a carbon dioxide, CO$_2$?
  b sulfuric acid, H$_2$SO$_4$?

11 Calculate the mass of:
  a 3 moles of carbon dioxide
  b 0.6 moles of carbon dioxide
  c 5 moles of sulfuric acid
  d 0.2 moles of sulfuric acid

12 What is the percentage of carbon in the following compounds:
  a CH$_4$?
  b CaCO$_3$?
  c Urea, molecular formula (NH$_2$)$_2$CO?

13 A compound consists of 75% sodium and 24% of oxygen. What is its empirical formula?

  A_r Na = 23.0, A_r O = 16.0

14 A gardener has to choose between three different substances to add nitrogen compounds to his garden. His choices are:
  • ammonium sulfate, (NH$_4$)$_2$SO$_4$
  • urea, (NH$_2$)$_2$CO
  • ammonium nitrate, NH$_4$NO$_3$

  Which contains the highest percentage of nitrogen?

Chemistry 2

Chapter 6 Mass, moles and yield

Finding the formula of red copper oxide

HOW SCIENCE WORKS / PRACTICAL SKILLS

As well as black copper oxide, there is a different oxide of copper that is red. After finding that the formula of black copper oxide is CuO, George and Meera decided to investigate the formula of red copper oxide. They took a weighed amount of red copper oxide and reduced it to copper (Figure 6.7).

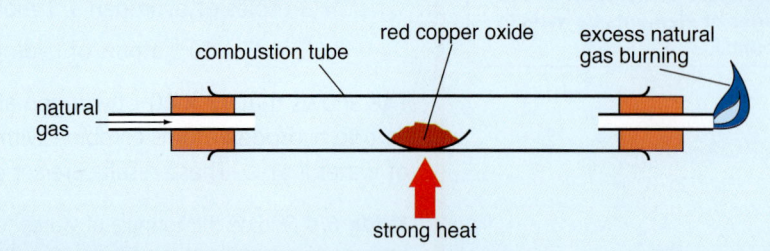

Figure 6.7 Reducing red copper oxide by heating in natural gas.

They carried out the experiment five times, starting with different amounts of red copper oxide. Their results are shown in Table 6.5.

Table 6.5

Experiment	Mass of red copper oxide in g	Mass of copper in g
1	1.43	1.27
2	2.10	1.87
3	2.72	2.54
4	3.55	3.15
5	4.29	3.81

1. Look at Figure 6.7. What safety precautions should George and Meera take during the experiment?

2. Write a word equation for the reduction of red copper oxide to copper using methane (CH_4) in natural gas. (*Hint:* The only solid product is copper.)

3. What steps could George and Meera take to ensure that all the copper oxide is reduced to copper?

4. Make a table similar to Table 6.5 but with three extra columns.
   a In the first of these extra columns, write the mass of oxygen in each of the samples of red copper oxide.
   b In the second extra column, calculate the number of moles of copper in the oxide. (A_r Cu = 63.5)
   c In the third extra column, calculate the number of moles of oxygen in the oxide. (A_r O = 16.0)

5. Use the results in the last two columns of your table to plot a graph of moles of copper (*y*-axis) against moles of oxygen (*x*-axis). Draw the line of best fit through the points on your graph.

6. Which of the points is anomalous and should be disregarded in drawing the line of best fit?

> 7 Look at your graph.
>   a Find the average value for the ratio moles of copper : moles of oxygen.
>   b How many moles of copper combine with one mole of oxygen in red copper oxide?
>   c What is the formula for red copper oxide?
> 8 How did George and Meera improve the **reproducibility** of their results?
> 9 Is their result for the formula of red copper oxide valid? State 'Yes' or 'No' and explain your opinion.
> 10 Use the results in your table to explain why a balance reading to only one decimal place would have been useless for this investigation.
> 11 a Name the independent variable and the dependent variable in this investigation.
>   b Are these categoric, continuous or ordered variables?

> Results are valid if they are **reproducible** and the measurements taken are affected by a single independent variable.

C2 6.4 Using masses in equations

Learning outcomes

- Be able to calculate masses of reactants and products involved in chemical changes.
- Be able to prove that mass is neither gained nor lost in a chemical change.
- Be able to calculate the amount of product obtained from a given amount of reactant.

Finding the masses of reactants and products in a reaction

If you want to bake a 1 kg cake, you need to know how much of each ingredient to use. So recipes give quantities that you could easily turn into masses. If you start with a 1 kg set of ingredients in the correct ratios, you should get 1 kg of cake.

Industrial chemists need to know how much product they can obtain from a given amount of starting material. It is vitally important to know the amounts of reactants needed for a chemical process and the amount of product that can be obtained.

To calculate masses of reactants and products, chemists use relative formula masses and equations. As an example, we can calculate how much iron we could obtain by reducing 1 kg (1000 g) of pure iron ore (containing iron(III) oxide) with carbon monoxide in a blast furnace.

The equation for the reaction is:

$$Fe_2O_3(s) + 3CO(g) \rightarrow 2Fe(l) + 3CO_2(g)$$

From this equation:

$$1 \text{ mole } Fe_2O_3 \rightarrow 2 \text{ moles of Fe atoms}$$

Using relative formula masses for Fe_2O_3 and 2Fe:

$$(2 \times 55.8) + (3 \times 16) \text{ g } Fe_2O_3 \rightarrow 2 \times 55.8 \text{ g Fe}$$
$$159.6 \text{ g } Fe_2O_3 \rightarrow 111.6 \text{ g Fe}$$
$$1.0 \text{ g } Fe_2O_3 \rightarrow \left(\frac{111.6}{159.6}\right) \text{g Fe}$$
$$= 0.7 \text{ g Fe}$$

Chemistry 2

Chapter 6 Mass, moles and yield

So, 1.0 kg Fe$_2$O$_3$ → 0.7 kg Fe. Therefore, the reduction of 1 kg of pure iron(III) oxide produces 0.7 kg of iron.

The amount of product actually made in a chemical process does not always match the theoretical (calculated) amount. You will learn more about this in Section 6.5.

Calculating masses
HOW SCIENCE WORKS — PRACTICAL SKILLS

1. Find the amount of water of crystallisation of copper sulfate crystals by heating a known mass of copper sulfate until the sample reaches constant mass. This happens when all the water incorporated in the crystals has been driven off.

2. Find the mass of anhydrous copper sulfate left. Calculate the number of moles this is. This will be the same number of moles as the original hydrated copper sulfate crystals according to the equation:

$$CuSO_4 \cdot nH_2O(s) \rightarrow CuSO_4(s) + nH_2O(l)$$

3. Calculate the formula mass of the original copper sulfate crystals and work out the empirical formula of the compound.

4. Find the mass of a quantity of calcium carbonate. Heat the mass until no more carbon dioxide is given off. Check this by heating and reweighing until the mass is constant to a reasonable degree of precision. Calculate the original number of moles of calcium carbonate and verify that this is equal the number of moles of calcium oxide in the product.

Use the equation:

$$CaCO_3(s) \rightarrow CaO(s) + CO_2(g)$$

5. Find the empirical formula of black copper oxide by heating a known mass of copper oxide in a stream of methane to reduce it to copper. The difference in the masses is the mass of oxygen originally combined with the copper.

Figure 6.8 This sulfur is being stored after mining. 32.1 g of sulfur can be reacted with oxygen and water to manufacture 98.1 g of sulfuric acid.

Test yourself

15 a Look at Figure 6.9. What elements are present in lead chromate besides lead?

b What is the relative formula mass of lead chromate?

c What is the percentage of lead in lead chromate?

(A_r Pb = 207, A_r Cr = 52, A_r O = 16)

Figure 6.9 The yellow substance in 'no parking' double lines is lead chromate, PbCrO$_4$.

C2 6.4 Using masses in equations

Test yourself

16 Copy and complete the following calculation to find how much lime (calcium oxide) can be obtained by heating 1 kg of pure limestone (calcium carbonate).

(A_r Ca = 40, A_r C = 12, A_r O = 16)

The equation for the reaction is:

calcium carbonate → ____ + ____

$CaCO_3$ → ____ + CO_2

1 mole $CaCO_3$ → ___ mole CaO

___ g $CaCO_3$ → ___ g CaO

1000 g $CaCO_3$ → ___ g CaO

17 When petrol (octane, C_8H_{18}) is burned, a lot of carbon dioxide is added to the atmosphere.

In the equation, C_8H_{18} burns to make $8CO_2$.
a What is the mass of 1 mole of octane?
b How many moles of carbon dioxide are produced in the combustion reaction?
c What is the mass of carbon dioxide produced?
d Where has the extra mass come from?

18 If 1 tonne (1000 kg) of aluminium oxide is reduced by electrolysis to aluminium, what mass of aluminium is produced?

$$2Al_2O_3(l) \rightarrow 4Al(l) + 3O_2(g)$$

Extracting tin from tinstone
HOW SCIENCE WORKS ACTIVITY

The production manager at a tin smelter has to meet certain production targets. The smelter must produce 595 tonnes of tin every month. In order to achieve this, the manager must calculate how much purified tinstone (tin oxide) he must buy to produce 595 tonnes of tin. To do this he uses the equation:

tin oxide + carbon (coke) → tin + carbon monoxide

$SnO_2(s)$ + 2C(s) → Sn(l) + 2CO(g)

The equation shows that 1 mole of SnO_2 produces 1 mole of Sn. The relative atomic masses of tin, carbon and oxygen are Sn = 119, C = 12, O = 16.

1 Copy and complete the following statement:

___ g SnO_2 produces ___ g Sn.

2 What mass of tinstone is needed to meet the production target of 595 tonnes of tin per month?

3 What mass of carbon (coke) is needed each month?

4 What concerns should the production manager have about the emissions from the smelter?

Chemistry 2

Chapter 6 Mass, moles and yield

5. The smelter normally operates all through the day and night. It can produce a maximum of 1 tonne of tin per hour. Is the production target possible? Explain your answer.

6. In the first 6 months of 2010, the amount of tin produced each month, to the nearest 5 tonnes, was 650, 630, 575, 585, 560 and 600.
   a What is the range in production over the 6 months?
   b What is the mean (average) production per month?
   c Has the monthly target been achieved on average over the first 6 months of 2010?
   d What should the production manager do to get a more reliable value for the average production per month?
   e What should the production manager do to get a more accurate value for the average production per month?

7. In 2010, tin from the smelter was sold at £10 000 per tonne. What is the value of the tin produced by the smelter in the first 6 months of 2010?

8. Why did the smelters have tall chimneys?

9. Why is tin no longer mined and produced in Cornwall?

C2 6.5 Reversible reactions and yield

Learning outcomes

- Know that in some chemical reactions, a reverse reaction turns products back into reactants.
- Know that an equilibrium is set up between the forward and reverse reactions.
- Know that in some chemical reactions, not all of the reactants will be converted into products.
- Know how chemical calculations can be used to calculate theoretical yield and percentage yield.

Reversible reactions

Chemical reactions are about change. A general way of summing up a chemical reaction is shown below:

$$\text{reactants} \xrightarrow[\text{conditions}]{\text{in certain}} \text{products}$$

Chemical changes start with 'reactants' as raw materials and turn these into 'products'. Chemical reactions can:
- make useful new materials — such as nylon or ammonia
- transfer stored energy — such as in a combustion reaction
- involve a useful process — such as when chlorine bleach makes a stained tea mug white again

Many chemical reactions are irreversible — they do not easily go back to the reactants once the products are formed. For example:

$$\text{magnesium} + \text{oxygen} \rightarrow \text{magnesium oxide}$$
$$2Mg(s) + O_2(g) \rightarrow 2MgO(s)$$

Some chemical reactions go backwards and forwards easily, depending on the conditions — these are reversible reactions. An example is the reaction between the colourless gases ammonia and hydrogen chloride. At room temperature, a white cloud of solid ammonium chloride forms (Figure 6.10):

$$\text{ammonia} + \text{hydrogen chloride} \rightarrow \text{ammonium chloride}$$
$$NH_3(g) + HCl(g) \rightarrow NH_4Cl(s)$$

C2 6.5 Reversible reactions and yield

Figure 6.10 Ammonium chloride is made when ammonia reacts with hydrogen chloride.

When ammonium chloride is heated, even gently, the compound decomposes to give the ammonia and hydrogen chloride again (Figure 6.11):

ammonium chloride → ammonia + hydrogen chloride
$$NH_4Cl(s) \rightarrow NH_3(g) + HCl(g)$$

A **reversible reaction** is usually shown as:

ammonium chloride ⇌ ammonia + hydrogen chloride
$$NH_4Cl(s) \rightleftharpoons NH_3(g) + HCl(g)$$

Factors affecting efficiency and outcome of reactions

Getting as much product as possible is important — the more product made, the greater the profit. Being able to control the speed of reactions is also important in industry — time is money. If a reaction needs a high temperature and/or pressure, then equipment and fuel costs could be high. Reducing the amount of energy input can save money on fuel costs.

To control chemical changes, you need to consider:
- how much of the reactants is turned into the useful product
- how to make the change happen at the most cost-effective rate
- how much energy needs to be transferred during the change

Some reasons why you cannot get all the product you expect from a reaction are given below:
- The reactants may not be pure.
- Some of the product may get lost in the separation process.
- Not all of the reactants may react, or they may react in an unexpected way.
- In reversible reactions, some of the products may change back to reactants.

> In a **reversible reaction** the product(s) may react or fall apart to reform the reactants.

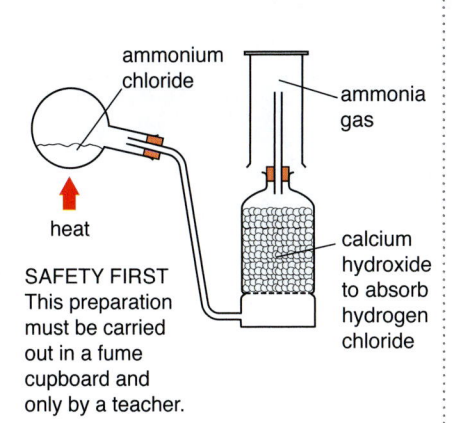

SAFETY FIRST This preparation must be carried out in a fume cupboard and only by a teacher.

Figure 6.11 The reverse reaction to that in Figure 6.10 — ammonium chloride decomposes, making ammonia gas and hydrogen chloride. This is one way of making ammonia.

Chemistry 2

Chapter 6 Mass, moles and yield

Atoms cannot be created or destroyed, and they always stay the same type of atom in a chemical reaction — but reactions do not usually convert all the atoms in the reactants into particles of the desired product. Some atoms stay unconverted or just get lost in the surroundings.

> **Test yourself**
>
> 19 Give four reasons why the amount of product in a chemical reaction could be less than expected.
>
> 20 A cook started with a total of 1.0 kg of ingredients (reactants). She found the cake that she made had a mass of 920 g. Explain what may have happened to the 'missing mass'.
>
> 21 Look at Figure 6.12. The only materials going into a diesel engine are hydrocarbon fuel vapour and air. But sometimes black, solid particles come out of the exhaust pipe.
>   a What do you think has happened to produce the carbon particles?
>   b Is this carbon one of the desired products of the reaction in the diesel engine?
>   c What effect does the production of carbon particles have on the energy output of the engine?
>
> **Figure 6.12** Black smoke in a diesel exhaust contains particles of carbon.

Yield of a chemical reaction

The amount of product obtained in a reaction is known as the **actual yield**. It is also possible to calculate the maximum possible amount of product in a reaction — this is called the **theoretical yield**.

For example, the mass of iron formed by reducing iron(III) oxide with carbon monoxide can be calculated from the masses of reactants if you know the relative formula masses and the equation:

$$Fe_2O_3(s) + 3CO(g) \rightarrow 2Fe(l) + 3CO_2(g)$$

> **Actual yield** is the amount of product obtained.
> **Theoretical yield** is the maximum amount of product that could theoretically be made from a mixture of reactants.

C2 6.5 Reversible reactions and yield

No atoms are gained or lost in a chemical reaction. The equation shows that the number of atoms of iron in the reactant (iron(III) oxide) balances the number of atoms of iron in the product. The theoretical yield of a reaction assumes that all the reactants are turned into products, and that all the products are successfully separated from the reaction mixture. However, chemists cannot always obtain the amount of product predicted from the equation and so the **percentage yield** is usually not 100%.

> **Percentage yield** is the actual amount of a product made expressed as a percentage of the theoretical ammount.

Worked example

Calcium hydroxide ($Ca(OH)_2$) is used as an alkali to neutralise acid soil. Calcium hydroxide can be made by heating calcium carbonate to make calcium oxide, which is then reacted with water:

$$CaCO_3(s) \rightarrow CaO(s) + CO_2(g)$$
calcium carbonate — calcium oxide (lime) — carbon dioxide

$$CaO(s) + H_2O(l) \rightarrow Ca(OH)_2(s)$$
calcium oxide — water — calcium hydroxide

a How much calcium hydroxide can, in theory, be produced from 200 g of calcium carbonate?

First, calculate the relative formula masses (A_r Ca = 40, A_r O = 16, A_r C = 12, A_r H = 1)

$$M_r \, CaCO_3 = 40 + 12 + (3 \times 16)$$
$$= 100$$
$$M_r \, Ca(OH)_2 = 40 + (2 \times (16 + 1))$$
$$= 74$$

From the equations, 1 mole of calcium carbonate yields 1 mole of calcium hydroxide.

Using the relative formula masses for calcium carbonate and calcium hydroxide, 100 g of calcium carbonate gives 74 g of calcium hydroxide, so 200 g of calcium carbonate should give:

$$200\,g \times \frac{74}{100} = 148\,g \text{ of calcium hydroxide}$$

b When the actual mass of calcium hydroxide produced in this reaction was reacted with hydrochloric acid, it neutralised the equivalent of 3.30 moles of hydrochloric acid. What was the percentage yield of the conversion of calcium carbonate to calcium hydroxide?

The equation for the reaction is:

$$Ca(OH)_2(s) + 2HCl(aq) \rightarrow CaCl_2(aq) + 2H_2O(l)$$
calcium hydroxide — hydrochloric acid — calcium chloride — water

From the equation, 1 mole of calcium hydroxide neutralises 2 moles of hydrochloric acid. 3.30 mol of acid were neutralised, so the solid added to the hydrochloric acid contained $\frac{3.30}{2}$ or 1.65 moles of calcium hydroxide.

What mass of reactant is this? Using the relative formula mass of calcium hydroxide:

$$1.65 \times 74\,g = 122\,g$$

From the answer to part (a), the theoretical yield is 148 g. The percentage yield is:

$$\frac{\text{actual yield}}{\text{theoretical yield}} \times 100\%$$

$$= \frac{122\,g}{148\,g} \times 100\%$$

$$= 82.4\%$$

Possible explanations for the yield being less than 100%:
- The starting material (marble chips) contained some impurity, such as sand.
- Some calcium hydroxide was lost in the dissolving, filtering and drying processes.
- The thermal decomposition was incomplete — not all the calcium carbonate decomposed.

Chemistry 2

Chapter 6 Mass, moles and yield

Another factor that can affect the yield is the fact that some chemical reactions are reversible. This means that the products can react to produce the original reactants. Such reactions can be represented by:

$$A + B \rightleftharpoons C + D$$

Observations and calculations

HOW SCIENCE WORKS — PRACTICAL SKILLS

1. Carry out a reversible reaction by heating ammonium chloride in a test tube. Make detailed observations of what you see.
2. Heat a known mass of magnesium ribbon in air (use a crucible and lid) until it reaches constant mass. The increase in mass is the amount of oxygen that combines with the magnesium. The product is magnesium oxide.

   Use the equation $2Mg(s) + O_2(g) \rightarrow 2MgO(s)$ to calculate the theoretical yield of the reaction. Use this to calculate the percentage yield of the reaction.
3. Repeat the procedure by using a known mass of wire wool (iron) and heating this in air. Again, calculate the theoretical yield and the percentage yield. Use the equation $4Fe(s) + 3O_2(g) \rightarrow 2Fe_2O_3(s)$

Figure 6.13 Pharmacists need to know exactly how much of a product they are making.

Boost your grade ✓

Use the balanced equation to work out the theoretical yield. This will be a mass. It could be greater than the reactant mass if, for example, gases are reacting.

Test yourself

22. One mole of sodium hydroxide (NaOH) is reacted with hydrochloric acid according to the equation below.

    $NaOH(aq) + HCl(aq) \rightarrow NaCl(aq) + H_2O(l)$

    What is the theoretical amount of sodium chloride that can be produced?
    $A_r Na = 23, A_r O = 16, A_r H = 1, A_r Cl = 35.5$

23. One mole of methane was burned in air. Some sooty deposits were noticed. The carbon dioxide produced was absorbed by a chemical which gained 32 g in mass.
    a What was the percentage yield of carbon dioxide?
    b Why was the percentage yield low?

C2 6.6 Instrumental methods of analysis

C2 6.6

Learning outcomes

- Know that most chemical analysis is carried out by computer-controlled instruments that can handle small sample sizes.

- Explain how chromatography can identify unknown materials including food additives.

- Know that the output of a chromatography column can be linked to a mass spectrometer to give a definite identification of the sample.

Figure 6.14 You might have seen analytical science in television shows such as *CSI*. Analysis is not done as quickly as on television, but the machines used in these programmes are reasonably accurate. Analysis is done using machines rather than old-fashioned glassware.

In Section 6.2 you learned about using an instrument called a mass spectrometer to measure the masses of atoms. Many other analytical machines are used by scientists. During the past 100 years, the work of analytical chemists has been revolutionised by the development of sophisticated measuring instruments.

Analytical chemists working in many different areas use these instruments. Those working in hospital laboratories can take a small specimen from a patient, analyse it and help doctors to diagnose illnesses. Researchers for drug companies can develop and analyse new medicines quickly and effectively. Environmental scientists can check the quality of the air we breathe and the water we drink. Forensic scientists can obtain evidence from minute fragments of material taken from the scene of a crime.

Modern instrumental methods of chemical analysis have taken over from traditional laboratory tests because they are:

- faster — many samples can be analysed in a short time. This is important when making rapid security checks at airports, and in industry when materials require regular, frequent monitoring.
- more sensitive — this allows analysis of tiny fragments of material and the detection of small amounts in a huge sample. This is important for drug testing in sport and in testing materials at the scene of a crime.
- more accurate — human error is reduced as a result of automation involving computer analysis and recording.

The development of modern instrumental methods could not have happened without equally rapid progress in technologies such as electronics and

Chemistry 2

Chapter 6 Mass, moles and yield

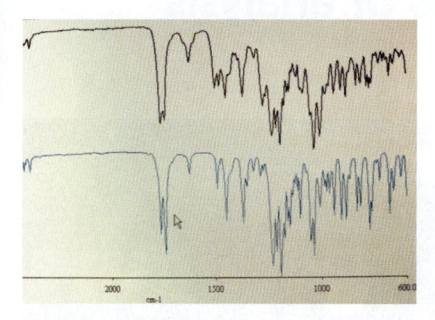

Figure 6.15 Using infrared spectroscopy, forensic scientists can check if a sample of material contains drugs such as heroin by comparing the sample's spectrum (black) with that of heroin (blue).

computing. Computers, printers and other electronic devices have speeded up the process of chemical analysis. For example, the analysis of an oil slick used to take weeks of painstaking laboratory experiments. Today, the same analysis can be carried out in a few hours. Computers have also speeded up the interpretation of complex data requiring detailed calculations and the search for information.

More about instrumental methods

The instruments used by analytical chemists can detect the presence of certain elements and compounds. Some instruments, such as emission and absorption spectroscopes, are suited to identifying elements. Other instruments, such as infrared and NMR spectrometers, are usually used to identify compounds. Mass spectrometers can be adapted to identify elements or compounds.

Table 6.6 gives more information about the instruments used, what they identify and a short explanation of how they work. You are not required to remember the details of how these instruments work.

Table 6.6 Instruments used in modern chemical analysis.

Instrument used	What does the instrument identify?	How does it work?	Example of the application
Mass spectrometer	Elements and compounds — from particles as small as atoms to complex molecules like proteins	Positive ions of elements and compounds pass through a magnetic field where they are deflected. Lighter particles are deflected by more than heavier particles. So, the mass of an ion can be determined by measuring how much it is deflected.	Checking that the additives present in food are permitted
Emission spectroscope — visible, infrared (IR) or ultraviolet (UV)	Elements	When substances are heated strongly, their electrons are excited to higher energy levels. The electrons soon release this energy as visible, IR or UV radiation. By identifying the wavelength or frequency of the radiation, it is possible to identify the elements in the material being analysed.	Analysing the composition of elements in the Sun and other stars
Absorption spectroscope — visible and UV	Elements	Strong visible and UV light pass through a sample. Electrons in the sample absorb radiation of a particular wavelength and are excited to higher energy levels. By comparing the incoming and outgoing radiation, it is possible to identify the elements in the sample.	Detecting elements in steel; monitoring pollution to identify heavy metals
Infrared spectrometer	Organic compounds — by identifying the bonds in compounds	These involve the same technique as absorption spectroscopes but with IR, which is absorbed by electrons in the bonds of molecules. Different bonds absorb energy at different wavelengths, so bonds such as C—H, C—O and C=C can be identified.	Identifying drugs in forensic testing
Nuclear magnetic resonance (NMR) spectrometer	Organic compounds — by identifying the different groups of atoms containing hydrogen	Nuclei of hydrogen atoms in organic compounds can act as small magnets. Radio waves of specific frequencies are absorbed by the nuclei of hydrogen atoms as they move into line with a strong magnetic field. The amount of energy absorbed corresponds to the number of hydrogen atoms in the same type of bond.	Analysing the molecular structure of new pharmaceutical products
Gas–liquid chromatograph	Volatile liquids	The liquid is allowed to vaporise into a stream of inert gas passing through a long, thin tube packed with porous material. The time it takes a vaporised substance to reach the end of the tube is used to identify it.	Analysing the different components in petrol

AQA GCSE Additional Science

C2 6.6 Instrumental methods of analysis

Figure 6.16 This prototype 'walk through' spectrometer can detect traces of explosives left in the air that passes over a person who has handled explosives. It is being tested in some airports.

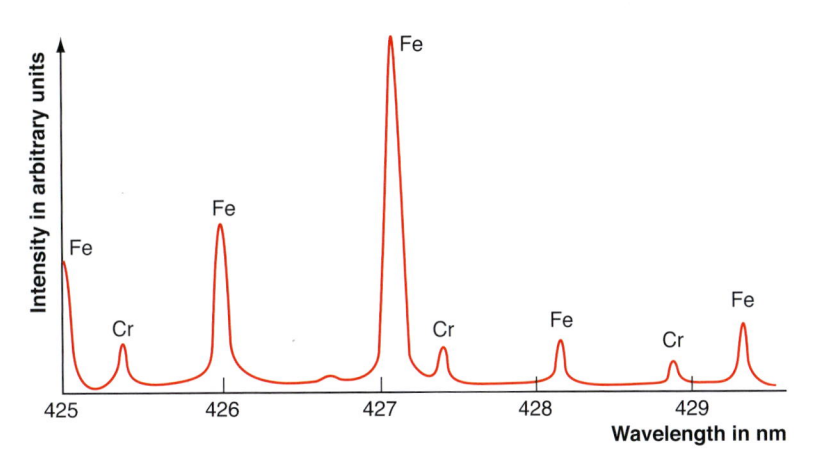

Figure 6.17 Absorption spectroscopy is used in the steel industry to monitor the amounts of different elements in steels to control an alloy's quality as it is being made. This printout shows the presence of chromium along with iron in a steel alloy.

Chromatography

You will have used chromatography to analyse food stuffs for colours or additives, but modern chromatography apparatus is not made of glass and paper, using liquid samples like you see in schools. They are complex computer-controlled machines where the sample is injected by syringe and passes through a long, thin tube filled with material on which the separation into different substances takes place.

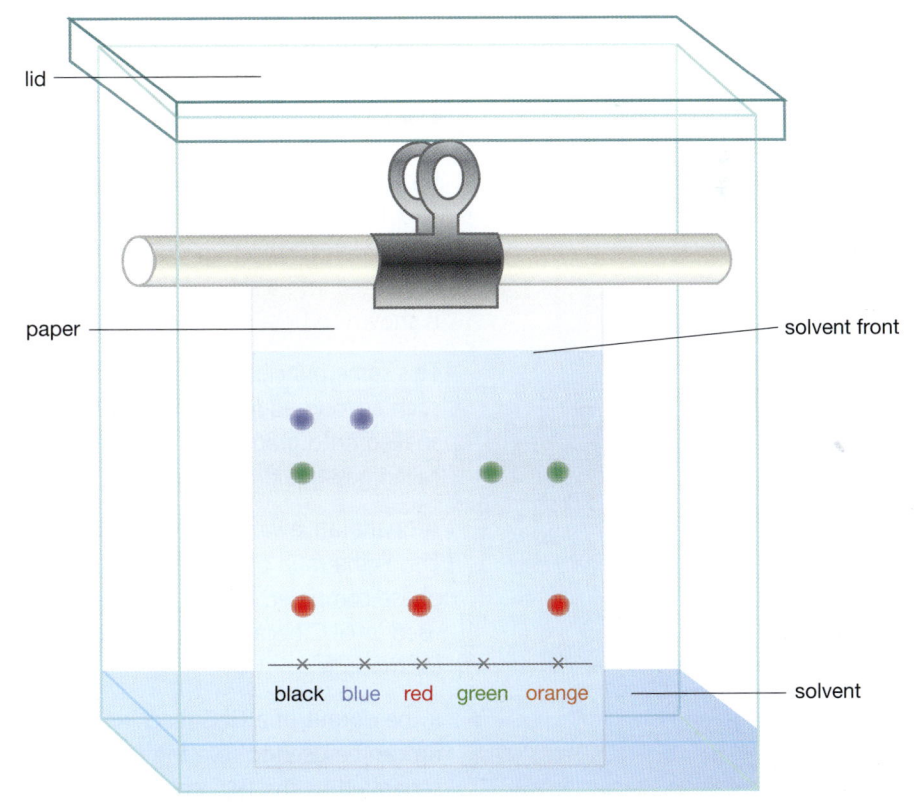

Figure 6.18 Sophisticated gas chromatography machines use the same principles as the simple separation of ink colours with a solvent and paper shown here.

Chemistry 2 143

Chapter 6 Mass, moles and yield

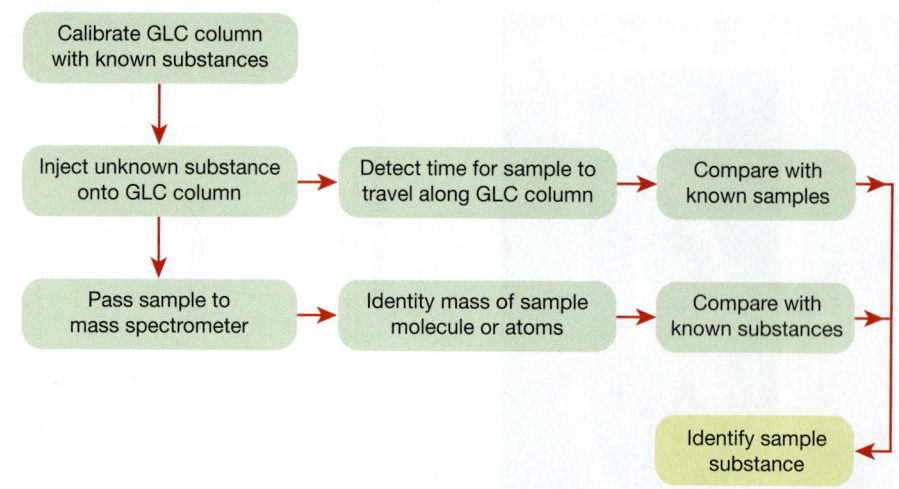

Figure 6.19 Flow chart for using a gas–liquid chromatograph to identify an unknown compound

Chromatography works because the molecules of substances have different amounts of attraction to other substances and also different solubilities in different solvents.

Figure 6.18 shows a simple chromatography set-up to separate and identify mixtures of different inks. The system consists of a stationary phase (the paper) and a mobile phase (the solvent that is allowed to soak up the paper from the bottom). Spots of the inks to be separated are spotted onto the bottom part of the paper above the solvent start line. The paper is then supported in the solvent tank and the solvent starts soaking upwards by capillary action. The distance the solvent and the ink spots have travelled is recorded as shown in Figure 6.18.

The inks are separated because their molecules stick to the paper with different strengths and because the inks have different solubilities in the solvent soaking up the paper. The red ink 'sticks' to the paper best and travels up only slowly, so it lags far behind the solvent front. The blue ink sticks to the paper least well and travels furthest up the paper just behind the solvent front. The black ink is separated into a series of colours that can be identified, and also the orange ink is shown to be a mixture of red and green inks.

This same principle works in separating many similar but different compounds such as food additives or different sugars. The stationary phase is often a liquid soaked onto granules in a thin tube, and the mobile phase can be a gas. This is called gas–liquid chromatography (GLC).

A quantitative value, called an R_f value, can be used to identify each compound. This is the ratio of how far a substance moves along the paper (or stationary) phase compared to how far the solvent (or mobile) phase moves. This can be used in GLC because known substances can be used to calibrate the apparatus (or machine).

If the output of a GLC machine is connected to a mass spectrometer, this set-up can be used to give the relative molecular mass of the substances being separated by the GLC column. The molecular mass is shown by a strong peak of the heaviest ion produced from each molecule.

AQA GCSE Additional Science

C2 6.6 Instrumental methods of analysis

Paper chromatography
HOW SCIENCE WORKS / PRACTICAL SKILLS

Separate a mixture of food colours by paper chromatography. Identify the original colours in the mixture. Use thin chromatography paper as the stationary phase and water as a mobile phase.

Work out the R_f values of the different food colours.

Methods of analysis
ICT

Create a presentation, video or podcast to explain what 'Instrumental methods of analysis' are and why they are important.

Test yourself

24 a State three advantages of using modern instruments to detect and analyse substances.
   b Name two kinds of instrument now used by analytical chemists.
   c Give two examples of the use of instrumental analysis.

25 In December 1978, four American and two Russian space probes landed on Venus. They had on board a range of instruments including mass spectrometers, spectroscopes and gas chromatograms. Because of weight limitations on the probe, the mass spectrometers had only low resolution. A particle of relative mass 64.0 was identified, but the resolution could not show whether the particle was SO_2 or S_2.
   a Why are instrumental methods of analysis important in space exploration?
   b Why could the mass spectrometers not show whether a particle on Venus was SO_2 or S_2? (A_r S = 32.0, A_r O = 16.0)
   c High-resolution mass spectrometers can measure relative masses to three decimal places. (A_r O = 15.995, A_r S = 31.972)
   Explain how a high-resolution mass spectrometer could determine whether a particle was SO_2 or S_2.

Chemistry 2

Chapter 6 Homework questions

Homework questions

1 Use the periodic table on page 283 for reference.
  a Copy and complete this table

Name	Symbol	Atomic number	Number of neutrons in nucleus	Average mass number (A_r)
Carbon				
	Ti			
		15		
Silver				
		53		
			18 or 20	35.5

  b Explain what the symbol $^{48}_{22}$Ti means.
  c What is an isotope and why is this important for the atomic mass (A_r) of chlorine?
  d Explain why the mass of electrons is not counted when working out the mass of an atom.

Use these atomic masses for the following questions:
A_rNa = 23.0 A_rO = 16.0 A_rN = 14.0 A_rK = 39.0
A_rS = 32.0 A_rCl = 35.5 A_rCu = 63.5 A_rC = 12.0
A_rPb = 207.0 A_rCr = 52.0 A_rCa = 40.0 A_rFe = 55.8
A_rH = 1.0

2 Calculate the formula mass (M_r) of:
  a sodium nitrate, $NaNO_3$
  b potassium sulfate, K_2SO_4
  c ammonium chloride, NH_4Cl
  d copper hydroxide, $Cu(OH)_2$
  e ammonium sulfate, $(NH_4)_2SO_4$
  f butane, C_4H_{10}

3 Calculate the mass of:
  a 1.0 mole of potassium atoms
  b 2.0 moles of oxygen molecules (O_2)
  c 12.0 moles of carbon atoms
  d 0.5 moles of chromium atoms
  e 0.01 moles of ammonia gas (NH_3)
  f 0.004 moles of calcium oxide (CaO)

4 How many moles in:
  a 80 g of calcium atoms?
  b 3.65 g of hydrochloric acid (HCl)?
  c 0.106 g of sodium carbonate (Na_2CO_3)?
  d 4.14 g of lead?

5 There is 22.2% of oxygen in an oxide of iron.
  a What is the composition by mass of 50 g of the iron oxide?
  b How many moles of each element are present in 50 g of the iron oxide?
  c What is the empirical formula of the iron oxide?

The iron oxide is reduced in a stream of carbon monoxide and 0.56 moles of iron metal are recovered.
  d What is the theoretical yield of this reduction reaction?
  e What is the percentage yield?

6 Chromium oxide can be reduced to chromium metal by heating it strongly with carbon powder. This was carried out for several samples of the oxide several times. The table shows the results.

Experiment	Mass of chromium oxide in g	Mass of chromium in the oxide in g
1	2.10	1.44
2	3.60	2.40
3	5.05	3.46
4	10.00	6.82
5	12.40	8.50

  a Why was the experiment carried out several times.
  b Plot a simple graph to compare the results. What does this tell you about the ratio of chromium to oxygen in the compound?
  c What does the graph tell you about the repeatability of the results?
  d What is the empirical formula of chromium oxide.
  e Do you think this evidence is valid, and how confident are you in the result you have calculated? Explain your answer.

AQA GCSE Additional Science

Chapter 6 Exam corner

Exam corner

Use these data:
A_r Na = 23.0 A_r O = 16.0 A_r Al = 27.0 A_r Ti = 48.0

a What is an isotope? (1 mark)

Student A
Radioactive material ✗

Examiner comment Some isotopes are radioactive, but this does not get the mark. Isotopes are about different numbers of neutrons in the nucleus of atoms of an element.

Student B
When two atoms of the same element have different numbers of neutrons in them, like carbon-12 contains 6 neutrons and carbon-14 contains 8 neutrons. ✓

Examiner comment Student B should have mentioned that it is different numbers of neutrons in the nucleus of atoms of the same element, but there is enough detail with the example for the mark.

b What is the percentage of oxygen in Al_2O_3? (2 marks)

Student A
60% is oxygen – three atoms out of 5. ✗✗

Examiner comment Student A has not calculated the formula mass and found what percentage oxygen is of this — the answer is 47.1%. No (correct) working, so no marks.

Student B
Al_2O_3: mass = (2 × 27) + (3 × 16) = 102

Percentage of oxygen $= \frac{3 \times 16}{102} \times 100\%$

$= \frac{48}{102} \times 100\%$

$= 47.1\%$ ✓✓

Examiner comment 1 mark for the correct formula mass, 1 mark for the correct proportion.

c 480 kg of titanium were obtained by reducing 1 tonne of titanium dioxide using this reaction: $TiO_2(s) + 4Na(l) \rightarrow Ti(s) + 2Na_2O(s)$
  (i) What is the theoretical yield from this reaction? (2 marks)
  (ii) What is the percentage yield in this case? (2 marks)

Student A
(i) Titanium is the product of the reaction. ✗

Examiner comment This is true, but not the answer required.

(ii) The percentage yield is $\frac{480}{1000} = 48\%$

   48% of the titanium ore is converted into metal ✗

Examiner comment Student A has not found the theoretical yield of titanium. If the student had stated the formula for finding the percentage yield, then 1 mark would have been awarded.

Student B
(i) M_r TiO_2 = 48 + (2 × 16) = 80

% titanium in ore = $\left(\frac{48}{80}\right) \times 100\% = 60\%$ ✓

Theoretical yield = 60% of 1000 kg (1 tonne) = 600 kg ✓

Examiner comment 1 mark for the correct % titanium in the ore, 1 mark for calculating the theoretical yield from 1000 kg.

(ii) Percentage yield $= \frac{\text{actual yield}}{\text{theoretical yield}} \times 100\%$ ✓

$= \frac{480}{600} \times 100\%$

$= 80\%$ ✓

Examiner comment 1 mark for the correct formula, 1 mark for calculating the correct % yield.

Chemistry 2

Chemistry 2

Chapter 7
Controlling rates of chemical reactions and energy transfer

Setting the scene

Cooking involves many chemical changes. The energy supplied by cooking starts chemical reactions. These reactions change the flavour and texture of food — making potatoes soft and turning meat brown and tender. If the temperature is too high, the reactions happen too fast and the meat and potatoes get browned on the outside, but remain cold and uncooked inside. Being able to control the speed of reactions is important when making new materials in industry too.

ICT

In this chapter you can learn to:
- use electronic sensors and data loggers
- create an animation that explains the collision theory model

Practical work

In this chapter you can learn to:
- draw conclusions about the effect of particle size on reaction rate
- identify controlling variables when investigating the decomposition of hydrogen peroxide
- draw a graph of experimental data and draw conclusions
- carefully measure temperature changes in reactions

C2 7.1

Learning outcome

Be able to measure the rate of a chemical reaction by measuring something that changes in the reactants or products per second of the time taken for the reaction.

How do we measure the rate of a chemical change?

Chemists need to control how fast a chemical change happens when they make new materials. This is so that the new material is produced in an economical way and in a suitable form to make it most useful.

How can you measure a rate of reaction?

A 'rate' is a speed. For example, your pulse rate is the number of heart beats in 1 minute. A **rate of reaction** is usually calculated as a measured quantity divided by the time taken.

Figure 7.1 Explosives can be used to demolish unwanted buildings. An explosion is a fast reaction — the reactants are used up quickly.

Figure 7.2 Sometimes you have to wait a long time to see the products form in a chemical reaction — rusting is a slow reaction.

Measuring the time taken for a reaction is straightforward. You can use a stopwatch or a wall clock — or a calendar for slow reactions. The quantity you measure in a chemical change can be:
- mass lost — if the reaction produces a gas
- time for all the solid reactants to be used up
- amount of unreacted solid left at any time
- temperature change produced — if energy is given out
- change in colour
- change in transparency of a solution
- change in viscosity — thickness or flow
- change in the output of a sensor, e.g. oxygen or carbon dioxide sensor with a data logger

The **rate of reaction** is the speed at which a reaction takes place. It can be measured as the rate of formation of a product or the rate of removal of a reactant.

Figure 7.3 When cooking an egg you can control the outcome carefully by controlling the temperature and how long you let the reaction continue.

Chemistry 2

Chapter 7 Controlling rates of chemical reactions and energy transfer

> **Boost your grade** ✓
>
> A rate of reaction must have units — for example, 'cm³ of gas given off per second' or 'grams of solid dissolved per minute'. However, you often just measure the time for a reaction to finish. In this case, the units would be 'minutes for mixture to stop fizzing' or 'seconds for the solid to dissolve'.

In a school laboratory, you usually measure rates of reaction by the amount of a reactant used up or the amount of a product formed:

$$\text{rate of reaction} = \frac{\text{amount of reactant used up}}{\text{time taken}}$$

or

$$\text{rate of reaction} = \frac{\text{amount of product formed}}{\text{time taken}}$$

For example, when magnesium reacts with an acid, hydrogen gas is formed.

Mg(s)	+	H$_2$SO$_4$(aq)	→	MgSO$_4$(aq)	+	H$_2$(g)
magnesium		sulfuric acid		magnesium sulfate		hydrogen

How could the rate of this reaction be measured? You could measure the volume of gas collected, or if the gas was allowed to escape, you could measure the mass of the flask at regular intervals (Figure 7.4).

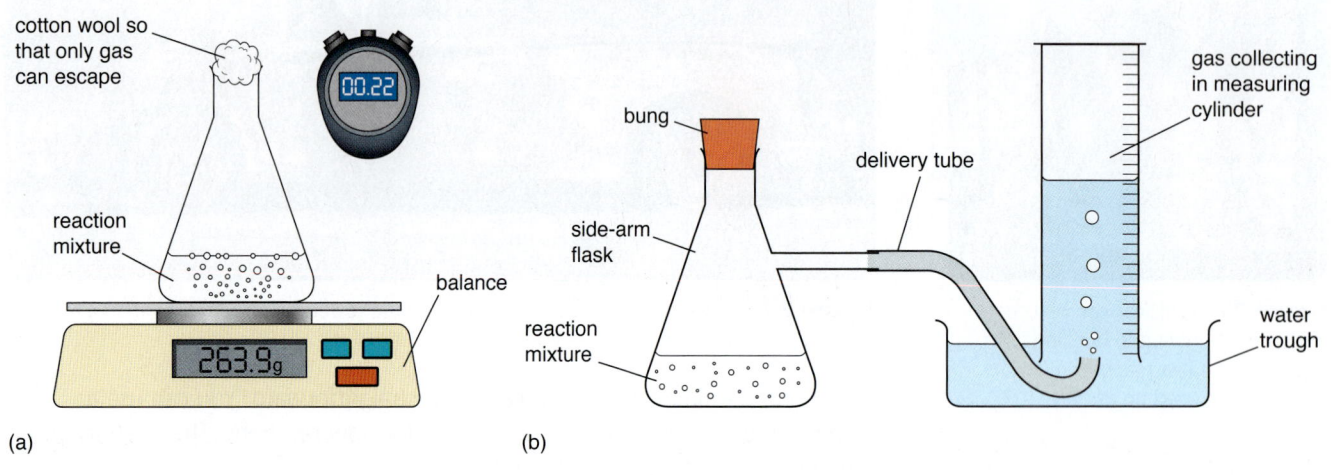

Figure 7.4 The rate of a reaction that produces a gas can be measured by: (a) recording the mass lost as the gas escapes and (b) collecting the gas produced and measuring its volume.

If you measure the time for the magnesium strip to completely dissolve, this does give you a way of judging the speed of the reaction, but this is not a rate because it does not measure 'something per second.'

Graphs of rates of reaction follow the same rules as the distance–time graphs you have studied (Figure 7.5). A steep graph line indicates a fast reaction. A horizontal graph line means that the reaction has stopped.

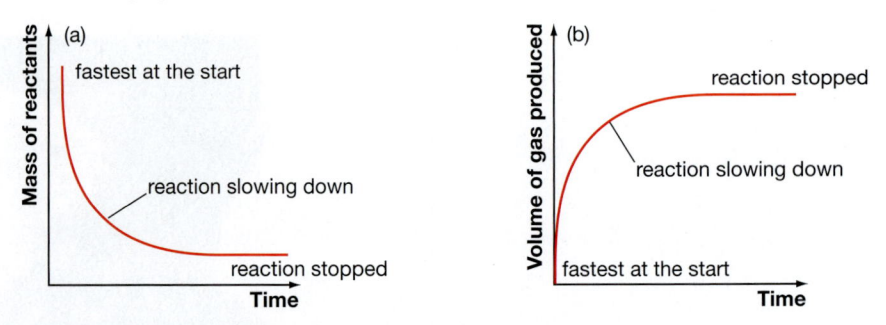

Figure 7.5 (a) The rate of a reaction can be measured by recording the change in mass over time. When the graph levels off, one of the reactants has been used up and the reaction stops. (b) This graph shows the volume of gas formed over time. The rate of reaction is highest when the gradient is steepest.

AQA GCSE Additional Science

C2 7.2 What makes a reaction happen, and how can we control it?

Test yourself

1. Give two methods for collecting data to measure the rate at which calcium carbonate reacts with hydrochloric acid to produce carbon dioxide.

2. You put a piece of sheet iron outside to react with the air and water in the environment.
   a. Is this a fast or a slow chemical change?
   b. What intervals between readings do you think you might use?
   c. For how long may you need to collect data?
   d. You decide to monitor the change by taking digital photos of the sheet of iron. How could you use this data to show the speed of the reaction? (Be creative!)

3. a. How long does it take to fry an egg?
   b. How long does it take to boil an egg?
   c. Does egg white (albumen) cook at the same rate as egg yolk?
   d. Explain how you decided your answer to part c.

4. What could you monitor or measure to find the rate of the following changes?
   a. A thin candle burning away.
   b. The change in the runniness of oil as it gets hotter.
   c. The amount of water boiled away from a glass beaker on a Bunsen burner per minute.
   d. The change in acidity of a bottle of milk as it is left to go sour

Investigating effect of temperature on reactions
HOW SCIENCE WORKS ACTIVITY

Glow sticks are used by anglers and emergency services as well as for amusement. How might you investigate the effect of temperature on the brightness and the length of time for which a glow stick works?

C2 7.2

Learning outcomes

- Be able to plot graphs of amount of products formed or reactants used over time in a reaction and to interpret the graphs in terms of the rate of the reaction.

- Be able to use the collision theory to model activation energy, and explain how increased energy or increased collision frequency changes the rates of reactions.

What makes a reaction happen, and how can we control it?

Collision theory model for chemical reactions

Most reactants are chemically stable. For a chemical change to happen, electrons have to be rattled free enough to be transferred or shared between atoms. For this to happen, the particles reacting have to collide — and collide quite hard.

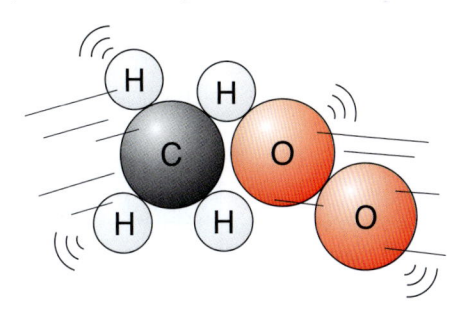

Figure 7.6 When two molecules meet, they react only if the collision transfers a sufficient amount of energy.

Observations show that most reactions are not instantaneous. The **collision theory** is a model that says that particles react only if the collision is hard

Chemistry 2

Chapter 7 Controlling rates of chemical reactions and energy transfer

> **Collision theory** states that reactant particles must collide in order for them to react chemically.
>
> The **activation energy** is the minimum energy needed for molecules to react when they collide.
>
> **Effective collisions** are collisions between atoms or molecules that successfully transfer enough energy for the atoms or molecules to react together.

enough for enough energy to be transferred to break some of the original bonds between atoms.

The minimum energy required for a reaction to happen is called its **activation energy**. The collision theory model can be summed up as follows:
- Particles can only react if they collide with each other.
- For particles to collide, they have to be moving about, so some of the particles must be in solution, a liquid or a gas.
- In solids, only the particles on the surface can react.
- Not all collisions result in a reaction. There has to be enough energy transferred when the particles collide for a reaction to happen — these are called the 'effective' collisions.

Making reactions go faster

There are two ways to make a reaction go faster (to increase the rate of reaction):
- increase the number of collisions
- make the collisions have more energy

It is obvious that increased temperature causes a faster reaction (increases the rate). This is because you get more collisions as the particles move faster. Other factors that can change the number of collisions, or change the number of effective collisions, are:
- the concentration of a solution of a reactant, or the pressure of a gas that is a reactant — both change the number of moving particles in a given volume of the reaction mixture
- changing the particle size of a solid reactant
- using a catalyst

The collision model explains how these factors can increase the number or energy of collisions.
- Increasing the temperature of the reactants makes the particles move about faster so they collide more often. The particles also collide with greater kinetic energy. Both factors increase the number of effective collisions.
- Increasing the concentration means that there are more particles in a certain volume of a solution. The particles are crowded together more closely, so there are more collisions.
- Increasing the pressure of a reacting gas is like increasing the concentration of a solution — there are more collisions.
- If a solid reactant is cut up into smaller pieces, there are more reactant particles exposed on the surface and therefore able to react (Figure 7.7).

eight cubes stacked together
surface area = 24 cm^2

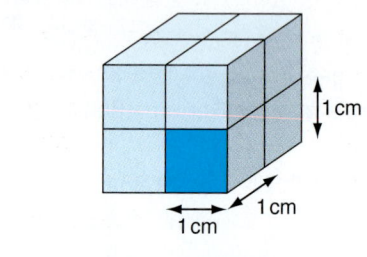

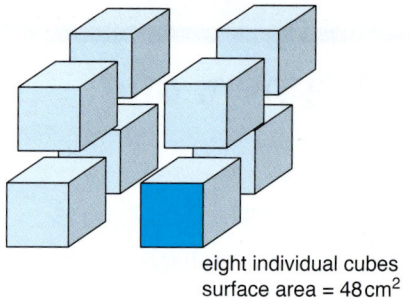

eight individual cubes
surface area = 48 cm^2

Figure 7.7 Making the pieces of reactant smaller increases the surface area and the number of particles exposed on the surface.

Investigating rates of reaction

HOW SCIENCE WORKS — PRACTICAL SKILLS

① Investigate the effect of temperature on rate of reaction by using the reaction between magnesium ribbon and dilute hydrochloric acid (0.5 mol/dm^3). Heat 25 cm^3 of the acid to a temperature of about 60°C, put a 2 cm strip of magnesium ribbon in the acid and record the time it takes to dissolve completely. Repeat this at several lower temperatures.

AQA GCSE Additional Science

C2 7.3 Predicting changes in rates of reaction

Test yourself

5. Explain why some collisions between reactants do not result in a chemical change.

6. Use the collision theory model to explain why temperature has a large effect on rates of reaction.

7. **a** Draw a particle picture to show the effect of changing the concentration of a solute reactant in a solution. Show a low concentration and a high concentration.
   b Draw a particle picture to show the effect of changing the pressure of a gas reactant. Show a low pressure and a high pressure.

❷ Investigate the effect of particle size/surface area of solid on rate of reaction by reacting an excess of marble chips (calcium carbonate, $CaCO_3$) with hydrochloric acid to produce carbon dioxide. Weigh out equal amounts of marble chips of different sizes — large chips, small chips and powdered chips. Put the marble into $50\,cm^3$ of $0.1\,mol/dm^3$ of acid in a $100\,cm^3$ conical flask placed on a top pan balance. Every 30 seconds record the loss in mass as the carbon dioxide is given off from the reacting mixture. Repeat this for each size of marble chips. Plot graphs of the resulting data. Devise a system using a data logger to record results and produce graphs.

❸ Sodium thiosulfate decomposes into sodium sulfate and sulfur powder when it is mixed with a small amount of an acid such as hydrochloric acid. As a result, the liquid mixture becomes cloudy with tiny particles of solid sulfur until you can no longer see through it. Use a series of dilutions of the $0.5\,mol/dm^3$ sodium thiosulfate solution provided to investigate the effect of concentration on rate of reaction. Devise a system using a data logger to record results and produce graphs.

❹ Use an electronic sensor and data-logging system to monitor the amount of different reactants over time. For example, you could measure the amount of carbon dioxide dissolved in water when elodea in the water is exposed to bright light. Or you could measure the amount of oxygen available in water as it becomes stagnant in a small experimental pond when nitrate is added to make the water plants grow.

C2 7.3 Predicting changes in rates of reaction

Learning outcomes

Know that the rate of a chemical reaction increases if the number of collisions increase or if the energy of the collisions increases.

Know that this can be achieved:
– by increasing the temperature
– by increasing the concentration of dissolved reactants or the pressure of gases
– if solid reactants are in smaller pieces

Figures 7.8–7.11 show how the rate of formation of products in a reaction is related to concentration, temperature and the particle size of the reactants.

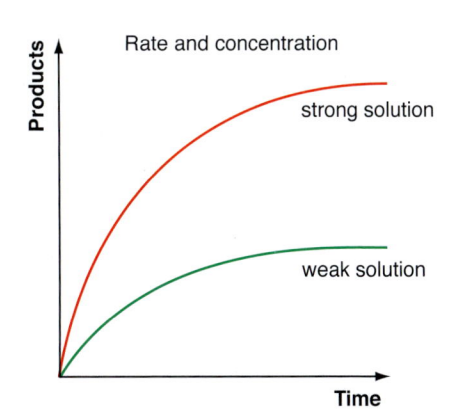

Figure 7.8 If a reactant in solution is diluted to half the concentration, this halves the number of collisions in a certain time.

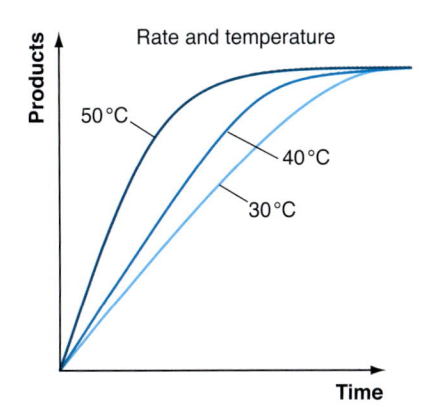

Figure 7.9 Each time a solution becomes 10°C hotter, the rate roughly doubles.

Chemistry 2 153

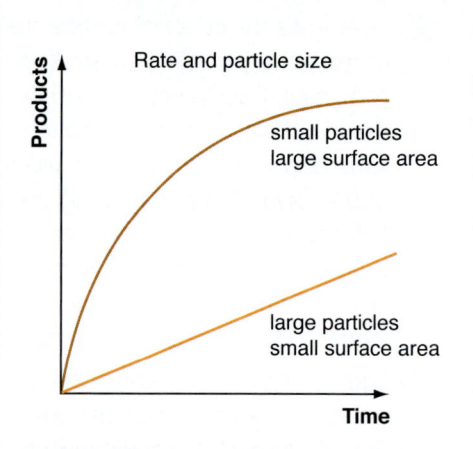

Chapter 7 Controlling rates of chemical reactions and energy transfer

Figure 7.10 If more surface area is exposed, the reaction rate increases.

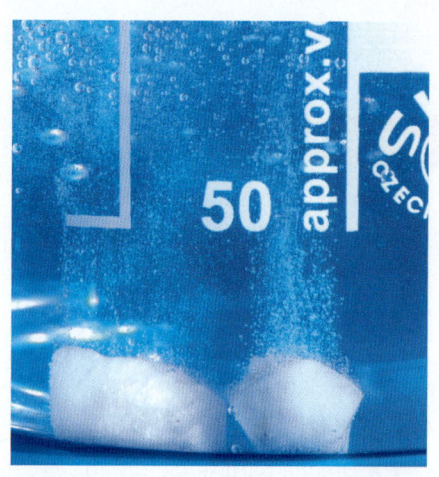

Figure 7.11 In an acid at 20°C not all the particles have the same energy — some are moving faster than others. Only the faster particles are moving fast enough when they collide with the marble chip to react and produce carbon dioxide gas. At a higher temperature many more of the acid particles have enough energy.

Boost your grade ✓

Remember that the mouse and crispy bar model is just a *thinking* model. It simply allows you to imagine how a reaction is affected by different factors.

A visual model for collision theory — blind mice and a crispy bar

Figures 7.12–7.17 show a visual model for magnesium reacting with hydrochloric acid. The acid particles (the mice) react with the magnesium (crispy bar) and gradually the magnesium is used up (the crispy bar is eaten and vanishes).

The blind mice are not clever. For a mouse to take a bite out of the crispy bar it has to bump its nose into the bar (this is like an effective collision needing the correct orientation of particles). If the mouse just brushes past the bar (unsuccessful collision) then no crispy bar gets eaten.

Figure 7.12 The acid particles (mice) will react with the magnesium (crispy bar) only if they collide directly with it. The inner crispies cannot get eaten.

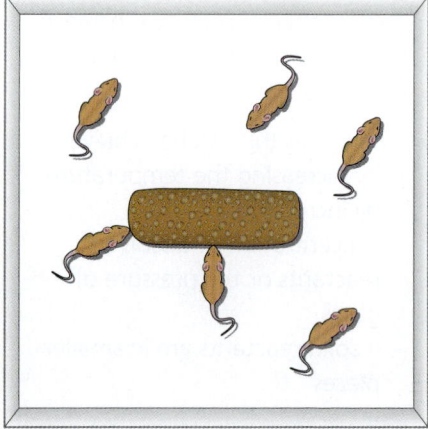

Figure 7.13 Slow-moving mice (low temperature) take all day to find and eat the crispy bar (slow reaction).

AQA GCSE Additional Science

C2 7.3 Predicting changes in rates of reaction

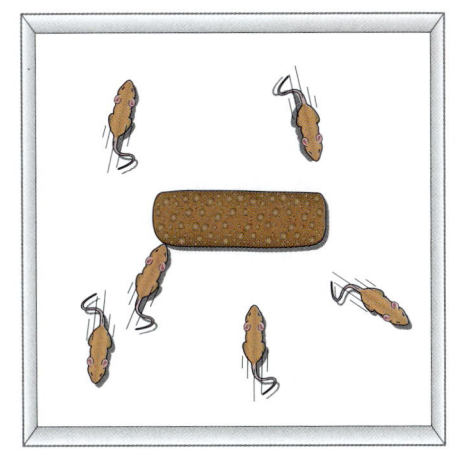

Figure 7.14 Fast-scurrying mice (higher temperature) eat the bar up much more quickly (faster reaction).

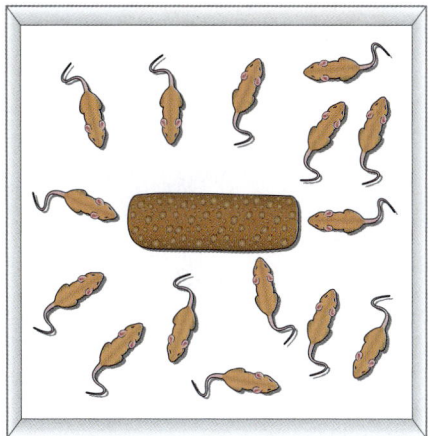

Figure 7.15 Lots more mice in the same space (higher concentration) eat the bar more quickly than a small number of mice.

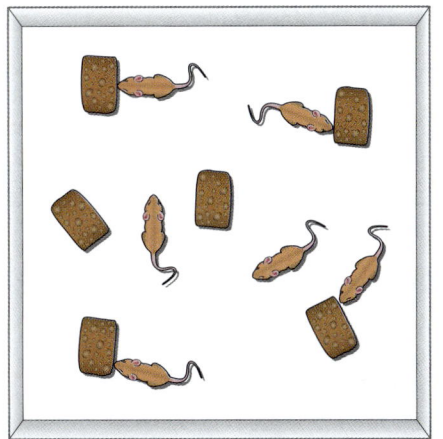

Figure 7.16 Even with a few mice, if the crispy bar is cut up into pieces, it is more likely that the mice will bump into a piece.

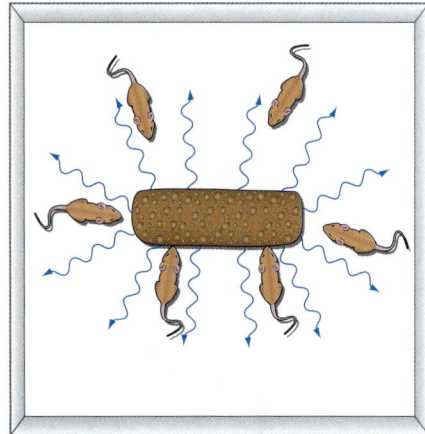

Figure 7.17 Explaining catalysts with this model is difficult. But if a vibration attracted the mice to the crispy bar, then it would get eaten much more quickly (faster reaction).

Modelling collision theory — ACTIVITY

How could you show the collision theory model using people to act as the particles? Describe a video you could make.

Explaining collision theory — ICT

Create an animation or presentation to explain how the collision theory model works.

Chemistry 2

Chapter 7 Controlling rates of chemical reactions and energy transfer

Figure 7.18 Cooking potatoes is a chemical change that sometimes takes ages, but sometimes only a minute or two. The cooking process breaks down the strong cell walls, making the food easier to digest. Cooking also releases the starch from inside the cells.

Test yourself

8. Explain why chips that are deep-fried in oil cook in 8 minutes, but boiling potatoes in water takes about 20 minutes.
9. If you cut the potatoes up smaller they cook faster. Crisps only take 20–30 seconds to deep-fry. Explain why this is so.
10. Explain why crinkle-cut chips cook faster than straight chips.
11. If the reaction between magnesium and hydrochloric acid is done in a beaker cooled in a freezer until the acid freezes, then the reaction stops. Explain this using the blind mice and crispy bar visual model.
12. If a crispy bar was broken up into individual crispies, how fast would the 'reaction' be in comparison to the other examples.

C2 7.4 Catalysts

Learning outcomes

- Know that if a catalyst is used, this increases the number of effective collisions.
- Understand that there are advantages and disadvantages of using catalysts in industrial processes.

Catalysts speed up, or catalyse, the rate of chemical reactions but are not used up in the course of the reaction. They are *not* one of the reactants — just an agent to speed things up. This is why the formula for a catalyst does not appear in the equation for the reaction.

Biological catalysts are called enzymes. The enzyme peroxidase catalyses the decomposition of hydrogen peroxide to water and oxygen. The reaction is also catalysed by manganese(IV) oxide.

A **catalyst** is a substance that speeds up the rate of a chemical change, without being used up itself. The same amount of catalyst will be present after the reaction has finished.

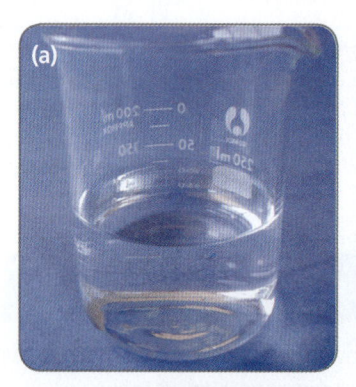

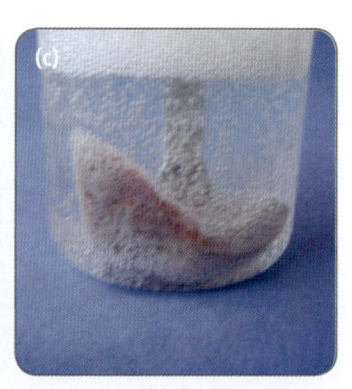

Figure 7.19 (a) Hydrogen peroxide (H_2O_2) decomposes slowly at room temperature to form water and oxygen. Adding a catalyst such as (b) manganese(IV) oxide or (c) peroxidase (an enzyme found in raw liver) makes the decomposition reaction much faster.

Test yourself

13. Name two catalysts that make hydrogen peroxide decompose more quickly.
14. Write a word equation for the decomposition of hydrogen peroxide.

C2 7.4 Catalysts

Test yourself

15 Suggest how you could measure the rate of decomposition of hydrogen peroxide with manganese(IV) oxide catalyst.

16 Sketch a graph that you would plot of your results for the decomposition of hydrogen peroxide with manganese(IV) oxide catalyst. Indicate the point at which the reaction stops.

17 When 1.0 mole of hydrogen peroxide decomposes, it forms 0.5 moles of oxygen.
 a What is the mass of 1.0 mole of hydrogen peroxide?
 b What is the mass of 0.5 moles of oxygen?

18 a Write a balanced symbol equation for the decomposition of hydrogen peroxide.
 b What mass of oxygen should be released from 1.0 mole of hydrogen peroxide by the end of the experiment? Suggest one reason why less gas than this might be collected.

Figure 7.20 Hydrogen peroxide is found in highlight kits and permanent hair dyes. It does not just bleach — the oxygen it releases combines with the dye molecules to produce different coloured products.

Catalysts are used to increase the rates of some important industrial processes. Without the catalysts, the reactions would happen only slowly, or would have to be carried out at a higher temperature. One disadvantage of using catalysts is that they need to be separated from the product after the reaction. This is easy if the catalyst is a solid and the reactants are gases or solutions, but could be difficult if both the catalyst and products are liquids or are in solution.

Different catalysts speed up different reactions, so generally each catalyst has only one use. Researchers are continually testing new catalysts for new industrial processes.

Investigation reaction rates

HOW SCIENCE WORKS / PRACTICAL SKILLS

1 The decomposition of hydrogen peroxide is catalysed by manganese(IV) oxide, MnO_2, or by peroxidase from living cells. Investigate the rate at which different substances will decompose '1 volume' hydrogen peroxide (made by diluting 20 volume hydrogen peroxide). Hydrogen peroxide has the formula H_2O_2.

2 Ignite a jet of hydrogen from a generator by using a small piece of platinum wire or mesh.

3 Crack liquid paraffin using broken porcelain as a high-temperature surface and as a catalyst.

Chemistry 2

Chapter 7 Controlling rates of chemical reactions and energy transfer

Table 7.1 Industrial reactions and catalysts.

Reactants	Catalyst	Product	Notes
Nitrogen and hydrogen	Iron mesh to give a large surface area	Ammonia	Haber process, manufacture of ammonia for fertilisers
Sulfur dioxide and oxygen	Granules of vanadium pentoxide	Sulfur trioxide	Contact process, manufacture of sulfuric acid
Hydrogen and unsaturated fats or oils	Fine nickel particles	Margarine	Nickel catalyses breakdown of carbon double bonds to make saturated fat, which is solid
Carbon monoxide and nitrogen oxides	Platinum and rhodium on a honeycomb metal support to increase surface area	Carbon dioxide and nitrogen (less polluting gases)	Used in catalytic converters in vehicle exhaust systems
Ethene and steam	Phosphoric acid absorbed on to silica (sand)	Ethanol	Very low yield reaction; gases are recycled to avoid waste
Cracking of hydrocarbons	Zeolite crystals	Shorter-chain hydrocarbons	The rate of cracking and the formation of diesel and petrol strongly depend on the presence of a catalyst
Ammonia and oxygen	Platinum mesh	Dinitrogen pentoxide	Used to make nitric acid for fertilisers and explosives

> **Test yourself**
>
> 19 Use Table 7.1 to describe three examples that show the importance of a large surface area when using a catalyst.
>
> 20 Describe a problem that can reduce the effectiveness of catalysts. Explain how you would overcome this problem.
>
> 21 Catalysts are used in industrial processes to reduce costs.
>   a Name two industrial processes where catalysts are used.
>   b Suggest how catalysts can reduce energy costs in industrial processes, and explain how this might work.
>   c Suggest how catalysts can increase the rate of industrial chemical reactions and explain how this might work.
>
> 22 Some catalysts are supported on an inert substance so they stay in one place. Other catalysts are mixed with the reacting mixture.
>   a Give one example of a catalyst from Table 7.1 that is supported on an inert substance.
>   b Margarine has nickel particles in it when it is manufactured. What has to happen to these nickel particles before the margarine can be sold?
>   c What does this processing do to the cost of making margarine?

Is there a cost-effective amount of a catalyst?

HOW SCIENCE WORKS / PRACTICAL SKILLS

Lyubov investigated this question as part of her GCSE practical work. She chose the decomposition of hydrogen peroxide (H_2O_2) using different masses of manganese(IV) oxide catalyst.

C2 7.4 Catalysts

$$2H_2O_2(aq) \xrightarrow{MnO_2 \text{ catalyst}} 2H_2O(l) + O_2(g)$$

She used 60 cm³ of 2 volume hydrogen peroxide solution in all her tests. All the tests were carried out at room temperature and atmospheric pressure. She used the same apparatus each time.

Note: hydrogen peroxide solutions do not have their concentration shown as molar concentrations. They have an indication of how much oxygen you can get from a certain volume of the solution. For example, 100 cm³ of '20 volume' hydrogen peroxide produces 100 cm³ × 20 = 2000 cm³ of oxygen gas.

Lyubov collected the results shown in Table 7.2. She needs to analyse the data to reach a conclusion.

Table 7.2

Amount of catalyst used	0 g	5 g	10 g	15 g	20 g	25 g
Time in seconds	Total volume of oxygen gas released in cm³					
0	0	0	0	0	0	0
10	0	5	12	20	22	24
20	0	10	20	37	40	43
30	0	15	30	55	57	60
40	1	20	39	72	74	76
50	1	25	47	86	87	88
70	1	35	65	110	110	110
100	2	48	90	115	115	115
150	2	70	118	120	120	120
200	2	92	120	120	120	120

1. Materials such as platinum are used as catalysts in a number of chemical processes. Explain why it is important to be cost-effective in the use of catalysts.

2. Draw the apparatus that Lyubov could have used to carry out the investigation described above.

3. Copy and complete Table 7.3 to summarise how Lyubov dealt with the possible factors that could affect this investigation.

4. a What is the maximum volume of oxygen produced in this experiment?
   b What is the longest time measured?
   c Use your answers to draw an *x*-axis scale (time) and a *y*-axis scale (oxygen gas given off) for a set of graphs.

5. Plot a line graph for each set of results. Plot them all on the same axes on one piece of graph paper.

6. Write a conclusion answering Lyubov's question. Suggest a quantitative answer for this set of reaction conditions.

Chemistry 2

Chapter 7 Controlling rates of chemical reactions and energy transfer

Table 7.3

Factor	Type of variable	Range chosen	Controlled by Lyubov
Type of catalyst used		Manganese(IV) oxide	
Temperature and pressure			
Concentration of solution		2 volume solution	Yes
Amount of catalyst			
Volume of oxygen given off			

C2 7.5 Energy change in reactions

Learning outcomes

- Know that energy can be transferred both to reactants and from reactants in chemical changes.
- Be able to describe exothermic and endothermic reactions.
- Know that in a reversible reaction the products can react to reform the reactants, so the reaction goes both ways.

When a chemical change happens, energy is transferred to or from the surroundings. When energy is given out by a chemical reaction, the energy often heats up the surroundings. Sometimes the energy is transferred as light (Figure 7.21) or sound (explosions). These energy changes are called **exothermic** reactions (Figure 7.22). Combustion is an example of an exothermic reaction between a fuel and oxygen in the air.

Figure 7.21 These wrist bands contain luminol, a chemical that gives out light energy when it reacts with another liquid in the band.

Self-warming cans heat their contents using an exothermic reaction. First you must press the bulb on the bottom of the can and wait 2 minutes. Water reacts with anhydrous calcium oxide in a sealed compartment inside the can. The reaction gives out energy. This energy is transferred to the liquid inside the can, causing it to heat up. A similar exothermic reaction is used in instant hot packs to keep you warm on a cold day.

Figure 7.22 An exothermic reaction transfers energy to the surroundings.

AQA GCSE Additional Science

C2 7.5 Energy change in reactions

> An **exothermic** reaction is a chemical change where energy is transferred to the surroundings, often causing heating.
>
> An **endothermic** reaction is a chemical change where energy has to be absorbed from the surroundings, often as a heat transfer that causes cooling of the surroundings.

Not all chemical reactions cause a temperature increase. Some chemical changes cool down the reactants and their surroundings. To do this the reaction takes in energy from the surroundings and transfers it to the reactants. This is called an **endothermic** reaction (Figure 7.23). Thermal decomposition is an example of an endothermic reaction, as energy must be transferred to the reactant before it breaks down.

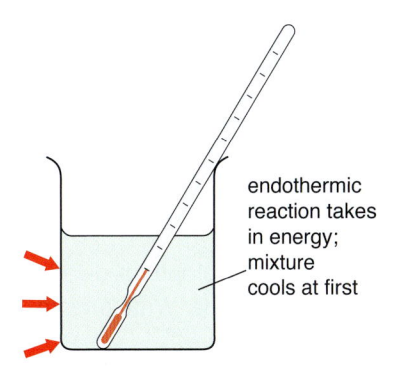

Figure 7.23 An endothermic reaction absorbs energy from the surroundings.

Figure 7.24 Twisting a cold pack starts a chemical reaction that makes the pack cold enough to numb the pain of a sports injury. Injuries such as sprains or bruises hurt because of the swelling, and cold helps to reduce swelling. The chemicals combine to take in energy, so the pack gets cold.

Measuring temperature changes

HOW SCIENCE WORKS — PRACTICAL SKILLS

1. Measure the energy transferred in the reactions listed in the table below. Carry out the reactions in a polystyrene cup supported in a glass beaker. Measure the temperature changes that take place during the reactions and record the highest temperature change reached. Keep the concentrations and quantities of solutions constant during the tests.

   Monitor these reactions using a temperature sensor. Copy and complete the following table.

Reaction	Temperature change	Explanation
25 cm³ of 2.0 mol/dm³ hydrochloric acid (irritant) with 25 cm³ of 2.0 mol/dm³ sodium hydroxide (corrosive)		
4 g sodium thiocyanate and 8 g barium hydroxide (corrosive) in 100 cm³ of water		
Zinc metal powder with copper sulfate solution		
Iron filings with copper sulfate solution		
Magnesium metal powder with copper sulfate solution		
Zinc metal powder with iron sulfate solution		
Magnesium metal powder with zinc sulfate solution		

2. Carry out a series of neutralisation reactions with different 2.0 mol/dm³ acids and alkalis.

   Prove that the only change that takes place in a neutralisation reaction is the formation of water molecules, so the size of the exothermic reaction is always the same:

   $$A^+(aq) + OH^-(aq) + H^+(aq) + B^-(aq) \rightarrow A^+(aq) + B^-(aq) + H_2O(l)$$
   alkali　　　　　　acid　　　　　　salt　　　water

3. Investigate dissolving the following in water:
   - ammonium nitrate
   - citric acid
   - anhydrous copper sulfate
   - sodium thiosulfate crystals
   - sodium hydrogencarbonate

Chemistry 2

Chapter 7 Controlling rates of chemical reactions and energy transfer

Take 0.1 mole of each compound and dissolve it in 100 cm³ of cold water. Measure the temperature changes and classify each reaction as endothermic or exothermic. Present the results in a table.

④ Investigate mixing 17 g of ammonium nitrate crystals and 32 g of barium hydroxide crystals. Stir the mixture and measure the change in temperature (and be careful of frostbite). Write a report and suggest an application for this reaction.

⑤ Investigate the maximum temperature change when instant hot and cold sports injury packs are used. Use a temperature sensor and data logger to record the changes. Write a consumer report deciding which is better for their purpose.

Test yourself

23 When would self-heating coffee or soup cans be useful?

24 Self-heating cans are heated using the reaction between calcium oxide and water. The only product of this reaction is calcium hydroxide. Write a word equation for the reaction.

25 What type of injury is treated with cold packs?

26 Suggest the materials you could put in an instant cold pack.

27 Copy and complete Table 7.4.

Table 7.4

Chemical change	Endothermic or exothermic?
Acid–alkali reaction	
Methane burning in a Bunsen burner	
Sodium reacting with water	
Calcium carbonate decomposing to calcium oxide and carbon dioxide	
Making aluminium metal	
Luminol reacting	

Designing a heat or cold pack

HOW SCIENCE WORKS ACTIVITY

Design and create a heat pack or a cold pack for treating sporting injuries.

What chemicals will you use, what quantities and what temperature change will they create? You will need to carry out the experiment to get some results.

Create a marketing campaign to scientifically promote your product.

AQA GCSE Additional Science

C2 7.6

Learning outcomes

- Know that in a reversible reaction the products can react to reform the reactants, so the reaction goes both ways.
- Know that in a reversible reaction if the forward reaction is endothermic, then the reverse reaction has to be exothermic.
- Know that the amounts of energy transferred in the forward and reverse reactions have to be equal and opposite.

Energy changes in reactions that go both ways

When blue crystals of hydrated copper sulfate are heated, water vapour is given off and white anhydrous copper sulfate is formed (Figure 7.25). The reaction is reversible, depending on how much water is present. The blue copper sulfate reforms when water is added to anhydrous copper sulfate. This colour change can be used as a test for water (Figure 7.26).

$$CuSO_4 \cdot 5H_2O(s) \underset{\text{exothermic}}{\overset{\text{endothermic}}{\rightleftharpoons}} CuSO_4(s) + 5H_2O(l)$$

hydrated copper sulfate (blue)   anhydrous copper sulfate (white)

Figure 7.25 Hydrated (blue) copper sulfate crystals (left) have a formula of $CuSO_4 \cdot 5H_2O$. They can be turned into anhydrous copper sulfate (right) on heating. Anhydrous copper sulfate ($CuSO_4$) has a slight tinge of colour due to water absorbed from the air.

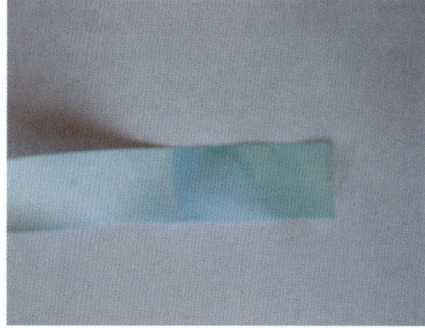

Figure 7.26 Testing for water. Filter paper is soaked in copper sulfate solution and dried in a hot oven. When the paper comes in contact with water, the anhydrous copper sulfate turns blue.

The thermal decomposition of hydrated copper sulfate is endothermic because heat is absorbed to break the water molecules out of the giant ionic lattice.

The reverse reaction is exothermic — when 5 g of anhydrous copper sulfate is dissolved in 50 g of water the temperature rises by 10 °C.

In a reversible reaction, if the forward reaction is endothermic, then the reverse reaction is exothermic. Exactly the same amount of energy is transferred in each case.

Figure 7.27 A reusable hand warmer has gives an exothermic reaction when it is used on a cold day. But if you heat the used hand warmer in hot water, you can put the energy back in. This is an endothermic reaction taking energy in. So overall the process is a reversible change.

Chemistry 2

Chapter 7 Controlling rates of chemical reactions and energy transfer

> **Copper sulfate crystals** — HOW SCIENCE WORKS ACTIVITY
>
> Carry out the dehydration and rehydration of copper sulfate crystals. Prove that the theory is correct and the change to anhydrous copper sulfate is reversible. Compare the reversible change of copper sulfate crystals with the similar reversible change of cobalt chloride crystals. Use a temperature sensor to monitor changes in temperature.
>
> Investigate the process of discharging a reusable instant heat pack and then recharging it to store energy in it again.

Test yourself

28 Describe how you would test a few drops of an unknown colourless liquid to see if it is water.

29 Explain why energy is needed to remove water of crystallisation from a giant ionic lattice.

30 Use the internet to research other chemical changes that are reversible. Also list some chemical changes that are almost impossible to reverse.

31 Use textbooks and the internet to find out about photosynthesis in plants and respiration in cells. Is this a reversible reaction?

Chapter 7 Homework questions

Homework questions

1
  a Give four examples of evidence you might see or measure to make you think that a chemical change is taking place.
  b Give four ways of measuring the progress of a reaction.
  c Put these chemical changes in order of rate of reaction. Put the fastest first.
  - digestion of food in the body
  - dye fading in the sunlight
  - baking a cake
  - egg frying
  - dynamite exploding
  - rusting of a car exhaust
  - concrete setting
  - firework rocket

  d Explain what you see happen when you fry an egg. How can you tell if an egg is cooked? How is this different from baking a cake and knowing if it is cooked?
  e Why does it take less time to fry chips than to boil potatoes until they are cooked?

2 Some marble chips were dropped into dilute hydrochloric acid and the gas produced was collected in a gas syringe.

The reaction is:

calcium carbonate + hydrochloric acid
  → calcium chloride solution
    + carbon dioxide gas + water

The amounts of gas collected were measured at regular intervals and they are listed in Table 7.5.
  a Draw the apparatus you could use to carry out this reaction.
  b Plot a suitable graph to show the progress of the reaction.
  c When was the reaction fastest? Mark this place 'fastest'.
  d When had the reaction stopped? Mark this place 'stopped'.
  e Sketch on your graph the curve you would expect if the acid used in the reaction was the same but hotter. Mark this line 'hotter'.
  f Carbon dioxide has a measurable mass. How could you follow the progress of the reaction without collecting the gas? Draw some apparatus you could use to do this.

Table 7.5

Time in s	Gas collected in cm^3
0	0.0
15	3.5
30	7.6
45	13.6
60	19.5
75	27.0
90	35.1
120	47.1
150	54.1
180	57.7
210	59.3
240	60.0
270	60.0

3
  a Explain what is meant by the 'collision theory' of reactions.
  b Explain why one of the reactants must be a gas or a liquid for a chemical reaction to happen.
  c Explain why not all collisions between particles result in a reaction. Use the idea of activation energy in your explanation.

4
  a Explain the meaning of the terms:
    (i) exothermic reaction
    (ii) endothermic reaction
  b Explain, with reference to the energy involved in bond making and bond breaking, why a reaction could be exothermic.
  c Explain, with reference to the energy involved in bond making and bond breaking, what activation energy is in this context.
  d Describe in detail, with a procedure and safety precautions, a simple demonstration you could do to show what is meant by an exothermic reaction.

Chemistry 2

Chapter 7 Exam corner

Exam corner

a Figure 7.28 shows a rusty exhaust pipe.
  (i) Car exhausts go rusty far faster than the rest of the car. Explain why this happens using the collision theory.
(4 marks)

Figure 7.28

Student A

The car exhaust is hot, reactions are faster when hotter. ✓

Examiner comment There are 4 marks for: hotter temperature particles move faster; so more collisions; reactant particles have more energy; so collisions are more effective. Only the first mark was scored, and there is no mention of collisions.

Student B

The exhaust and the air in contact with it is at a higher temperature. ✓ So this results in more collisions with more energy in them. ✓ Both these factors increase the rate of reaction. ✓

Examiner comment The 'particles move faster' mark was not scored.

  (ii) Metal parts of cars can corrode quickly if car battery acid is spilled on them, but do not corrode so fast if the acid is diluted with water as soon as it is spilled. Explain this effect using the collision theory.
(4 marks)

Student A

If you dilute battery acid it reacts more slowly. ✓

Examiner comment The 4 marks are for: more concentrated acid has more particles per unit of volume; so there are more collisions; dilute with water there are fewer particles per unit volume; so fewer collisions.
Student A has scored the 'dilute battery acid' mark — just about.

Student B

The battery acid has lots of acid particles per litre ✓ so there are likely to be more collisions ✓ of the acid with the metal. When you add water the acid is diluted ✓ and the number of collisions is hugely reduced. ✓

Examiner comment This is a complete answer.

b When hydrochloric acid is neutralised by sodium hydroxide, the reaction is described as exothermic.
Explain exactly what is meant by an 'exothermic' reaction.
(3 marks)

Student A

Exothermic means gets hot. ✓

Examiner comment The marks are for: stored chemical energy is transferred; to the surroundings; causes heating or other energy transfer. Student A scored the last mark point for 'gets hot'. Such an answer is too short for 3 marks.

Student B

An exothermic reaction results in some of the chemical energy ✓ in the reactants being transferred to the surroundings ✓ causing a rise in temperature ✓ or emission of light or other energy transfer.

Examiner comment This full answer scores all the points.

166 AQA GCSE Additional Science

Chemistry 2

Chapter 8
Using ions in solutions

ICT

In this chapter you can learn to:
- use the internet to find out how to grow crystals
- use electronic sensors to record data
- use multimedia to highlight the importance of recycling aluminium

Setting the scene

Common salt is used for more than adding flavour to chips. It is a raw material for many useful products, such as bleach, soap, washing soda, caustic soda (sodium hydroxide), water softeners, chlorine, PVC, baking soda and glass. Salt is also used for preserving meat (and Egyptian mummies), making freezing mixtures, dyeing and removing traces of water from aviation fuel after it is purified. Every cell of your body contains salt, a total of 250 g per person. Salt in body fluids is essential for muscle action — too little and you get cramp. For many of these uses and products, a solution of salt is needed. If salt did not dissolve, none of these uses would be possible.

Practical work

In this chapter you can learn to:
- use a burette to measure a volume of liquid precisely
- make crystals using different techniques
- make careful observations of electrolysis products

Chapter 8 Using ions in solutions

C2 8.1

Learning outcomes

- Know that when an ionic substance is melted or dissolved in water, the ions are free to move about in the liquid or solution.

- Know that salt solutions can be crystallised to produce a solid salt.

What does melting and dissolving do to ions?

In chemical equations, you have already learned that (s) means solid, (l) means liquid, (g) means gas and that (aq) implies an aqueous solution.

Ions are charged particles. They are formed when some elements react — for example, when a metal and a non-metal combine. The atoms react and electrons are transferred, leaving atoms, or groups of atoms, with extra electrons or fewer electrons. This makes the particles more stable because they have a noble gas electron configuration. Ions can be made from a single atom (e.g. a magnesium ion, Mg^{2+}) or from a group of atoms (e.g. a nitrate ion, NO_3^-).

Pure ionic substances are all solids at room temperature. There are strong electrostatic bonds holding the ions in a giant lattice (Section 5.1 and giant ionic structures in Section 5.4). It takes a lot of energy to break up a giant lattice of ions (Figure 8.1). When an ionic substance melts, the ions are free to move around in the resulting liquid. Ions can also be freed up from a lattice arrangement by dissolving the solid in water (Figure 8.2). When soluble ionic solids are put into water, the water molecules can get between the ions and separate them.

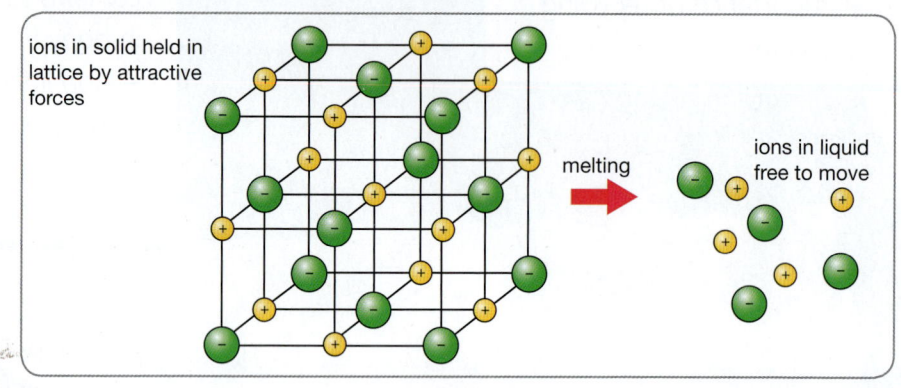

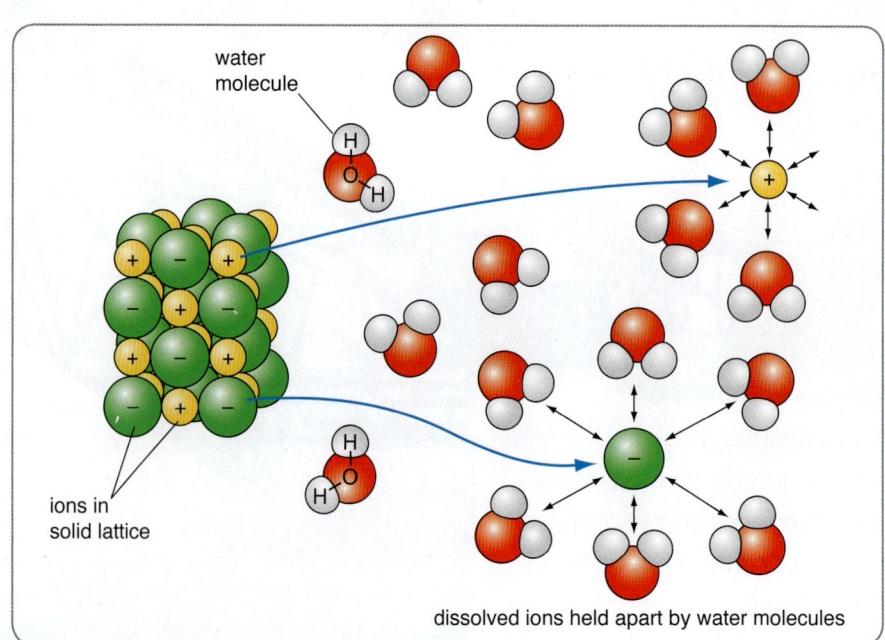

Figure 8.1 Melting an ionic compound means that the ions are free to move.

Figure 8.2 A model for dissolving an ionic substance. In a solution of an ionic compound the positive and negative ions are free to move about separately, kept apart by water molecules round the ions.

168 AQA GCSE Additional Science

C2 8.2 How do acids and alkalis react?

Figure 8.3 The regular stacking of ions in a crystal makes a structure with sharp points and straight sides.

A **crystal** is a regular solid lattice that has sharp edges and flat sides.

Investigating conductivity
HOW SCIENCE WORKS — PRACTICAL SKILLS

Using a bulb, power supply, two wires and two electrodes, investigate what happens to the conductivity of solid salt (sodium chloride) when water is added to it. Explain your observations.

Crystallisation

If the water, or other solvent, in a solution is allowed to evaporate, the solution becomes more concentrated. When this happens there may not be enough solvent to hold the ions in solution. The positive and negative ions come together to form a solid. If the solvent evaporates slowly, the ionic solid forms large, regular **crystals** as each ion takes its place in an extended giant structure or lattice of ions.

> **Test yourself**
>
> 1 Use the data in Table 5.2 on page 96 to write the formulae of:
>   a sodium nitrate
>   b aluminium chloride
>   c magnesium sulfate
>   d ammonium carbonate
>   e calcium hydroxide
> 2 What exactly is NaCl(aq)?
> 3 Draw particle pictures showing the difference between a solid salt, a molten salt and a dissolved salt?
> 4 Why do all ionic compounds have a relatively high melting point?
> 5 Draw a diagram to show one layer of atoms in a crystal of sodium chloride.

HOW SCIENCE WORKS — ICT

Research the best crystals to grow and make a presentation to your teacher to convince him/her to give you the chemicals needed to grow them.

C2 8.2 How do acids and alkalis react?

Learning outcomes

- Know that acids can be neutralised by bases.
- Know that soluble bases are called alkalis.
- Know that an acid and a base react to make a salt and water only.

Oxygen is reactive. It oxidises both metal elements and non-metal elements. Several non-metal oxides form **acids** when they dissolve in water. Insoluble non-metal oxides can have acid-like properties as well.

Soluble metal oxides are called **alkalis** — that is where most alkaline solutions come from. Insoluble metal oxides also react like alkalis but they are called **bases**. This is a precise scientific term for a substance that neutralises an acid to make a salt, and has nothing in common with other meanings of the word 'base'.

A salt is an ionic compound. Most salts consist of positively charged metal ions and negatively charged non-metal ions. Salts form when an acid neutralises a base. This is also different from the common meaning of the word 'salt'.

Chemistry 2

Chapter 8 Using ions in solutions

> An **acid** is a substance that releases H⁺ ions to make an acidic solution with a pH less than 7.
>
> An **alkali** is a substance that releases OH⁻ ions to make an alkaline solution with a pH greater than 7.
>
> A **base** is a substance that neutralises an acid to make a salt and water.

Acid solutions all contain hydrogen ions. Hydrogen ions (H⁺) are the ions responsible for all acid reactions. Alkali solutions all contain hydroxide ions (OH⁻). These are the ions responsible for all alkali reactions.

When an acid or an alkali dissolves in water, the substance splits up to make hydrogen ions or hydroxide ions respectively — the other dissolved ion is a 'spectator ion'.

$$HCl(aq) \rightarrow H^+(aq) + Cl^-(aq)$$
$$NaOH(aq) \rightarrow Na^+(aq) + OH^-(aq)$$

All acids react in similar ways. This is because only the hydrogen ion from the acid reacts. The alkali reactions of the hydroxide ion are all the same for a similar reason. The spectator ions (sodium ions and chloride ions in this case) take no part in acid or alkali reactions.

Explaining neutralisation

In a neutralisation reaction between an acid and an alkali, hydrogen ions react with hydroxide ions to produce water:

$$H^+(aq) + OH^-(aq) \rightarrow H_2O(l)$$

The spectator ions stay dissolved in the solution.

$$H^+(aq) + NO_3^-(aq) + Na^+(aq) + OH^-(aq) \rightarrow Na^+(aq) + NO_3^-(aq) + H_2O(l)$$
nitric acid   sodium hydroxide   sodium nitrate   water

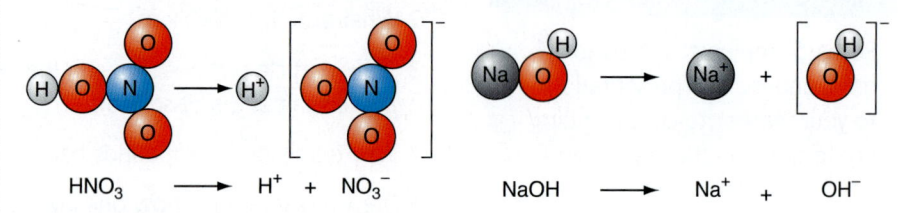

Figure 8.4 Nitric acid HNO₃ splits into H⁺ (hydrogen ions) and NO₃⁻ (nitrate ions).

Figure 8.5 Sodium hydroxide NaOH splits into Na⁺ (sodium ions) and OH⁻ (hydroxide ions).

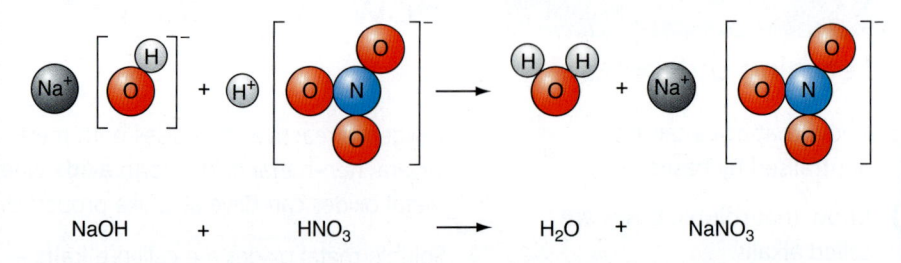

NaOH + HNO₃ ⟶ H₂O + NaNO₃

Figure 8.6 A neutralisation reaction. When sodium hydroxide (an alkali) reacts with nitric acid, the H⁺ and OH⁻ ions react to form water, but the Na⁺ and NO₃⁻ ions stay dissolved and unchanged.

Note that if we remove the spectator ions from the picture, then all neutralisation equations become the same:

$$H^+(aq) + OH^-(aq) \rightarrow H_2O(l)$$

C2 8.2 How do acids and alkalis react?

This means that the only change happening in any acid–alkali reaction is hydrogen ions joining with hydroxide ions to make water molecules. The spectator ions are 'left over' — they form the salt (in solution) produced by the reaction.

pH scale

The pH scale is used to measure the acidity or alkalinity of solutions. It is a measure of hydrogen ion concentration.
- pH 7 is the value for neutral solutions — 1 in every 10^7 water molecules splits to make an H^+ ion.
- pH 1 is strong acid — 1 water molecule in 10 splits to make an H^+ ion.
- pH 14 is a strong alkali — low concentration of H^+ ions.

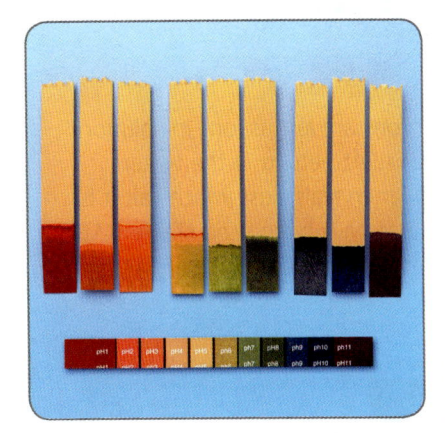

Figure 8.7 Universal indicator is a mixture of substances that responds to the pH of a solution by turning different colours — this range of colours matches the pH scale.

Investigating pH
HOW SCIENCE WORKS — PRACTICAL SKILLS

Investigate the pH of several household chemicals and classify them as acid, alkali or neutral.

Compare the colours of universal indicator in different strengths of acid and alkaline solutions.

Use a pH sensor and/or a conductivity sensor to monitor an acid–alkali neutralisation.

Wasp and bee stings
HOW SCIENCE WORKS — ICT

Wasp and bee stings have different pH values. Find out which is alkaline and which is acidic. Research which common household substances could be used to treat these stings and decide whether these treatments would work when judged scientifically.

Test yourself

6 a Draw a particle picture (like Figure 8.6) of hydrochloric acid being neutralised by potassium hydroxide.

7 Write an ion equation (like the sodium hydroxide/nitric acid example above) and name the salt produced for:
  a hydrochloric acid reacting with lithium hydroxide (LiOH)
  b sulfuric acid (H_2SO_4) reacting with sodium hydroxide
  c hydrochloric acid reacting with calcium hydroxide ($Ca(OH)_2$)

8 Estimate the pH of:
  a a weak acid
  b a moderate, but not strong alkali
  c a solution that is just on the acid side of neutral

9 a What substances dissolve in water to produce acids?
  b What substances dissolve in water to produce alkalis?

10 a 'Only the hydrogen ion matters in an acid reaction.' Explain what is meant by this statement.
   b What is the corresponding ion in an alkali reaction?
   c What new molecule is produced in neutralisation reactions?
   d What is meant by a 'spectator ion'?

Chemistry 2

Chapter 8 Using ions in solutions

C2 8.3

Learning outcomes

- Be able to predict which salt is produced in a reaction.
- Be able to explain how ammonium salts are produced and describe their use as fertilisers.

What salts can be made with an acid and an alkali?

When an acid is neutralised, all the spectator ions are left as ions in solution. If all the water is then evaporated, a solid salt is left behind. If the water evaporates slowly, the solid salt forms as regular crystals. Just which salt it is depends on the acid and the alkali (or base) that have been used (Table 8.1). For example, when potassium hydroxide and ethanoic acid react with each other, potassium ethanoate is made.

Table 8.1 Salts made from acids/alkalis.

Acid or alkali	Salt made
Nitric acid	Nitrates
Hydrochloric acid	Chlorides
Sulfuric acid	Sulfates
Ethanoic acid (vinegar)	Ethanoates
Sodium hydroxide	Sodium salts
Potassium hydroxide	Potassium salts
Ammonia solution (alkali)	Ammonium salts

Using indicators to prepare a salt

Acid and alkali solutions are usually colourless, and when you are making a salt you need to determine when the solution is exactly neutral. In a neutral solution, only the dissolved salt will be present with no excess acid or alkali.

You could use a pH meter to test when the solution is neutral. You could also add the acid or alkali solution a little at a time to the other, and check the pH after each addition by dropping a sample onto indicator paper (Figure 8.8).

Another method is to leave the indicator in the reaction flask as the solutions are mixed and react. The indicator is removed before crystallising the salt by adding

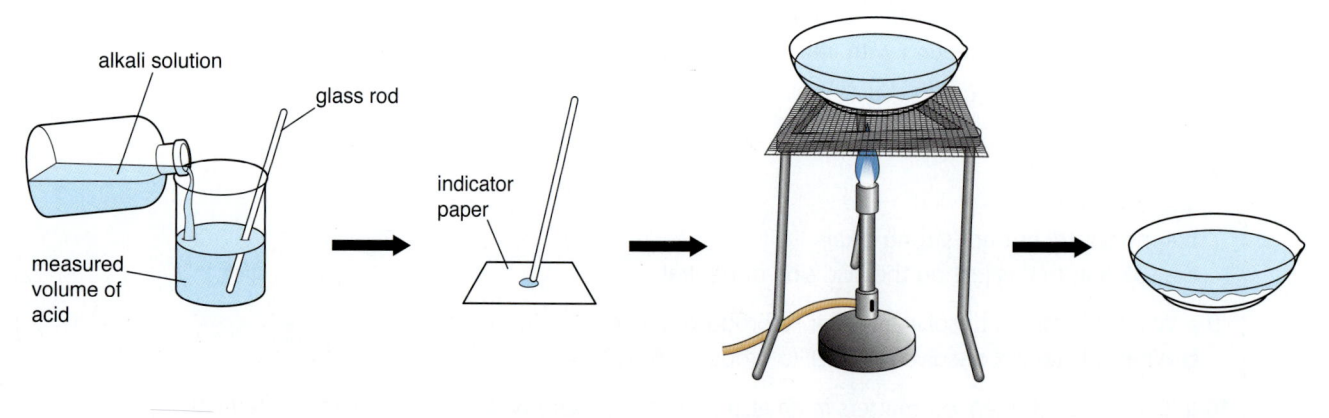

1. Add alkali a little at a time to the acid. Stir well.
   (alkali solution, glass rod, measured volume of acid)

2. Check pH after each addition.
   (indicator paper)

3. When the acid is just neutralised, evaporate the salt solution to crystallising point.

4. Set aside to cool and crystallise.

Figure 8.8 Preparing a pure soluble salt by neutralisation.

172 AQA GCSE Additional Science

C2 8.3 What salts can be made with an acid and an alkali?

two spatulas of charcoal and heating. The charcoal absorbs the indicator and can be filtered off, leaving a colourless solution.

Ammonia

Ammonia (NH_3) is a smelly gas. It is soluble in water — 1 dm^3 (1 litre) of water dissolves 1500 dm^3 (1500 litres) of ammonia gas. When ammonia dissolves in water, some ammonia molecules react with the water to form an ammonium ion (NH_4^+) and a hydroxide ion (OH^-) to give an alkaline solution:

$$NH_3(aq) + H_2O(l) \rightarrow NH_4^+(aq) + OH^-(aq)$$

Ammonium salts are important constituents of fertilisers. The ammonia provides the source of nitrogen for plants to make proteins.

Figure 8.9 Measuring the volume of acid accurately for a neutralisation reaction

Growing crystals
HOW SCIENCE WORKS / PRACTICAL SKILLS

1. Investigate making sodium chloride crystals using a technique like the one shown in Figure 8.8. Make the crystals as big as possible.
2. Ammonium sulfate is important as a nitrogen-based fertiliser. Make ammonium sulfate crystals by adding a small excess of ammonia solution to 0.1 mol/dm^3 sulfuric acid.

Use a burette and a standardised solution (0.1 mol/dm^3) of hydrochloric acid to find out how much standard acid it takes to neutralise an alkaline solution of unknown strength.

Test yourself

11 Copy and complete Table 8.2.

Table 8.2

Name of acid	Name of alkali	Name of salt	Formula of salt
Sulfuric acid	Sodium hydroxide		
Hydrochloric acid		Lithium chloride	
		Calcium nitrate	
			$MgCl_2$
			NH_4NO_3

12 Draw a flow chart for a method of producing clean, dry salt crystals by neutralising sulfuric acid with potassium hydroxide solution, using universal indicator.

13 Figure 8.9 shows a device called a burette being used. Describe how to set up and use a burette, with safety precautions, to measure the amount of a standard acid solution required to neutralise an alkali solution of unknown strength.

Chapter 8 Using ions in solutions

C2 8.4

Learning outcome

Be able to carry out methods to make a named salt.

What other ways are there of making salts?

Alkalis are soluble bases. Don't confuse the chemical word 'base' with other meanings of the word — for example in sports, in apparatus, in mountain climbing, in 'back to base' and so on. In chemistry, a **base** is a metal oxide or hydroxide. If a base dissolves in water, it is an alkali, but most bases are insoluble.

Acid plus insoluble base forms a soluble salt

$$2(H^+ + Cl^-)(aq) + CuO(s) \rightarrow (Cu^{2+} + 2Cl^-)(aq) + H_2O(l)$$
hydrochloric acid — copper oxide — copper chloride — water

> A **base** is a compound that neutralises an acid by forming water molecules as one of the products.

To make the salt copper chloride, add copper oxide to hydrochloric acid until no more of the black solid will react. The oxygen in the oxide reacts with hydrogen ions in the acid to form water. The spectator ions are left in solution to make copper chloride (Figure 8.10).

When all the acid is neutralised, filter off the remaining (excess) copper oxide and then heat the mixture gently for a few minutes. This evaporates the water from the copper chloride solution until it starts to crystallise.

Many metal carbonates react in a similar way to metal oxides. Carbon dioxide is an additional product because the carbonate reacts with the hydrogen ions from the acid:

$$2(H^+ + Cl^-)(aq) + MgCO_3(s) \rightarrow (Mg^{2+} + 2Cl^-)(aq) + H_2O(l) + CO_2(g)$$
hydrochloric acid — magnesium carbonate — magnesium chloride — water — carbon dioxide

Figure 8.10 Copper oxide dissolves in hydrochloric acid to give a blue solution of copper chloride. Unreacted copper oxide has settled to the bottom.

Acid plus metal forms a soluble salt plus hydrogen

Many reactive metals form salts when reacted with acids. The reaction is a **displacement reaction**, not a neutralisation reaction. The reaction does not work for the less reactive metals such as copper and gold.

> A **displacement reaction** is a reaction in which a more reactive element displaces a less reactive one from one of its compounds.

Only the metals above hydrogen in the reactivity series (see Table 8.3) react with an acid to make a salt. Aluminium does not make salts with acids readily because there is a hard aluminium oxide layer on the surface of aluminium, and this resists acid attack.

To make a salt from a metal plus an acid, add an excess of metal. Leave the reaction for some time, but don't heat, because hydrogen gas forms an explosive mixture with air. Once the reaction has finished, filter off the unreacted metal to leave the salt solution.

For example, you can make zinc sulfate by adding zinc to sulfuric acid:

$$Zn(s) + H_2SO_4(aq) \rightarrow ZnSO_4(aq) + H_2(g)$$
zinc — sulfuric acid — zinc sulfate — hydrogen

Figure 8.11 Soluble zinc sulfate is formed when zinc reacts with sulfuric acid — the fizz is hydrogen being released.

AQA GCSE Additional Science

C2 8.5 Precipitation

Table 8.3 Reactivity series for metals.

| Potassium |
| Sodium |
| Lithium |
| Calcium |
| Magnesium |
| Aluminium |
| (Carbon) |
| Zinc |
| Iron |
| Lead |
| (Hydrogen) |
| Copper |
| Silver |
| Gold |

Making salts

HOW SCIENCE WORKS / PRACTICAL SKILLS

Use the method of acid plus excess of base and filter to make:
- copper sulfate crystals from copper oxide
- copper sulfate crystals from copper carbonate
- magnesium sulfate crystals from magnesium oxide
- copper chloride crystals from copper oxide
- zinc nitrate crystals from zinc oxide

Use conductivity sensors to monitor changes in conductivity of a solution during a precipitation reaction. Use conductivity to determine the end point of the reaction.

Test yourself

14 What is the difference between a base and an alkali?

15 Magnesium chloride can be made by reacting magnesium with hydrochloric acid.
  a What safety precautions do you need to take?
  b Why would this method not work for copper chloride?
  c Write a step-by-step plan for making magnesium chloride crystals by this method, detailing what you would expect to see at each stage.

16 Write a step-by-step plan for making salt crystals from zinc oxide and nitric acid.

17 a Explain how you would know that a certain volume of sulfuric acid had been completely neutralised by copper carbonate.
  b What salt is made in this reaction?
  c Write a balanced symbol equation for the reaction between sulfuric acid and copper carbonate.

C2 8.5 Precipitation

Learning outcome

Be able to explain how precipitation reactions can be used to make salts.

Insoluble salts

When two ionic solutions are mixed, a **precipitate** can sometimes form. The precipitate is an insoluble salt and can be filtered off. The precipitate can be washed with pure water to remove any residual salt solution, and dried to give a pure, dry insoluble salt.

If two solutions are mixed and one pair of ions reacts to form an insoluble substance as a solid, this is an example of **precipitation**. For example:

$$BaCl_2(aq) + Na_2SO_4(aq) \rightarrow BaSO_4(s) + 2NaCl(aq)$$

barium chloride solution + sodium sulfate solution → barium sulfate solid + sodium chloride solution

> A **precipitation reaction** is a reaction that produces an insoluble solid when solutions are mixed. The solid produced is called a **precipitate**.

Chemistry 2

Chapter 8 Using ions in solutions

The solid barium sulfate precipitates out (Figure 8.12), leaving the spectator sodium ions and chloride ions still in solution — remember, '(aq)' means aqueous or in solution.

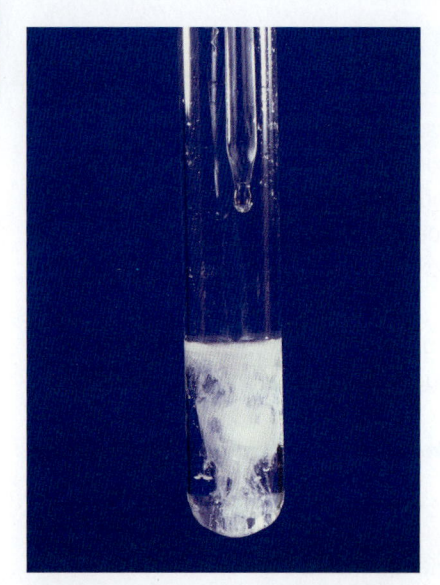

Figure 8.12 The colourless solutions in the pipette and in the test tube are two separate mixtures, one of barium and chloride ions and the other of sodium and sulfate ions. When the solutions are mixed, the barium and sulfate ions react together to form a white precipitate.

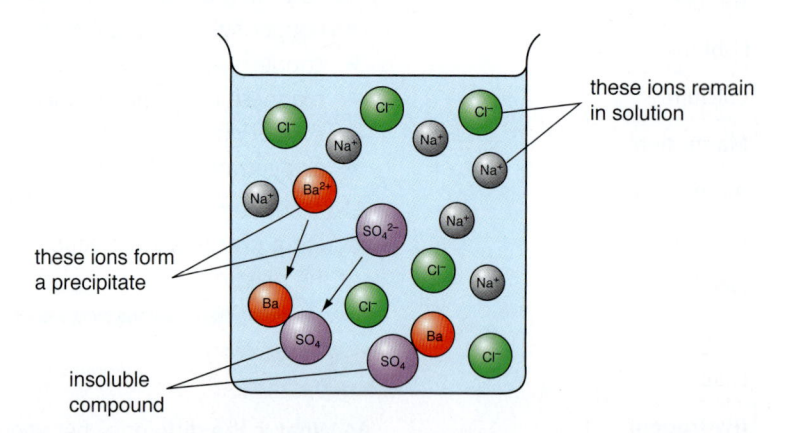

Figure 8.13 When these two ionic solutions are mixed, one pair of ions forms a solid precipitate that can be removed from the solution, but the other ions remain in solution.

Uses of precipitation reactions

Precipitation can be used to remove selected ions from a solution. The solution is mixed with another solution containing ions that form an insoluble precipitate with the unwanted ions in the first solution.

Iron and some other metal ions in drinking water are not harmful to human health but can be a nuisance. Dissolved iron forms brown stains on sinks, and darkens cups when combined with tannin in tea. It also discolours vegetables during cooking. These iron ions can be removed in water treatment plants by adding calcium hydroxide solution. The hydroxide ions react with the iron ions in solution to form insoluble iron hydroxide. Filters remove the precipitated particles.

The same precipitation reaction can be used to remove soluble metal ions from mine discharges and industrial effluent. Dissolved heavy metal ions, such as cadmium and lead, are precipitated out as metal hydroxides when the effluent is mixed with calcium hydroxide solution.

Phosphate fertilisers can contaminate rivers and groundwater and cause eutrophication by excess weed growth. Adding a metal salt (such as aluminium sulfate) to the waste water precipitates the phosphate ions as aluminium phosphate.

Precipitation reactions are used in forensic analysis to test if a sample contains particular ions, for example those from poisons or explosives residues. Barium chloride solution is used as a test for dissolved sulfate ions.

Precipitation reactions   ICT

Create a presentation explaining the uses and benefits of precipitation reactions.

C2 8.6 Electrolysis

> **Precipitation reactions** HOW SCIENCE WORKS / PRACTICAL SKILLS
>
> 1. Make pure, clean lead iodide by mixing lead nitrate solution and potassium iodide solution. Write an equation and draw a particle diagram for the precipitation reaction. What salt is also produced?
> 2. Repeat the activity for barium sulfate by mixing solutions of sodium sulfate and barium chloride.

Test yourself

18. Explain why dissolved metal ions can form a solid when you mix a colourless sample solution with another colourless solution.
19. Draw a diagram to represent the ions and molecules in a solution of magnesium chloride ($MgCl_2$).
20. Use the idea of ions and lattices to explain why large crystals of sodium chloride take a long time to dissolve.
21. Table 8.4 gives some data about solutions and precipitates.

    What is the white precipitate in each case? Explain your answers.

    Table 8.4 Solutions and precipitates.

Solutions mixed	Colour of precipitate
Barium chloride and sodium sulfate	White
Carbon dioxide and calcium hydroxide solution	White
Silver nitrate and calcium chloride	White

C2 8.6 Electrolysis

Learning outcomes

- Know that the electrolysis of ionic liquids results in the decomposition of compounds into elements.
- Know that half-equations are used to represent the reactions occurring at the electrodes.
- Know that electrolysis can be explained in terms of oxidation and reduction (electron gain and loss).

When an ionic substance is molten or dissolved in water, any ions in it are free to move about. In **electrolysis**, an electric potential difference is applied to the liquid by two **electrodes**. The ions move towards the electrodes with the opposite charge. Chemical changes take place when the ions reach the electrodes because the ions gain electrons at the **cathode** (negative electrode) and lose electrons at the **anode** (positive electrode). The energy transferred by the electricity can re-form elements from ions and break a compound down into its elements.

A good example is the electrolysis of molten lead bromide (Figure 8.15). The lead bromide has to be molten so that its ions can move freely.

Chemistry 2

Chapter 8 Using ions in solutions

Figure 8.14 Electrolysis does many jobs in science — this shows rows of electrolysis cells being used to electrolyse brine.

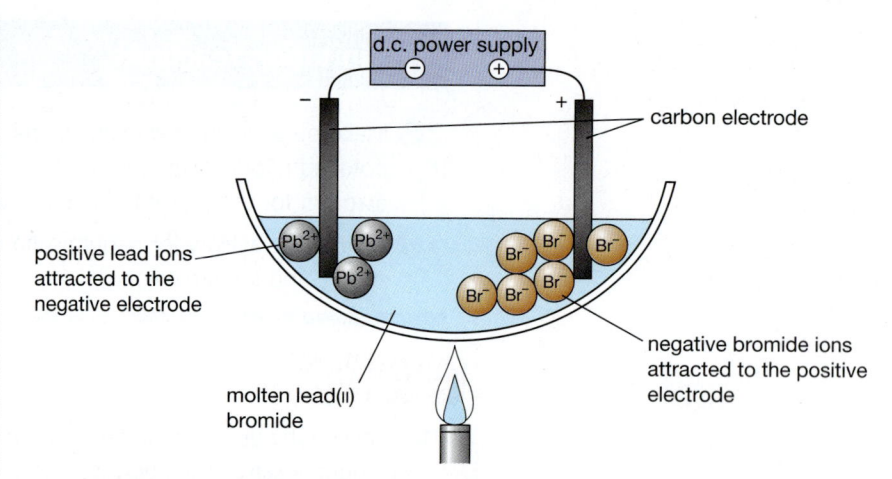

Figure 8.15 Lead bromide can be split into its elements, lead and bromine, by electrolysis.

Lead ions (Pb^{2+}) are attracted to the cathode (negative electrode). The lead ions gain electrons from the cathode so they become lead metal, molten at this temperature:

> lead ions + electrons → lead
> (molten metal)

Bromide ions (Br^-) are attracted to the anode (positive electrode), which pulls electrons away from the bromide ions forming bromine gas:

> bromide ions − electrons → bromine
> (gas)

If a salt can exist in a molten state, then it can be decomposed into its elements like this. However, some salts decompose when they are heated and cannot be electrolysed.

Electrolysis forms the basis of the production of most of the reactive metals, such as sodium and aluminium, from their molten metal ores.

Electrolysis equations

The reactions at the electrodes can be represented by **half-equations**. For example, in the electrolysis of molten lead bromide described above, we have:

> cathode: $Pb^{2+}(l) + 2e^- \rightarrow Pb(l)$
> anode: $2Br^-(l) - 2e^- \rightarrow Br_2(g)$

These are called ionic half-equations because each equation represents only half of the overall electrolysis reaction. Half-equations must obey the same rules as normal chemical equations. Each half-equation must be balanced in terms of atoms and charges, and the loss and gain of electrons in each pair of half-equations must match.

Electrolysis as oxidation and reduction

When oxygen reacts with a metal and oxidises it, the oxygen atoms remove electrons from the metal atoms to make negatively charged oxide ions (O^{2-}). The metal atoms become positively charged ions. When any other non-metal reacts with a metal, electrons are also transferred from the metal to the non-metal. This process is also called **oxidation**.

Electrolysis is the process of splitting a chemical compound into its elements by passing an electric current through a molten liquid or aqueous solution containing ions (the **electrolyte**).

An **electrolyte** is a liquid chemical that carries the electrical current and undergoes a chemical change.

An **electrode** is the physical connection between an electric circuit and an electrolyte. It is usually made from a material that won't corrode, such as a carbon rod or platinum foil.

The **anode** is the positive electrode (the positively charged connection between the electrolyte and the electric circuit).

The **cathode** is the negative electrode (the negatively charged connection between the electrolyte and the electric circuit).

A **half-equation** describes the reaction that takes place at one of the electrodes in an electrolysis cell. This is only half of the total process because another, different, reaction takes place at the other electrode.

AQA GCSE Additional Science

C2 8.6 Electrolysis

Oxidation is the loss of electrons — in electrolysis this happens at the positive electrode when a negatively charged ion loses electrons.

Reduction is the gain of electrons — in electrolysis this happens at the negative electrode when a positively charged ion gains electrons.

Boost your grade ✓
Use 'OIL RIG' to help you remember that '**O**xidation **I**s **L**oss' and '**R**eduction **I**s **G**ain'.

Reduction of metal oxides with carbon or hydrogen is the reverse process. The carbon atoms pull oxygen away from less reactive metal ions and give the electrons back, so the ions become metal atoms again.

We can define oxidation and reduction as follows:
- Oxidation Is Loss (of electrons)
- Reduction Is Gain (of electrons)

This gives us the mnemonic 'OIL RIG'.

In electrolysis, the anode takes electrons away — so the anode oxidises materials. The cathode gives electrons to materials — so the cathode reduces them.

Electrolysis of solutions

Salt solutions can be electrolysed in a similar way to molten ionic compounds. The ions in solution are attracted to the electrode with the opposite charge, where they are turned into atoms of the element.

Negatively charged ions (e.g. Cl^-) lose electrons (oxidation) and positively charged ions (e.g. Cu^{2+}) gain electrons (reduction). However, the presence of water in the solution, and the hydrogen ions and hydroxide ions that are always present in water, means that a different substance can sometimes be produced at the electrodes.

With some solutions, hydrogen ions (H^+) in the water gain electrons from the cathode to form molecules of hydrogen gas (H_2). For example, when sodium chloride solution is electrolysed, hydrogen is formed at the cathode.

The negative ions from the salt (Cl^-) lose electrons to the anode and form molecules of chlorine gas (Cl_2). The positive ions from the salt (Na^+) stay in solution. Hydroxide ions (OH^-) from the water also stay in solution. This means the original sodium chloride solution forms three products — hydrogen, chlorine and sodium hydroxide solution.

The half-equations for the reactions at the electrodes are:

$$\text{cathode: } 2H^+(aq) + 2e^- \rightarrow H_2(g)$$
$$\text{anode: } 2Cl^-(aq) \rightarrow Cl_2(g) + 2e^-$$

Mixtures of ions in solution

For solutions containing a mixture of ions, the products formed depend on the reactivity of the elements concerned. If a mixture of positive metal ions is present, then the ions of the element with the *lowest* reactivity form atoms first (the reactivity series for metals is given in Table 8.3). For example, in a mixture of magnesium chloride and hydrochloric acid, hydrogen is produced and not magnesium. In a mixture of iron sulfate and copper sulfate, copper is produced and not iron.

The negative ions in salt solutions are often simple halide ions (e.g. Cl^-, Br^-). The reactivity series for the halogens determines which product forms in the electrolysis of solutions with a mixture of halide ions. For example, in a mixed solution of sodium bromide and sodium chloride, bromine is less reactive and so bromine gas is produced at the anode.

Chemistry 2

Chapter 8 Using ions in solutions

Modelling electrolysis
ACTIVITY

Using models that you make out of paper or Plasticine, work through an electrolysis process showing the movement of electrons and the formation of ions, along with what happens at the anode and cathode.

Electrolysis reactions
HOW SCIENCE WORKS **PRACTICAL SKILLS**

1. Carry out the electrolysis of molten lead bromide (in a fume cupboard).
2. Carry out the electrolysis of solutions of these compounds using carbon electrodes:
   - copper chloride
   - sodium chloride
   - zinc bromide
   - zinc iodide
3. Electrolyse acidified water. Test the gases given off at each electrode to identify what they are. If appropriate, use a Hoffmann voltameter.

Test yourself

22. Explain what these terms mean:
    a electrolyte
    b electrolysis
    c anode
    d cathode

23. a Write a plan for a procedure to turn copper chloride crystals into copper metal and chlorine using electrolysis.
    b What safety precautions should you take?
    c Draw a diagram of the apparatus you would use.

24. Copy and complete Table 8.5.

    Table 8.5

Liquid being electrolysed	Product at cathode	Product at anode
Molten lead bromide		
Molten sodium oxide		
Copper chloride solution		
Potassium iodide solution		
A mixture of sodium iodide solution and copper chloride solution		

C2 8.7 Uses of electrolysis

Learning outcomes

- Know that the electrolysis of sodium chloride solution produces important substances for the chemical industry.
- Know that copper can be purified by electrolysis.
- Know that aluminium is manufactured by electrolysis.
- Know that several analytical and research techniques rely on electrolysis.

Electrolysis of water

Pure water is a poor conductor. If a little acid is added to the water, it conducts electricity. This means that acidified water can be electrolysed. This reaction breaks down water into hydrogen and oxygen. It is the basis of making hydrogen for use in fuel cells for electric vehicles.

$$\text{cathode: } 4H^+(aq) + 4e^- \rightarrow 2H_2(g)$$
$$\text{anode: } 4OH^-(aq) - 4e^- \rightarrow O_2(g) + 2H_2O(l)$$

Making aluminium metal

Aluminium is too high in the reactivity series to extract it from its ore by reduction with carbon. Most reactive metals are produced by electrolysis of a molten compound.

Aluminium metal is produced by the electrolysis of molten aluminium oxide (obtained from bauxite ore). The aluminium oxide has a high melting point and is dissolved in molten sodium aluminium fluoride (cryolite) to make it melt at a lower temperature (Figure 8.16). Only aluminium and no sodium is produced during electrolysis because sodium is higher in the reactivity series.

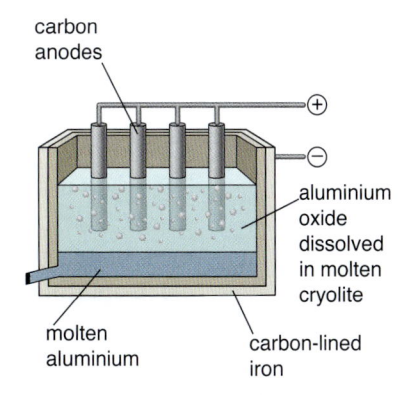

Figure 8.16 Aluminium oxide is decomposed by electrolysis in this industrial cell. The molten aluminium is siphoned from the bottom of the cell.

The oxygen produced reacts with the carbon electrode burning it away to make carbon dioxide:

$$\text{cathode: } 4Al^{3+}(l) + 12e^- \rightarrow 4Al(l)$$
$$\text{anode: } 6O^{2-}(l) - 12e^- \rightarrow 3O_2(g)$$
$$\text{then } 3C(s) + 3O_2(g) \rightarrow 3CO_2(g)$$

Electrolysis of sodium chloride solution — the chlor–alkali process

The electrolysis of salt solution produces some important chemicals. Concentrated brine (salt solution) is used to produce chlorine, sodium hydroxide (a common alkali solution, also known as caustic soda) and hydrogen gas. Hydrogen gas is a by-product of the process.

Chemistry 2

Chapter 8 Using ions in solutions

Sodium hydroxide is used to manufacture soap, and chlorine is used to make bleach and plastics. Some of the many industrial uses of salt and its products are described on the first page of this chapter.

Sodium chloride solution is electrolysed in a cell where a membrane separates the anode chamber from the cathode chamber (Figure 8.17). This method is called the chlor–alkali process. The porous membrane prevents the mixing of brine on one side with the sodium hydroxide formed on the other side.

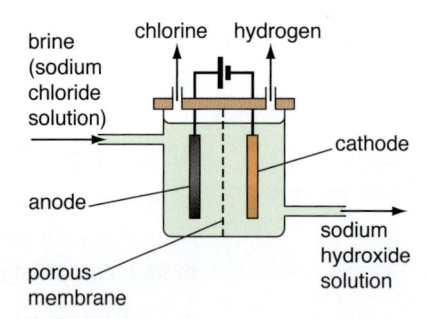

Figure 8.17 Cells like this membrane cell are used in a chlor–alkali plant.

At the anode, chlorine is produced:

$$2Cl^-(aq) - 2e^- \rightarrow Cl_2(g)$$

At the cathode, water is reduced to hydrogen gas releasing hydroxide ions into solution:

$$2H_2O(l) + 2e^- \rightarrow H_2(g) + 2OH^-(aq)$$

Electroplating and other uses

Electrolysis is also used to plate relatively unreactive metals onto more reactive ones. Chromium, nickel or tin can be deposited on steel materials to protect them from corrosion, or just for decorative purposes. There is a huge **electroplating** industry worldwide. Gold and silver are plated onto cutlery and tableware, copper is plated onto objects to preserve them. Even plastic objects can be plated with metal in a process invented by the eminent scientist Richard Feynman.

> **Electroplating** is the process of depositing metal on a material using electrolysis.

Figure 8.18 Chromium plating prevents the rusting of the steel.

C2 8.7 Uses of electrolysis

Test yourself

25 Name three products that can be made by the electrolysis of sodium chloride solution.

26 Write ionic half-equations for the reactions at the anode and cathode of the membrane cell shown in Figure 8.17.

27 a Electrical energy is needed for electrolysis. What other costs are involved in the manufacture of aluminium?
   b Explain why producing aluminium by electrolysis is not as sustainable as recycling aluminium.

Using electricity to make chemical changes
HOW SCIENCE WORKS / PRACTICAL SKILLS

1. Electroplate copper foil with nickel metal using nickel sulfate solution.
2. Investigate the movement of purple manganate ions in potassium permanganate by putting a crystal of potassium manganate on filter paper soaked in sodium chloride solution and apply a potential difference of about 4V across it.
3. Copper chromate solution is made from blue copper(II) ions and yellow chromate ions, so it appears green in solution. Put some dilute acid in a U-tube. Pipette dense copper chromate solution into the bottom of the U-tube so that the acid gets displaced into the vertical arms. Make up the copper chromate with lots of urea in it to make it dense. Put the electrodes in the colourless acid in the arms and investigate the movement of the coloured copper and chromate ions.

Recycling aluminium
ICT

Create a campaign to recycle aluminium, explaining to the public why this is important.

Chemistry 2

Chapter 8 Homework questions

Homework questions

1. 'Acid solutions are about H⁺(aq) ions; alkali solutions contain many OH⁻(aq) ions.'
   a. Explain, with examples, what the statement above means.
   b. What ions do the following substances split up into in solution:
      (i) sodium hydroxide?
      (ii) sulfuric acid?
      (iii) hydrochloric acid?
      (iv) calcium hydroxide?
   c. Write balanced symbol equations for the following neutralisation reactions:
      (i) sodium hydroxide plus sulfuric acid
      (ii) hydrochloric acid plus calcium hydroxide
   d. Explain what the term 'spectator ion' means.
   e. Rewrite both equations in part **c**, but remove the spectator ions on each side.
   f. (i) Explain what you need to do to find out if an acid solution has exactly neutralised an alkali solution.
      (ii) Explain what pH is and how it will change as an acid solution is neutralised by an alkali solution.

2. 'Common salt is only one of many salts — all with different colours and uses.'
   a. Explain what a scientist means by a 'salt' with reference to how one is made and its structure when it is a solid.
   b. (i) Write a detailed procedure, with safety precautions, for an experiment to prepare a sample of potassium chloride crystals from hydrochloric acid and potassium hydroxide solution.
      (ii) Include a balanced symbol equation.
   c. An alkali is just a soluble example of a chemical substance called a 'base'.
      (i) Explain what is meant by a 'base,' and why alkalis are special examples of a base.
      (ii) Why is it easy to confuse people when talking about a 'chemical base'?
   d. (i) Explain how to neutralise hydrochloric acid exactly with a base such as copper oxide.
      (ii) What would you see happen during the neutralisation?
      (iii) Write a balanced symbol equation for the reaction.
      (iv) Explain why an indicator solution is not needed in this procedure.

3. Salts will conduct electricity, but only when they are in liquid form.
   a. (i) What is the name for the process of passing electricity through liquids?
      (ii) Why do solid salts not conduct electricity?
      (iii) Explain the **two** ways that a solid salt can be made liquid.
      (iv) What happens to the chemicals in the liquids that carry the electricity? Give examples.
      (v) Explain why pure water does *not* conduct electricity.
   b. (i) Draw a labelled diagram of the apparatus you would use for passing an electric current through copper chloride solution.
      (ii) What are the products of this process? Write ionic half-equations for the reactions happening at the electrodes.
      (iii) How would the products of the process be different if copper sulfate solution was used?
   c. (i) Draw a labelled diagram for passing an electric current through molten tin bromide ($SnBr_2$; melting point 215°C).
      (ii) Explain how you would carry out this experiment, including all the safety precautions you would take and what hazards you would expect.
      (iii) What are the products of this process? Write ionic half-equations for the reactions happening at the electrodes.
   d. Aluminium metal is made by passing electricity through aluminium oxide (Al_2O_3; bauxite) that has been dissolved in molten calcium fluoride.
      (i) Why is aluminium manufactured in this way and not by smelting with carbon (coke), the traditional method of metal extraction?
      (ii) Write ionic half-equations for the reactions happening at the electrodes.
      (iii) Why is aluminium metal and not calcium metal produced in this process? Both ions are present in the electrolyte.
      (iv) Give social, economic and environmental reasons for why aluminium is a useful material.

AQA GCSE Additional Science

Chapter 8 Exam corner

Exam corner

a(i) Exactly what would you see happen if you added universal indicator to an acid solution? *(2 marks)*

Student A

(i) It would go red ✓

Examiner comment The required mark points are: it would turn from green or yellow–green; to red — so only 1 mark is scored.

Student B

(i) The universal indicator is a green solution ✓ or yellow–green paper. The indicator turns orange in a weak acid or red in a strong acid. ✓

Examiner comment Full marks — more detail is included than was required.

(ii) What are the products of the reaction between sulfuric acid and sodium hydroxide? *(2 marks)*

Student A

(ii) Sodium sulfate ✓

Examiner comment The products are sodium sulfate *and* water. If there is an excess of acid then sodium hydrogensulfate can be produced.

Student B

(ii) The salt sodium sulfate is produced ✓ and water molecules are made ✓ in the neutralisation reaction. Sodium ions and sulfate ions are spectator ions in solution. If there is an excess of acid then sodium hydrogensulfate can be produced.

Examiner comment A long answer but only the first sentence was required.

(iii) Write a balanced chemical equation for the reaction between sulfuric acid and sodium hydroxide. *(2 marks)*

Student A

(iii) sodium hydroxide + sulfuric acid → sodium sulfate + water ✗

Examiner comment This is a word equation, not a balanced symbol equation.

Student B

(iii) $2NaOH(aq) + H_2SO_4(aq) \rightarrow Na_2SO_4(aq) + H_2O(l)$ ✓✓

Examiner comment Full marks are awarded for this complete answer.

(iv) What would you see if the sodium hydroxide in part (ii) was contaminated with sodium carbonate solution? *(2 marks)*

Student A

(iv) The sodium hydroxide would be cloudy. ✗

Examiner comment Incorrect. Sodium carbonate is soluble and could be in the solution. The mark points are for: there would be a gas given off; the neutralisation and indicator colour change would be the same.

Student B

(iv) The reaction between the carbonate and acid would produce a fizz reaction. ✓ But the neutralisation and indicator colour change would be the same because sodium carbonate is a base. ✓

Examiner comment Full marks for a complete answer.

Chemistry 2

Chemistry 2

Exam questions

Chapter 5

1 This question is about lithium oxide and how it is formed from atoms of lithium and oxygen.

 a Copy and complete the following sentences by choosing the most suitable words from the box below to fill in the blank spaces.

**atoms   covalent   giant   hard   high
ionic   ions   low   simple   soft**

Lithium oxide is a substance with ____ bonding. It has a ____ structure composed of ____ . Strong forces between particles in the structure of lithium oxide make it very ____ with a ____ melting point.
(5 marks)

 b The electron structure of a lithium atom is 2,1 and the electron structure of an oxygen atom is 2,6. Describe the changes in electron structure of the lithium atom and the oxygen atom when lithium oxide forms. Use diagrams to do this. **(4 marks)**

 c Explain why the formula of lithium oxide is Li$_2$O. **(2 marks)**

2 Table 5A shows some properties of diamond and graphite.

Table 5A

Diamond	Graphite
Colourless crystals	Shiny black solid
Hardest natural substance	Soft and slippery

 a Why would you expect diamond and graphite to have the same properties? **(1 mark)**
 b Why do diamond and graphite have different properties? **(1 mark)**
 c Explain why graphite is soft and slippery. **(2 marks)**
 d Explain why diamond is hard. **(2 marks)**
 e Diamond does not conduct electricity but graphite does.
    Explain why graphite conducts electricity. **(3 marks)**

3 Substance X melts at a high temperature and the liquid that forms conducts electricity.
 a Which of the following could X be?
    aluminium chloride, cellulose, titanium, polythene, bronze, carbon disulfide **(3 marks)**
 b Explain your answers to part **a**. **(6 marks)**

4 Polythene is one of the most common plastics used today. Table 5B shows some of the properties of the plastic.

Table 5B

Property	LD polyethene	HD polyethene
Recycling symbol	4	2
Density	0.92 g/cm^3	0.95 g/cm^3
Max. temperature for use	80°C	110°C
Flexibility	Very flexible	Quite stiff
Amount of side chain branching	Considerable	Hardly any
Tensile strength	Lower	Higher
Resilience	Higher	Lower

 a What is the name of the monomer unit that is joined together to make polyethene? **(1 mark)**
 b Polyethene can be made in two ways that produce two different types of the material. Low-density polythene is very flexible; high-density polythene is much stiffer. Suggest two applications for each of the types of polythene. **(2 marks)**
 c Polythene is a thermosoftening plastic and, therefore, goes soft when heated. Why would you expect a thermosetting plastic to have a higher maximum use temperature? **(3 marks)**

5 Nanomaterials contain nanoparticles. Nanoparticles are manufactured particles with sizes measured in nanometres.

Clear sun cream contains nanoparticles of titanium dioxide, which allows the cream to be clear instead of the usual creamy white lotion. A health group is campaigning against the use of clear suncream because they think it could be dangerous. Why could this be the case? **(4 marks)**

186 AQA GCSE Additional Science

Chemistry 2

Chapter 6

1 Many processed foods contain chemicals as additives.
   a Sodium carbonate is used as an acidity regulator in some tinned foods. Sodium carbonate contains sodium ions (Na$^+$) and carbonate ions (CO$_3^{2-}$). Which one of the following is the formula of sodium carbonate? *(1 mark)*

   A NaCO$_3$ B Na$_2$CO$_3$ C Na(CO$_3$)$_2$ D Na$_3$(CO$_3$)$_2$

   b The electron structure of a sodium atom can be written as 2,8,1.
   Which one of the following is the electron structure of a sodium ion? *(1 mark)*

   A 2,2   B 2,8,2   C 2,8   D 2,8,8

   c Calcium chloride is used as a firming agent in some tinned foods. Calcium chloride contains calcium ions (Ca^{2+}) and chloride ions (Cl$^-$). Which one of the following is the formula of calcium chloride? *(1 mark)*

   A CaCl   B Ca$_2$Cl   C Ca$_2$Cl$_2$ D CaCl$_2$

   d Copy and complete the following sentence. *(2 marks)*

   A calcium ion (Ca^{2+}) is formed when a calcium atom _____ two _____ .

2 If a person is hypocalaemic, the amount of calcium ions in their blood is low. Doctors can prescribe calcium carbonate tablets.
   a What elements are present in calcium carbonate? *(3 marks)*
   b What is the chemical formula of calcium carbonate? *(2 marks)*
   c Calculate the relative formula mass (M_r) of calcium carbonate. *(2 marks)*
   d Calculate the percentage of carbon in calcium carbonate. *(2 marks)*

3 Potassium chloride can be mixed with sodium chloride to produce a 'reduced sodium' alternative to salt.
   a What are the chemical formulae of sodium chloride and potassium chloride? *(2 marks)*
   b What are the electronic structures of sodium and potassium? *(2 marks)*
   c How many protons, neutrons and electrons are there in a chlorine atom? *(3 marks)*
   d How does a chloride ion differ from a chlorine atom? *(2 marks)*

4 Read the newspaper cutting below, and then answer the questions that follow.

> ### Toxic gas scare at local factory
> Workers were forced to flee from the 'Cupromets' factory in Woodside Avenue last Tuesday. The factory was evacuated when fumes of highly toxic nitrogen dioxide filled the building. The gas was produced when concentrated nitric acid spilled onto thin copper pipes.

   a What does 'toxic' mean? *(1 mark)*
   b Explain why the reaction in the factory was more dangerous because the acid spilled onto thin copper pipes rather than thick copper bars. *(2 marks)*
   c Water was sprayed onto the copper and nitric acid to slow down the reaction. Why would this slow the reaction down? *(2 marks)*
   d Copy and balance this equation: *(2 marks)*

   Cu + __HNO$_3$ → Cu(NO$_3$)$_2$ + __H$_2$O + 2NO$_2$

   e Using this equation for the reaction, calculate the mass of nitrogen dioxide produced when 254 g of copper reacts completely with concentrated nitric acid. *(2 marks)*

   (Cu = 63.5, N = 14.0, O = 16.0)

5 a What is the mass, in grams, of 1 mole of NaF? *(1 mark)*
   b How many moles is 12 g of NaF? *(1 mark)*

Chapter 7

1 Two clear solutions, X and Y, are mixed in a conical flask. They gradually go cloudy and eventually a cross under the conical flask can no longer be seen. The time for the cross to disappear is measured for different temperatures of the reacting solutions — these are noted in Table 7A.

Chemistry 2

Table 7A

Temperature in °C	Time taken in s
22	327
27	260
35	198
43	194
50	118
58	100

a Draw a line graph of the data. Plot temperature on the x-axis (from 20°C to 60°C) and time on the y-axis (from 0 to 360 seconds). **(6 marks)**
b Identify the one anomalous result. **(1 mark)**
c Use your graph to find how long it takes for the cross to become obscured at 30°C. **(2 marks)**
d Use collision theory to explain the shape of the graph. **(5 marks)**

2 In the Contact process, sulfur dioxide reacts with oxygen to form sulfur trioxide in an exothermic reaction. The sulfur trioxide is then reacted with water to make sulfuric acid. Vanadium pentoxide is used as the catalyst:

$$2SO_2(g) + O_2(g) \rightleftharpoons 2SO_3(g)$$
sulfur dioxide   oxygen   sulfur trioxide

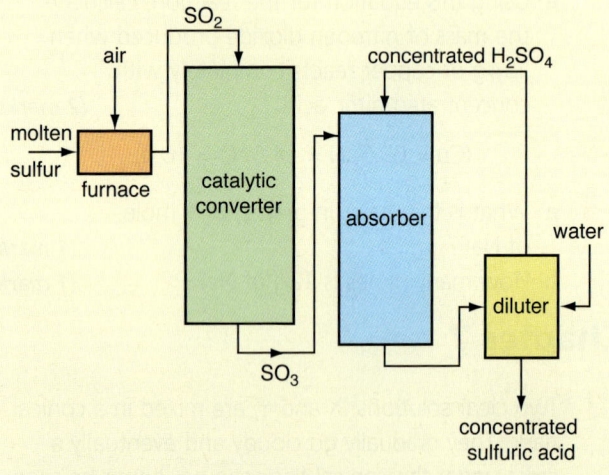

Figure 7A Flow diagram for making sulfuric acid.

a What is the purpose of the catalyst? **(2 marks)**
b Why is no extra catalyst being added in the flow diagram in Figure 7A to keep the process working? **(2 marks)**
c Explain how the best use is made of energy in the process. **(2 marks)**
d Suggest why it is important to make the best use of energy and materials in an industrial process. **(2 marks)**

3 For the process described in question 2, what is:
a the effect of pressure on the equilibrium mixture? **(2 marks)**
b the effect of temperature on the equilibrium mixture? **(2 marks)**
c the effect of removing sulfur trioxide from the equilibrium mixture? **(2 marks)**

4 Ethanoic acid is used to make plastics and the synthetic fibre called rayon. Before 1970, most ethanoic acid was made by heating butane in air with a catalyst:

$$2C_4H_{10} + 5O_2 \rightarrow 4CH_3COOH + 2H_2O$$
butane   oxygen   ethanoic acid   water

Several side products may also form, including butanone and formic acid. Formic acid is toxic.
a Suggest one reason why the yield of ethanoic acid is never 100%. **(1 mark)**

Most ethanoic acid is now made by reacting methanol with carbon monoxide in the presence of a catalyst:

$$CH_3OH + CO \rightleftharpoons CH_3COOH$$
methanol   carbon monoxide   ethanoic acid

Methanol and butane are both obtained from natural gas, but methanol is cheaper to obtain than butane. Methanol can also be obtained from wood or sewage.

b Explain why making ethanoic acid from methanol is more sustainable than making ethanoic acid from butane. **(3 marks)**

5 Instant cold packs can be used to treat sports injuries. The bag contains a pack of water and a pack of ammonium nitrate. When the bag is activated, the two chemicals mix and the solution becomes cold.
a What is the name of this type of reaction? **(1 mark)**
b Describe the energy changes that take place in this type of reaction, draw a graph to help explain. **(4 marks)**

AQA GCSE Additional Science

Chemistry 2

Chapter 8

1. Write balanced half-equations for the electrolysis of these liquids:
   a acidified water, H_2O *(2 marks)*
   b molten zinc chloride, $ZnCl_2$ *(2 marks)*
   c concentrated sodium chloride solution, NaCl(aq) *(2 marks)*
   d dilute copper sulfate solution, $CuSO_4$(aq) *(2 marks)*

2. Making salts is about neutralising acids. How would you make sure that all the acid had been exactly neutralised in each of the following reactions?
   a an acid plus insoluble metal carbonate *(2 marks)*
   b an acid plus metal oxide *(2 marks)*
   c an acid plus alkali *(2 marks)*

3. Oxidation is loss; reduction is gain.
   a Explain how the oxidation of a metal fits in with the definition of oxidation above. *(4 marks)*
   b How can this principle for defining oxidation and reduction be applied to electrolysis? *(4 marks)*

4. The apparatus in Figure 8A is used in industry to purify copper. The anode is impure copper, and the cathode is a sheet of pure copper. The electrolyte is copper(II) sulfate solution.

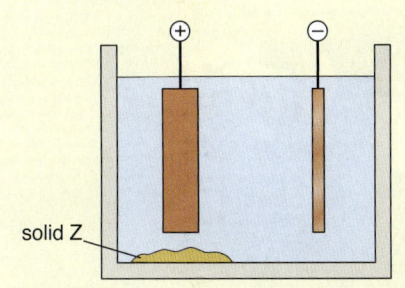

   Figure 8A

   a Explain why the anode is made of a large block of impure copper. *(2 marks)*
   b Explain why the cathode is made of a thin sheet of pure copper. *(2 marks)*
   c What is the difference in the properties of pure and impure copper? *(1 mark)*
   d What is the solid Z that accumulates below the anodes? *(2 marks)*
   e What are the main differences between this process and the electrolysis of copper(II) sulfate solution with carbon electrodes? *(4 marks)*

5. Briefly suggest what chemicals are needed for making each of the salts in Table 8A. *(10 marks)*

Table 8A

Salt	Function	Formula
Washing soda	Prevents scum from soap	Na_2CO_3
Epsom salts	Helps digestive system	$MgSO_4$
Moss killer for lawns	Gardening aid	$FeSO_4$
Road salt	Prevents water freezing	NaCl
Fish tank cleaner	Stops growth of fungi	$CuCl_2$

Physics 2

Chapter 9
Forces and their effects

ICT

In this chapter you can learn to:
- use digital photography to indicate forces in real situations
- use motion-tracking software to analyse extreme acceleration
- use multimedia to show what you know about stopping distances

Setting the scene

Before the parachute opens, the skydiver reaches her terminal velocity — a steady speed of about 60 m/s. Would the skydiver reach a faster terminal velocity if she were to jump from an aeroplane flying at a higher altitude?

Practical work

In this chapter you can learn to:
- consider the reliability of results when using a homemade spring balance
- control variables while investigating acceleration
- plot graphs to display distance–time data
- make predictions about reaction times and test ideas
- plan an investigation including a risk assessment
- take repeat readings, calculate the mean and identify anomalous results

P2 9.1 Effects of forces on motion

Learning outcomes

- Know that a force can change the motion of an object.
- Know that when the resultant force on an object is zero, the object remains in its state of motion.
- Know that when the resultant force on an object is not zero, the object accelerates in the direction of the resultant force.

Boost your grade ✓

Remember, the two forces always act on different objects — this is why they do not cancel out.

In diagrams, an arrow is used to represent a force. The length of the arrow represents the size of the force — the direction of the arrow shows the direction of the force.

To make a supermarket trolley move is simple — you push and the trolley moves forward; you pull and the trolley moves backward. But whatever the direction of the force you exert, the trolley always exerts an equal sized force on you in the opposite direction. This is because forces always come in pairs. You cannot have one force without another.

Whenever two objects interact, the forces they exert on each other are always equal in size and opposite in direction.

Figure 9.1 When you push on the trolley, it pushes equally hard on you.

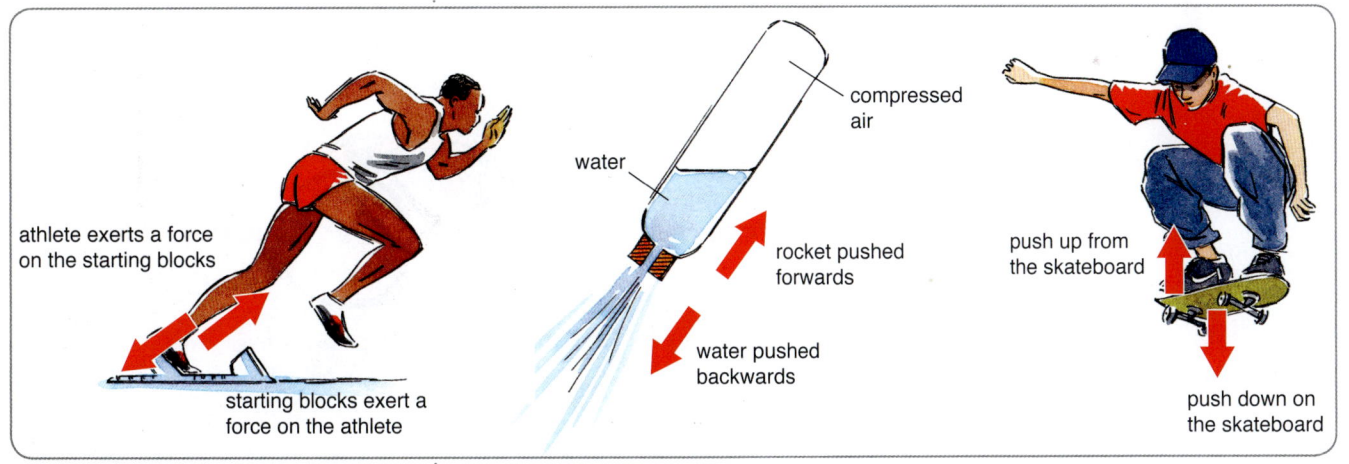

Figure 9.2 When two objects interact, there are always two forces. One force acts on one of the objects — the other force, which is equal in size, acts in the opposite direction on the second object.

The **resultant force** is a single force that has the same effect on an object as all the original forces acting together.

What happens when more than one force acts on the same object?

Each diagram in Figure 9.2 shows two forces. But only one force is shown acting on each different object. This would be unusual. In most situations, there will always be more than one force acting on a single object.

To find out what happens to an object when more than one force acts on it, we imagine all the individual forces being replaced by a single force that has the same effect. This is called the **resultant force**. The size and direction of the resultant force determine how, or even if, the object moves.

Physics 2   191

Chapter 9 Forces and their effects

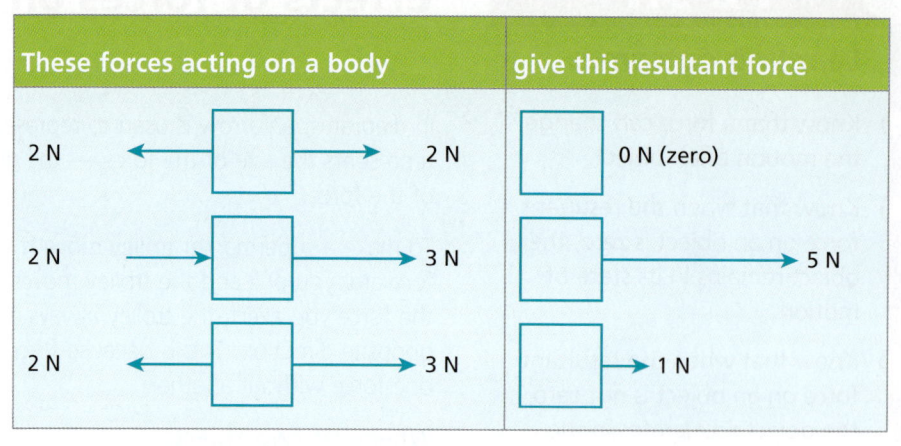

These forces acting on a body	give this resultant force
2 N ← □ → 2 N	□ 0 N (zero)
2 N → □ → 3 N	□ → 5 N
2 N ← □ → 3 N	□ → 1 N

Table 9.1 Two forces acting in the same direction add to give the resultant force. Two forces acting in opposite directions subtract to give the resultant force.

What happens when the forces are balanced?

Figure 9.3 shows the forces acting on a toy bird. The weight of the toy causes the spring to stretch. The spring stretches until the upward force in the spring (tension) equals the downward weight of the toy. The forces are now balanced — in effect, the forces have cancelled each other out. So, the resultant force is zero and the toy is stationary.

If the resultant force on an object is zero, the movement of the object does not change — if the object is stationary, it remains stationary. But what if the object is moving? If the forces are balanced so that the resultant force is zero, then the object keeps moving at the same speed and in the same direction (Figure 9.4).

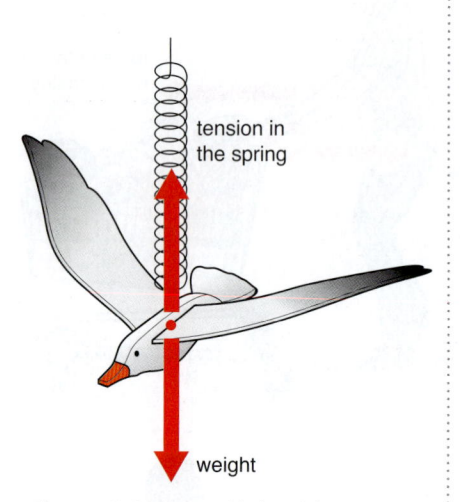

Figure 9.3 The toy bird weighs 3 newtons. What is the tension in the spring when the toy is stationary?

Figure 9.4 The cyclist is moving at a constant speed and in a straight line. This means that there is no change in the movement of the cyclist. What is the resultant force on the cyclist?

If the resultant force acting on an object is zero, the forces are balanced and the object does not accelerate. It remains at rest, or continues at the same speed and in the same direction.

Air resistance (or drag) is a force of friction. The direction of air resistance is always opposite to the direction of the moving object. Most of the frictional forces opposing the motion of a vehicle are caused by air resistance.

Figure 9.5 When the driving force forwards is balanced by the frictional forces backwards, the car travels at a constant speed.

What happens when the forces are unbalanced?

In a tug of war, when both teams pull equally hard the resultant force is zero and the rope does not move. But when one team starts to pull with a larger force the rope moves (Figure 9.6) — the resultant force is no longer zero.

AQA GCSE Physics

P2 9.1 Effects of forces on motion

Figure 9.6 The team pulling with the biggest force wins — the rope and everyone in both teams start to move in the direction of the resultant force.

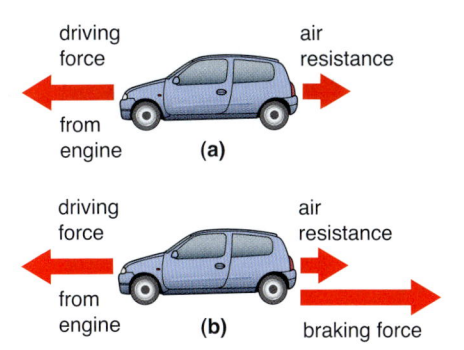

Figure 9.7 (a) The car accelerates in the direction of the resultant force. (b) The car decelerates in the direction of the resultant force.

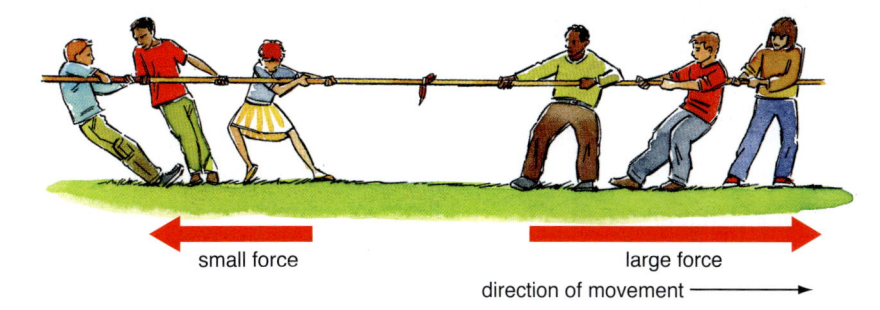

When the forces acting on an object are **unbalanced**, the resultant force can make an object speed up, slow down or change direction. This means that the resultant force makes the object **accelerate**.

An engine produces the driving force needed to move a car forwards. If the car is stationary, pushing the accelerator pedal produces a resultant force forwards. The car then accelerates in the direction of the resultant force (Figure 9.7).

When the resultant force on a stationary or moving object is not zero, the object accelerates in the direction of the resultant force.

If the car engine stops or the driver puts the brakes on, the car decelerates (slows down). This happens because the frictional forces backwards are now larger than the engine force forwards. The resultant force is in the opposite direction to the way the car is moving.

When an object is moving in the opposite direction to the resultant force, it decelerates.

Test yourself

1. Copy and complete the following sentences.
   a. When a book rests on a shelf its _____ pushes down on the shelf. The shelf pushes _____ on the book, with the same _____ force.
   b. The resultant of two forces is zero when the two forces are _____ in size and _____ in direction.

2. Tom and Sam are pulling on the opposite ends of a rope. Tom pulls with a force of 70 N and Sam pulls with a force of 100 N.
   a. What is the resultant force?
   b. Jim now joins in pulling the rope. The new resultant force is zero. What force does Jim pull with — and does he help Tom or Sam?

3. Look at the forces acting on the moving boat in Figure 9.8.
   a. Which two forces are always equal? Explain the reason for your choice.
   b. Describe the motion of the boat when:
      (i) $F_2 = F_3$
      (ii) F_2 is larger than F_3

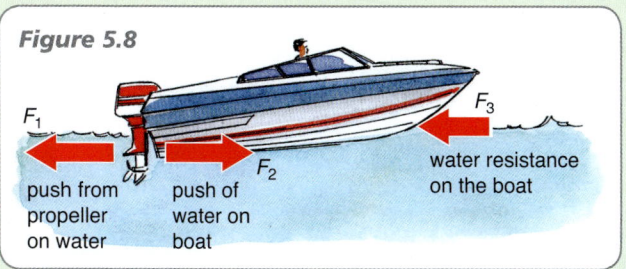

Figure 5.8

Forces in action   ICT

Have a photograph taken of you involved in some activity or sport. Using suitable software, add the force arrows to your photo showing all the forces acting. Show also which ones are balanced and which are not.

Physics 2

Chapter 9 Forces and their effects

P2 9.2 Elastic potential energy

Learning outcomes

- Know that a force can change the shape of an object.
- Understand that an elastic object stores elastic potential energy when it is stretched or squashed.
- Understand that the extension of a spring is directly proportional to the force applied, provided the spring is not overstretched.

Some objects break easily — a small force is all that is needed to cause permanent damage. If you apply a force to an elastic object it may change shape, but not permanently. When the force is taken away the object goes back to the way it was.

To stretch, squash, twist or bend an elastic object, work must be done and energy transferred to the object. The energy stored in an elastic object when work is done to change its shape is called **elastic potential energy**. There is more information about work and energy transfer in Section 10.1.

Figure 9.9 Bending the ruler and letting go makes it vibrate up and down. As long as the force is not too big, the ruler pings back into shape when the force is removed — the ruler is an elastic object.

> **Elastic potential energy** is the energy stored in an elastic object when work is done to temporarily change the shape of the object.

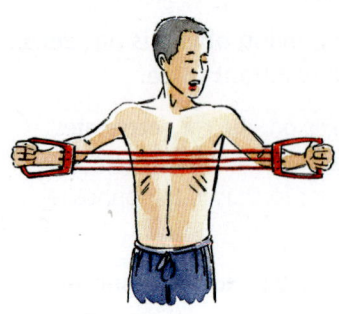

Figure 9.10 Using a chest expander involves doing work to stretch the springs. Doing work transfers chemical energy from the person's muscles to elastic potential energy, which is stored in the stretched springs.

Stretching a spring

The spring in Figure 9.11 is stretched by hanging masses from one end. Each 100 gram mass exerts a force of 1 newton on the spring. Measuring the extension of the spring each time a mass is added gives the data needed to plot a graph (Figure 9.12).

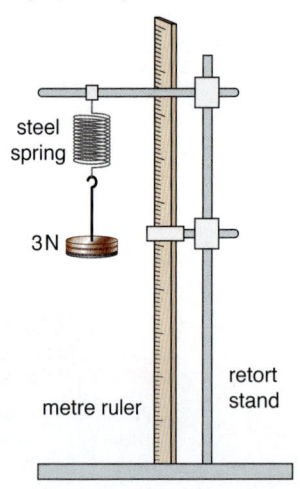

Figure 9.11 How much a spring stretches depends on the force applied to the spring. The difference between the stretched and unstretched lengths of the spring is called the **extension**.

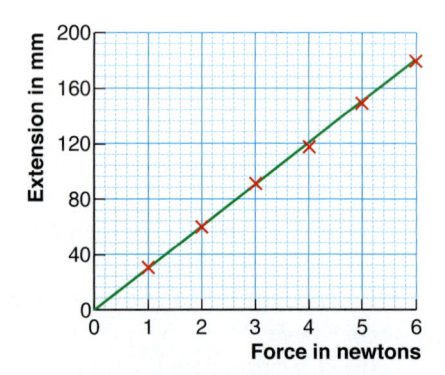

Figure 9.12 The graph gives a straight line through the origin. This means that the force applied to the spring and the extension of the spring are directly proportional.

To check the reliability of the data, the extension could also be measured as the masses are removed one by one. Provided the spring has not been overstretched

AQA GCSE Physics

P2 9.2 Elastic potential energy

Figure 9.13 The graph shows the data from an experiment with a steel spring. Why does the graph not give a totally straight line?

(gone past its **limit of proportionality**) the two sets of data should be the same.

If too much force is applied to a spring, it overstretches (Figure 9.13). When the force is removed, the spring does not go back to its original length. The spring has gone past its limit of proportionality.

The force exerted on a spring and the extension of the spring are linked by the equation:

$$F = k \times e$$

force = spring constant × extension
(newtons, N) (newtons per metre, N/m) (metres, m)

The equation works provided the spring has not gone past its limit of proportionality.

Worked example

George uses a spring to weigh the fish he has just caught. The spring stretches 8 cm. George knows that the spring constant is 300 N/m. How much does the fish weigh?

8 cm = 0.08 m

$F = k \times e$
$F = 300 \times 0.08$
$F = 24\,N$

Boost your grade ✓

Check that the units given in a question match. If, for example, one quantity involves metres and another centimetres then one of them must be changed.

Test yourself

4 Each of the objects in Figure 9.14 has had its shape changed.

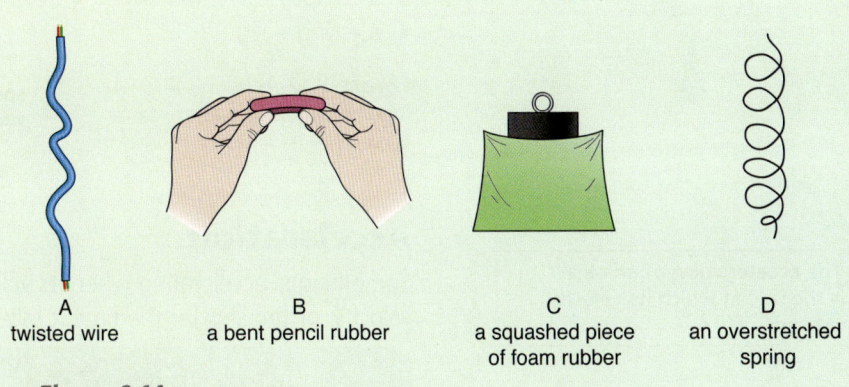

A twisted wire
B a bent pencil rubber
C a squashed piece of foam rubber
D an overstretched spring

Figure 9.14

Which of the objects has stored elastic potential energy? Explain the reason for each choice.

5 When a force of 2000 N is applied to a strong steel spring, the spring compresses by 4 cm. Calculate, in N/m, the spring constant for the spring.

6 A stone hung from the spring shown in Figure 9.11 causes the spring to extend by 75 mm. Use the graph (Figure 9.12) to find the mass of the stone in grams.

HOW SCIENCE WORKS — PRACTICAL SKILLS

If you were given a spring, a set of masses, a ruler and some card, how would you make a spring balance to find the mass of an apple?

Physics 2

Chapter 9 Forces and their effects

P2 9.3 Speed, velocity and acceleration

Learning outcomes

- Know that velocity involves both speed and direction.
- Know that acceleration is the rate of change of velocity.
- Know that an object that changes speed or direction is accelerating.

Figure 9.15 The average speed of a snail is just 0.0005 m/s. How far would a snail go in 1 hour?

> The **velocity** of an object is its speed in a given direction.

> The **acceleration** of an object is the rate at which its velocity changes.

Figure 9.18 This fairground ride takes the children round at a constant speed, but the children are accelerating. This is because they are constantly changing direction, which means their velocity is changing. Acceleration is to do with change in velocity — not just change in speed.

Speed and velocity

When you go somewhere it's not just speed that is important — direction also counts. Figure 9.16 shows two routes that Stacey can take to school.

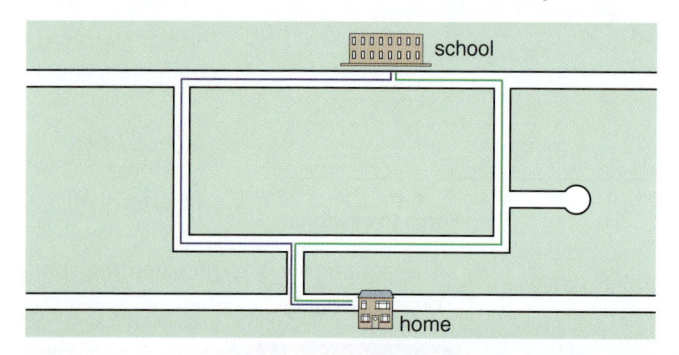

Figure 9.16 Stacey's two routes to school — same speed but different velocities.

The two routes are the same distance and take Stacey the same time. This means that Stacey goes at the same speed on both routes. But the two routes involve travelling in different directions, and each time Stacey changes direction her **velocity** changes. This is because velocity includes speed *and* direction.

To write the velocity of an object, you need to give both its speed and its direction. Direction can be shown in different ways. You can use:

- an arrow
- a compass direction
- a + or a − sign

Figure 9.17 Although the two joggers are moving at the same speed they have different velocities. Why?

speed = 3 m/s
velocity = 3 m/s ←
or 3 m/s west
or −3 m/s

speed = 3 m/s
velocity = 3 m/s →
or 3 m/s east
or +3 m/s

Acceleration

Something is accelerating when its velocity is changing. The bigger the change in velocity, or the shorter the time it takes, the larger the **acceleration**.

196 AQA GCSE Physics

P2 9.3 Speed, velocity and acceleration

Worked example

At the start of a 100 metre race, a runner accelerates to 12 m/s in 2 seconds. What is the acceleration of the runner?

Starting velocity = 0 m/s; final velocity = 12 m/s; time taken = 2 s.

$a = \dfrac{v - u}{t}$

$a = \dfrac{12 - 0}{2}$

$a = 6 \text{ m/s}^2$

Acceleration can be calculated using this equation:

$$a = \dfrac{v - u}{t}$$

$$\text{acceleration} = \dfrac{\text{final velocity} - \text{initial velocity}}{\text{time taken}}$$

where acceleration is in metres per second squared, m/s²
initial and final velocity are in metres per second, m/s
time is in seconds, s

Deceleration

An object that is slowing down is decelerating — the object has a negative acceleration. Putting the brakes on makes a car decelerate and slow down. If it decelerates at 3 m/s², then its velocity goes down by 3 m/s every second.

Figure 9.19 On landing, a jet fighter uses its brakes and a parachute to produce a large deceleration.

Test yourself

7 Figure 9.20 shows the speed and direction of four animals.

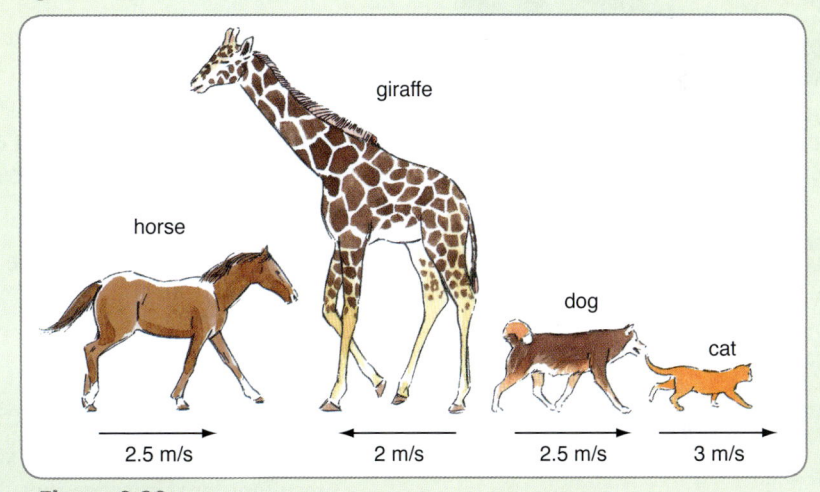

Figure 9.20

horse 2.5 m/s → , giraffe ← 2 m/s, dog 2.5 m/s →, cat 3 m/s →

Which animals have the same velocity? Explain the reason for your answer.

8 A falcon hovering in the sky sees its prey and dives downwards. The falcon reaches a velocity of 9 m/s in just 1.5 s. Calculate its acceleration.

9 A motorbike decelerates from 20 m/s to rest (0 m/s) in 4 seconds. Calculate the deceleration of the motorbike.

10 Explain the following statement:

'A model train running at a constant speed around a circular track is accelerating.'

Motion-tracking software

HOW SCIENCE WORKS — ICT

There is free motion-tracking software available on the internet. If possible (and ask your teacher's permission), download it and analyse videos from file-sharing websites of extreme acceleration — drag racers, jets etc. — to determine their acceleration.

Physics 2

Chapter 9   Forces and their effects

P2 9.4   Force, mass and acceleration

Learning outcomes

- Know that the acceleration of an object depends on the resultant force acting on the object and on the object's mass.
- Be able to use the equation acceleration = resultant force ÷ mass.

If you have ever tried to push a broken-down car you will know that it is not easy. If you are the only person pushing, the car accelerates and moves slowly. The more people that push, the bigger the resultant force on the car and the greater the acceleration (Figure 9.21).

The bigger the resultant force on an object, the greater the acceleration of the object.

Figure 9.21 The acceleration of the car depends on the resultant force applied to the car.

Common sense would tell us that it is much more difficult to give a big van the same acceleration as a car. A much larger force is needed — this is because the mass of the van is far bigger than the mass of the car. **Pushing a larger mass with the same force gives a smaller acceleration**. Therefore, the bigger the mass of an object, the bigger the resultant force needed to make it accelerate.

Figure 9.22 A large resultant force acting on the small mass of the drag car gives it massive acceleration.

The acceleration of an object can be calculated using this equation:

$$a = \frac{F}{m}$$

$$\text{acceleration} = \frac{\text{resultant force}}{\text{mass}}$$

where acceleration is in metres per second squared, m/s^2
resultant force is in newtons, N
mass is in kilograms, kg

Worked example

A resultant force of 14 N acts on a 2 kg ball. Calculate the acceleration of the ball.

$$a = \frac{F}{m}$$

$$a = \frac{14}{2}$$

$$a = 7 \text{ m/s}^2$$

We can rearrange this equation so that we can calculate the resultant force needed to give a mass a certain acceleration:

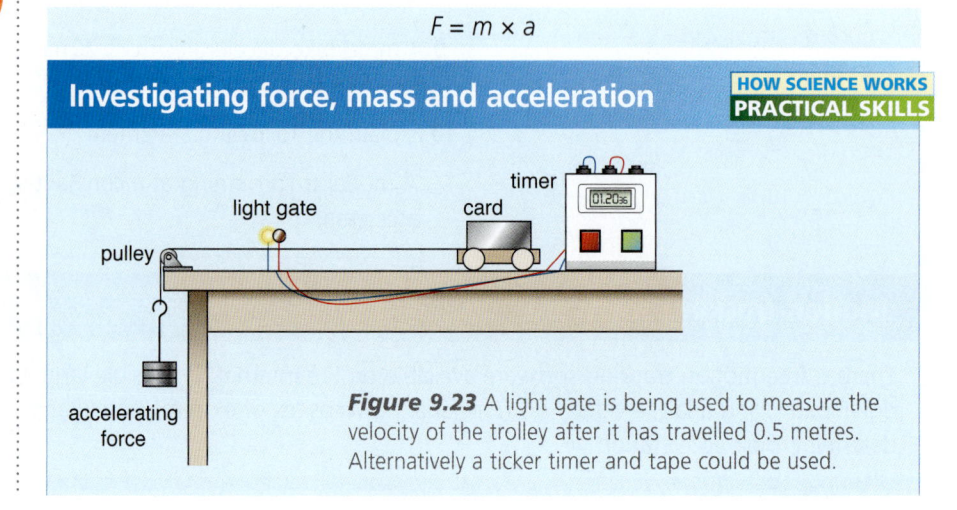

$$F = m \times a$$

Boost your grade ✓

You need to remember that it is the resultant force that causes acceleration, but you don't need to write the word 'resultant' in the equation. If you write the equation using symbols, make sure they are all correct.

Investigating force, mass and acceleration   HOW SCIENCE WORKS   PRACTICAL SKILLS

Figure 9.23 A light gate is being used to measure the velocity of the trolley after it has travelled 0.5 metres. Alternatively a ticker timer and tape could be used.

198   AQA GCSE Physics

P2 9.5 Using a distance–time graph to describe the motion of an object

The accelerating force is caused by the falling mass. (A mass of 100 g tied to the string exerts a force of 1 N.)

Use a constant force to accelerate one trolley. Time how long it takes the trolley to travel 0.5 m to the light gate. The velocity of the trolley is calculated using the length of the card and the time taken for the card to pass through the light gate.

Now calculate the acceleration of the trolley. Remember that the starting velocity of the trolley was 0 m/s.

Repeat this with a range of accelerating forces.

Now keep the accelerating force constant but change the mass of the trolley (by stacking one trolley on top of another).

In this investigation the mass is controlled when the accelerating force is changed. The accelerating force is controlled when the mass is changed. Why is this important?

Test yourself

11 What force is needed to make a ball of mass 0.5 kg accelerate at:
   a 3 m/s^2?
   b 2 m/s^2?
   c 0.4 m/s^2?

12 The part of the space shuttle that returns to Earth is called the orbiter. It has a mass of 78 000 kg and lands at a speed of 100 m/s. After touchdown, it takes 50 s to decelerate and come to a halt.
   a Calculate the deceleration of the orbiter.
   b Calculate the force needed to bring the orbiter to a halt.

13 Gurpal is cycling along a flat road when he stops pedalling. Gurpal's speed drops from 8 m/s to 5 m/s in 6 s.
   a Calculate Gurpal's deceleration.
   b The mass of Gurpal and his bicycle is 90 kg. Calculate the resistance force slowing Gurpal down.

P2 9.5 Using a distance–time graph to describe the motion of an object

Learning outcomes

- Understand that the gradient of a distance–time graph represents the speed of an object.
- Be able to calculate the speed of an object from the gradient of a distance–time graph.

A **distance–time graph** can be used to describe and calculate the speed of an object.

Figure 9.24 shows the distance–time graph for a car. The car travels the same distance, 600 metres, every minute. So the car is moving at a steady speed of 10 metres per second (m/s). This is the same as the gradient of the line.

Figure 9.24 The straight line shows that the car is moving at a constant speed.

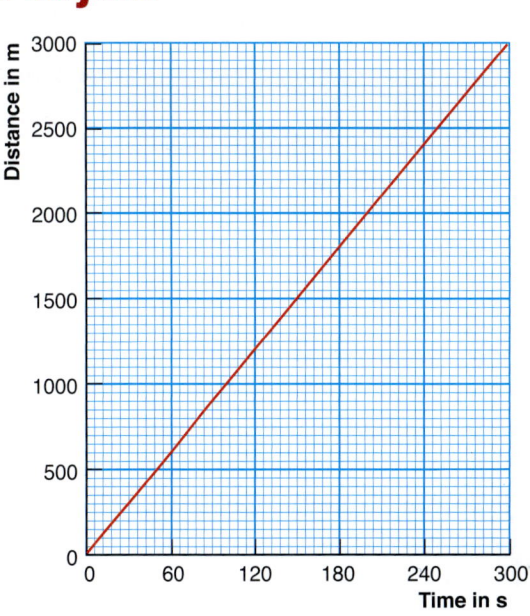

Physics 2

Chapter 9 Forces and their effects

> The **gradient** of a distance–time graph gives the speed of the object.

When a distance–time graph gives a straight line, the object is moving at a constant (or zero) speed.

Steve and his horse, Rocket, take part in a long-distance ride. Figure 9.25 shows the distance–time graph for the ride. The graph has been divided into three parts:

- Part A — Steve and Rocket move at a constant speed. They travel 16 kilometres in 2 hours. The slope of the line equals 8 km/h, which is their average speed.
- Part B — Steve and Rocket have stopped to take a rest. The distance moved does not change and the slope of the line is zero.
- Part C — Steve and Rocket move at a steady speed. They travel 24 kilometres in 2 hours. The slope of the line is 12 km/h. This means their speed was the fastest during this part of the ride.

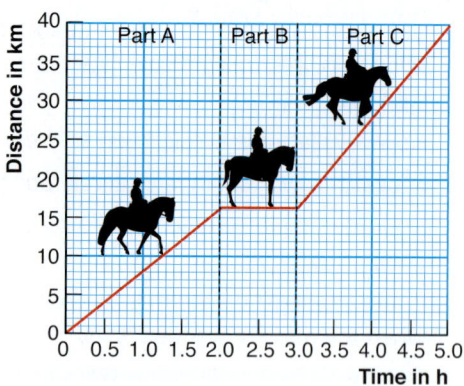

Figure 9.25 On a distance–time graph, the steeper the line, the faster the speed.

When an object is stationary (not moving), the line on a distance–time graph is horizontal.

Test yourself

14 Sketch a distance–time graph for a car travelling 90 metres at a constant 15 m/s

15 Curtis cycles to school. Figure 9.26 shows the distance–time graph for his journey.
   a How long did Curtis stop at the traffic lights?
   b During which part of the journey was Curtis cycling the fastest?

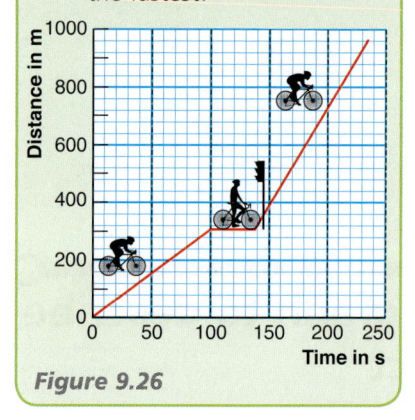

Figure 9.26

Plotting distance–time graphs

HOW SCIENCE WORKS · PRACTICAL SKILLS

In a safe place outside, mark every 5 metres along a line 50 metres long.

Obtain the data needed to plot distance–time graphs for a student:
- walking at their average pace
- walking at their fastest pace

How many times will you repeat the measurements? Compare the graphs for different students. How can you tell from the graphs who were the students that kept a constant speed? Who had the slowest average speed and who had the fastest speed?

P2 9.6 Using a velocity–time graph to describe the motion of an object

Learning outcomes

- Know that the gradient of a velocity–time graph represents acceleration.
- Be able to calculate the acceleration of an object from the gradient of a velocity–time graph.
- Be able to calculate the distance travelled by an object from a velocity–time graph.

A **velocity–time graph** can be used to describe the velocity and acceleration of an object.

Boost your grade ✓

Although a velocity–time graph may look similar to a distance–time graph, they mean different things. When answering a question, always look at the axes and make sure you know which type of graph has been drawn.

AQA GCSE Physics

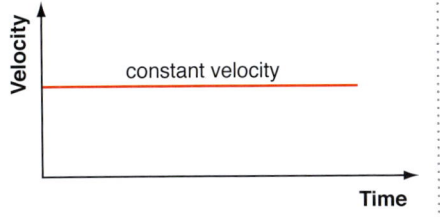

P2 9.6 Using a velocity–time graph to describe the motion of an object

Figure 9.27 The velocity–time graph for something moving with a constant velocity is a horizontal line.

Figure 9.27 shows the velocity–time graph for a car moving at a constant speed along a straight road. This means that the car is moving with a constant velocity.

Figure 9.28 shows the velocity–time graph for a car accelerating along a straight road. The gradient of the line shows how quickly the velocity is changing — this means that the gradient shows the acceleration of the car. A straight line means the acceleration is constant.

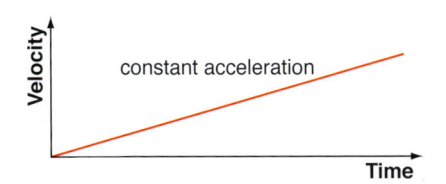

Figure 9.28 The gradient of a velocity–time graph gives acceleration — a steeper gradient represents a greater acceleration.

> The **gradient** of a line in a velocity–time graph gives the acceleration of the object.

Calculating the distance travelled

Figure 9.29 shows Ravi's velocity as he cycles downhill on a straight road. How far does Ravi cycle in 10 seconds?

distance = average velocity × time
average velocity = $\frac{1}{2}$ × 20

So, the distance cycled is:
($\frac{1}{2}$ × 20) × 10 = 100 m

Note that this distance that Ravi travelled equals:
$\frac{1}{2}$ × final velocity × time =
$\frac{1}{2}$ × XY × OY

This is the shaded area below the graph line.

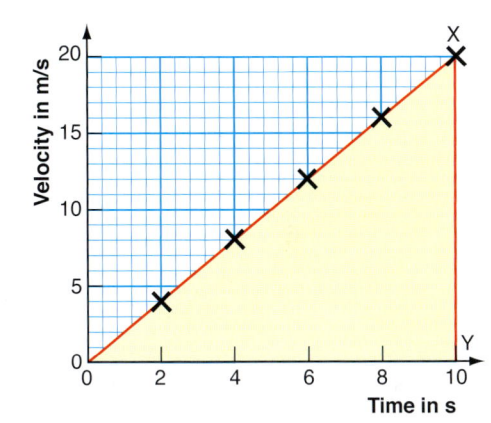

Figure 9.29 The straight line means that Ravi has a constant acceleration.

The area below the graph line in a velocity–time graph represents distance travelled.

Worked example

Figure 9.30 shows the velocity–time graph for a lift moving between two floors in a hotel. The graph has been divided into two parts:
- Part A — the lift leaves the first floor. The velocity is increasing at a constant rate; the lift has a constant acceleration.
- Part B — the velocity is decreasing at a constant rate (negative gradient); the lift has a constant deceleration. It then reaches the second floor.

To find the acceleration of the lift in part A, we work out the gradient of the graph line.

gradient = $\frac{LM}{KM}$ = $\frac{6}{3}$ = 2 m/s^2

You should be able to use the same method to show that the deceleration of the lift in part B is 1.2 m/s^2.

To find the distance the lift moves while it is accelerating, we work out the area under the line in part A:

area = $\frac{1}{2}$ × 6 × 3
area = 9 m

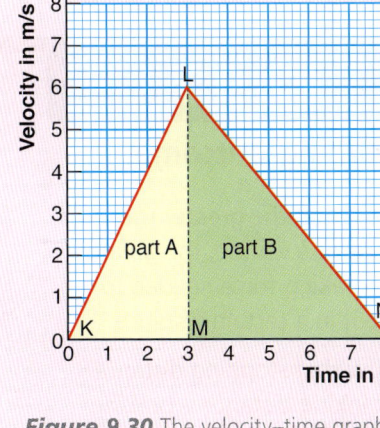

Figure 9.30 The velocity–time graph for a lift.

You should be able to use the same method to show that the distance between the two floors in the hotel is 24 m.

Physics 2

Chapter 9 Forces and their effects

> **Plotting the velocity–time graph for a trolley**
> **HOW SCIENCE WORKS — PRACTICAL SKILLS**
>
> Use either a ticker timer and tape, or a motion sensor linked to a computer, to plot the velocity–time graph for a trolley accelerating down a slope. Use a short runway propped up at one end by books to form the slope.

Test yourself

16 Figure 9.32 shows the velocity–time graph for a cheetah.
   a Calculate the acceleration of this cheetah.
   b Calculate the distance travelled by this cheetah in 10 seconds.

Figure 9.31 Cheetahs are capable of reaching 28 m/s in only 3 seconds. What acceleration is a cheetah capable of achieving?

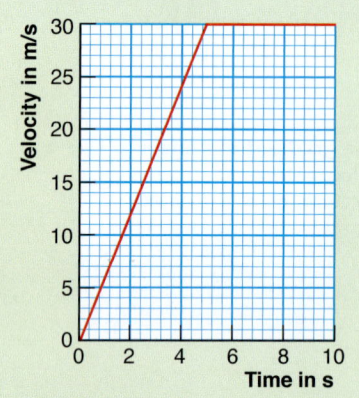

Figure 9.32 The velocity–time graph for a cheetah taking it easy.

17 Table 9.2 gives some data for a speed skier travelling in a straight line down a steep slope.
   a Draw a velocity–time graph for the skier.
   b Describe the motion of the skier.
   c Calculate the acceleration of the skier during the first 4 seconds.
   d Calculate the acceleration of the skier during the last 4 seconds.
   e Calculate the distance travelled by the skier in the final 4 seconds.

Time in s	0	2	4	6	8	10	12	14
Velocity in m/s	0	12	24	32	38	40	40	40

Table 9.2

P2 9.7 Stopping safely

Learning outcomes

Know that the greater the speed of a vehicle, the greater the braking force needed to stop it in a certain distance.

Know that stopping distance = thinking distance + braking distance.

Friction forces decelerate a moving car and bring it to a stop. In an emergency, the driver must apply the brakes as soon as possible. But the average driver has a reaction time of about three-quarters of a second. This is the time taken by the driver to react and move their foot from the accelerator pedal to the brake pedal. During this time the car carries on moving at a steady speed. The distance the car moves during the reaction time is called the **thinking distance**. Once the brakes have been applied the car slows down and comes to a stop. The distance the car moves once the brakes are put on is called the **braking distance**.

The total **stopping distance** of a vehicle (Figure 9.33) is made up of two parts — the thinking distance and the braking distance.

AQA GCSE Physics

P2 9.7 Stopping safely

> The **thinking distance** is how far a vehicle travels during the driver's reaction time, before the brakes are applied.
>
> The **braking distance** is how far a vehicle travels before stopping, once the brakes have been applied.
>
> **stopping distance** = thinking distance + braking distance

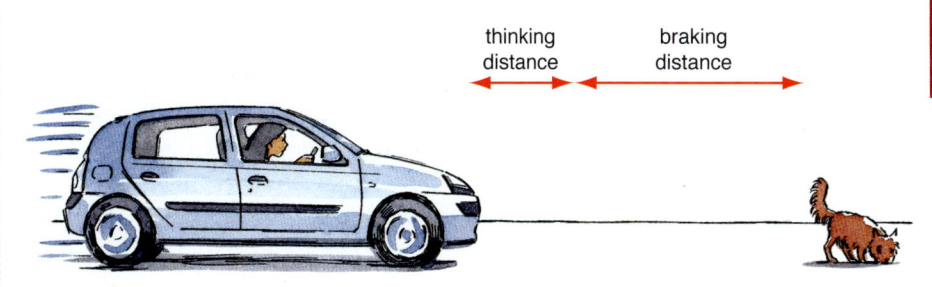

Figure 9.33 Thinking distance + braking distance = stopping distance.

Speed affects both the thinking distance and the braking distance. Figure 9.34 shows the stopping distances for a typical car at different speeds. The force applied by the brakes is the same for each of the speeds. As the vehicle goes faster, a greater braking force is needed to make the vehicle stop in the same distance. But beware — too great a braking force can make a vehicle skid.

The stopping distances given in Figure 9.34 are average values for a family car in good conditions on a dry road. Under different conditions the actual stopping distance can be much longer.

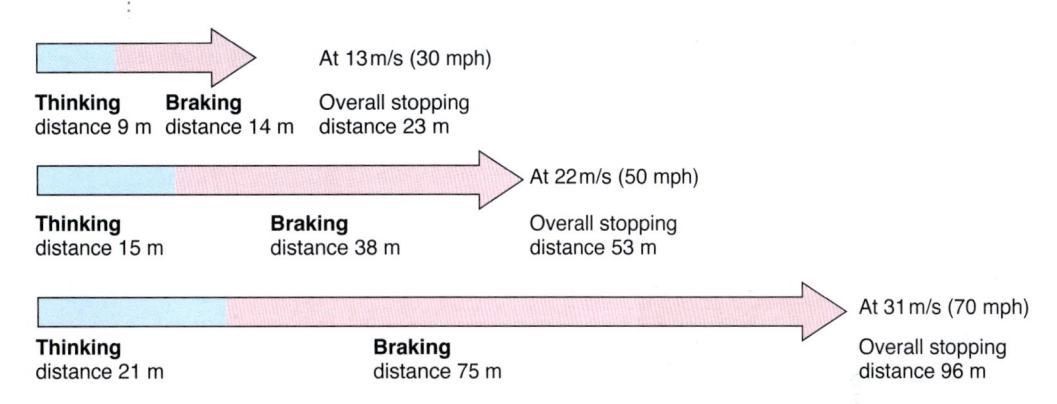

Figure 9.34 For a certain size braking force, the faster the vehicle, the greater the stopping distance.

At 13 m/s (30 mph)
Thinking distance 9 m   Braking distance 14 m   Overall stopping distance 23 m

At 22 m/s (50 mph)
Thinking distance 15 m   Braking distance 38 m   Overall stopping distance 53 m

At 31 m/s (70 mph)
Thinking distance 21 m   Braking distance 75 m   Overall stopping distance 96 m

Drivers' reactions may be slower than usual — the longer the reaction time, the longer the thinking distance. Drivers' reactions are much slower if they:
- have been drinking alcohol
- have been taking certain types of drug
- are tired

It is not just illegal drugs that can slow reactions. Anyone taking a medicine and intending to drive should always check the label.

A driver's ability to react may also be affected by distractions, such as talking on a mobile phone.

Braking distance can be affected by several factors as well as speed:
- weather conditions — in wet or icy conditions the friction between the car tyres and the road is reduced; this increases the braking distance
- the vehicle being poorly maintained — worn brakes or worn tyres increase the braking distance
- the road surface — a rough surface increases the friction between the road and the tyres, reducing the braking distance; you may have noticed that high-friction road surfaces are often used on the approaches to busy junctions and roundabouts

Boost your grade ✓

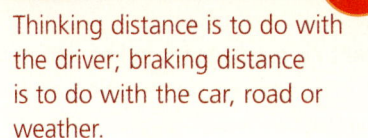

Thinking distance is to do with the driver; braking distance is to do with the car, road or weather.

Physics 2   203

Chapter 9 Forces and their effects

Reaction and distraction
HOW SCIENCE WORKS — PRACTICAL SKILLS

Hold a ruler between the fingers of a partner's open hand. Without giving any warning, let go of the ruler. Your partner should close their fingers as soon as they see the ruler fall.

Try it again, but this time with your partner listening to music on an MP3 player. What do you predict will happen?

If there is a difference in reaction time, how could you check that it was due to the music and not some other factor?

Test yourself

18 A sign beside a motorway reads 'Take a break — tiredness kills'. Explain the meaning of this message.

19 Explain why people should not drive a car after drinking alcohol.

20 How could the road surface near a busy junction be changed to help to reduce the braking distance of a vehicle approaching the junction?

Braking and kinetic energy

The work done by a braking force reduces a vehicle's kinetic energy, causing the vehicle to slow down and stop. As the kinetic energy of the vehicle goes down, the temperature of the brakes goes up. This is why friction brakes, when applied, always get hot. (Section 10.2 has more detail about work and kinetic energy.)

Braking distances
ICT

Create a multimedia resource to prepare a young driver for the theory part of the driving test. You need to explain stopping distances and the factors that affect them.

P2 9.8 Forces and terminal velocity

Learning outcomes

- Know that the faster an object moves through a fluid, the greater the frictional forces acting on it.
- Understand that when the resultant force on a falling object is zero the object falls at its terminal velocity.

A falling object accelerates because of the force of gravity. If there were no frictional forces (air resistance), a ball dropped out of a window would accelerate at $10\,\text{m/s}^2$. This means that the velocity would increase by $10\,\text{m/s}$ every second.

What happens when frictional forces act against gravity?

Air resistance usually acts on a falling object. This has the effect of reducing the resultant force on the object.

Joss is a skydiver. She understands how important air resistance is in keeping her alive. At the moment when Joss jumps from a plane, the only force acting on her is her weight. So she starts to accelerate downwards, as shown in the first part of Figure 9.35.

As soon as Joss starts to fall, air resistance begins to act upwards on her. Because her weight is bigger than the air resistance, she still accelerates towards the ground, but her acceleration decreases (the gradient of the graph decreases).

As Joss's speed increases, air resistance increases until it equals her weight. At this point the resultant force on Joss is zero and she stops accelerating, but she does not stop falling. She carries on falling but at a constant speed. The velocity–time graph is a straight line — Joss is falling at her **terminal velocity**.

Terminal velocity is the constant velocity reached by a moving object when the resultant force acting on it (in the direction it is moving) is zero.

P2 9.8 Forces and terminal velocity

Boost your grade ✓

You need to be able to apply the idea that a change in surface area changes terminal velocity to other situations. For example, a cyclist in a race reduces surface area to increase terminal velocity.

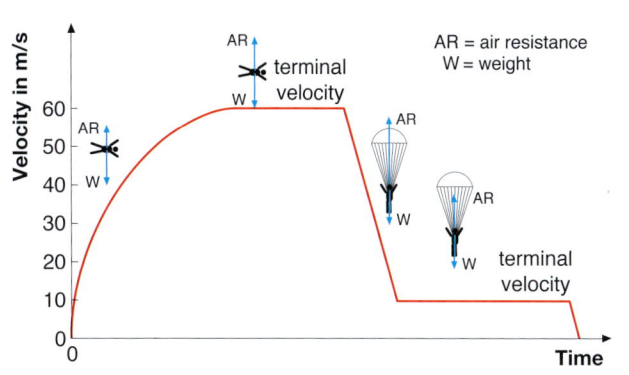

Figure 9.35 The graph shows how the velocity of Joss changes from the moment she jumps from the plane until the moment she lands and stops moving.

When Joss opens her parachute, the increased surface area causes an increase in the air resistance. This produces a resultant force upwards — the opposite direction to her motion — and so Joss decelerates and slows down. As she slows down, air resistance decreases until, once again, the two forces, weight and air resistance, become balanced. Joss then falls with a new, slower terminal velocity until she lands safely on the ground.

You could use apparatus like that in Figure 9.36 to find out whether the terminal velocity of a ball-bearing depends on its diameter. Just time different sized ball-bearings over the final 20 cm and then use the equation:

$$\text{speed (m/s)} = \frac{\text{distance (m)}}{\text{time (s)}}$$

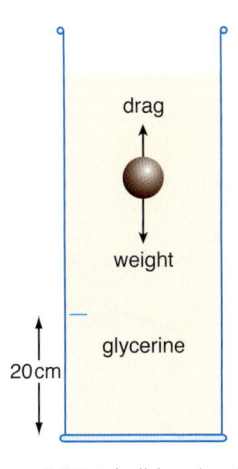

Figure 9.36 A ball-bearing falling through a thick liquid (like glycerine), reaches its terminal velocity before an identical ball-bearing falling through the air. Which is greater, the drag force in the glycerine or air resistance?

Figure 9.37 The top speed of a vehicle is affected by the power of its engine and the vehicle's body shape. Increasing the power of the engine increases the top speed. Making the body streamlined reduces air resistance and increases the top speed.

Terminal velocity and surface area
HOW SCIENCE WORKS — PRACTICAL SKILLS

Figure 9.38 shows a model parachute made from a sheet of plastic.

1. Plan an experiment to find out if the terminal velocity of the Plasticine figure depends on the area of the parachute. Include a risk assessment in your plan.
2. What measurements will you make?
3. How will you know that the Plasticine figure has reached its terminal velocity?
4. Which one of the following is the independent variable in this experiment?
   - distance the parachute falls
   - area of the parachute
   - weight of the Plasticine figure
   - time taken to fall
5. What are the control variables in this experiment?

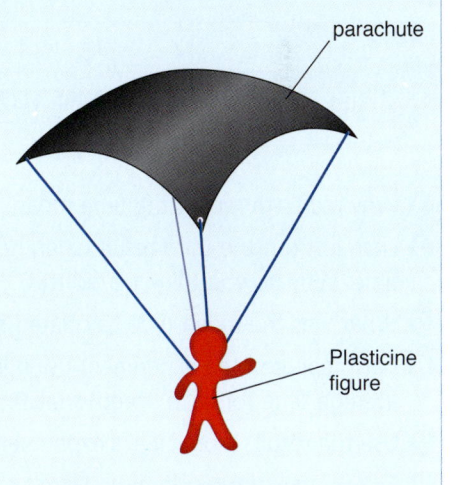

Figure 9.38 When the parachute is dropped, the Plasticine figure soon reaches its terminal velocity.

Physics 2 — 205

Chapter 9 Forces and their effects

Investigating the acceleration of a falling object

HOW SCIENCE WORKS
PRACTICAL SKILLS

Liam has read the story about Galileo's famous experiment with falling objects. Galileo believed that gravity makes all falling objects accelerate at the same rate. But most people had seen with their own eyes that different-sized objects took different times to fall the same distance, so they did not believe Galileo.

To show that his idea or **hypothesis** was right, Galileo is supposed to have dropped a small iron ball and a large iron ball from the top of the Leaning Tower of Pisa, in Italy. Reports on the experiment say that the two objects hit the ground at almost, but not quite, the same time.

> A **hypothesis** is an idea based on scientific theory. A hypothesis states what is expected to happen. It can then be used to make predictions that can be tested.

1 Before doing his experiment, Galileo had a theory about gravity and made a prediction.
   a What was Galileo's theory about gravity?
   b What did Galileo predict would happen in his experiment?

2 What was the observation that made people say that Galileo's theory was wrong?

Liam decided to investigate the acceleration of falling objects, to see if Galileo's prediction was right.

Liam started with a small sheet of lead. As soon as he let go, the sheet passed through the top light gate and the electronic timer started. The timer stopped when the sheet passed through the bottom light gate. Liam repeated this four times. The four timer readings are shown in Table 9.3.

Result	1	2	3	4
Time in s	0.326	0.330	0.465	0.319

Table 9.3

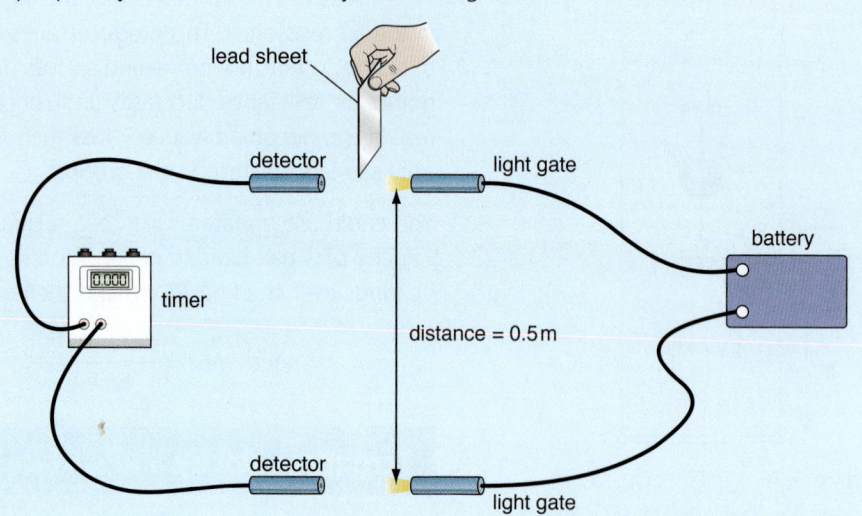

Figure 9.39 The apparatus used by Liam to measure the acceleration of a falling object.

3 Why did Liam repeat the time measurement?

4 Liam left out the third result when he calculated the mean (average) time. Why do you think he did this?

5 What was the mean time calculated by Liam?

6 Liam repeated the experiment using lead sheets of different sizes. The average time for each sheet to fall 0.5 m is given in Table 9.4

Sheet	Size	Mean time in seconds
A	Small	0.320
B	Medium	0.325
C	Large	0.322

Table 9.4

7 Use the results from Liam's experiment to draw a bar chart.

8 Explain why the results from Liam's experiment agree with Galileo's ideas about gravity.

9 When Liam timed a polystyrene ball falling through 0.5 m he got a value considerably longer than the value for any of the lead sheets. Why was this?

AQA GCSE Physics

P2 9.8 Forces and terminal velocity

Mass and weight

Weight is the force which gravity exerts on a mass. If you want to know your weight, you will probably stand on some bathroom scales. But bathroom scales usually give mass in kilograms and not weight, which is measured in newtons.

However, more mass does mean more weight, because there is more mass for the force of gravity to pull on. On Earth, gravity pulls on every 1 kilogram of mass with a force of about 10 newtons. This is called the gravitational field strength (g):

$$g = 10 \, N/kg$$

Sakhib, who has a mass of 50 kilograms, weighs:

$$50 \times 10 = 500 \text{ newtons}$$

To calculate weight use this equation:

$$W = m \times g$$

weight = mass × gravitational field strength
(newtons, N) (kilograms, kg) (newtons per kilogram, N/kg)

> **Worked example**
>
> An astronaut has a mass of 85 kg. Calculate the weight of the astronaut on the Moon.
>
> The gravitational field strength on the Moon = 1.6 N/kg.
>
> $W = m \times g$
> $W = 85 \times 1.6$
> $W = 136 \, N$

Figure 9.40 To measure your weight the bathroom scales must measure in newtons.

> **Test yourself**
>
> 21 Explain why a skydiver wearing loose clothing and spreading out her arms and legs has a slower terminal velocity than a skydiver curled up into a ball.
>
> 22 Explain why a ball-bearing falling through oil has a much slower terminal velocity than an identical ball-bearing falling through the air.
>
> 23 Kathy, Jon and Cary go on a diet. The table below shows the mass of each person at the start and end of the month.
>
> | Person | Mass at start of the month in kg | Mass at end of the month in kg |
> |---|---|---|
> | Kathy | 72.0 | 70.5 |
> | Jon | 115.0 | 110.5 |
> | Cary | 68.0 | 66.0 |
>
> How much weight did each person lose in 1 month?
>
> 24 When he was asked his weight, Gorese said it was 45 kg. Why is Gorese wrong?

Physics 2

Chapter 9 Homework questions

Homework questions

1. Josh knows that when he kicks a ball into the air, the force of gravity pulls the ball back down. Because forces come in pairs, he thinks that if the ball is pulled down by the Earth, then the ball should pull the Earth up. But he has never noticed this. Explain why.

2. Four forces act on a submerged submarine (Figure 9.41).
   a The submarine is cruising with a constant velocity and at a constant depth. Write down the connection between weight and upthrust; and between drag and thrust.
   b Explain what happens if the weight of the submarine is increased by filling the buoyancy tanks with water.

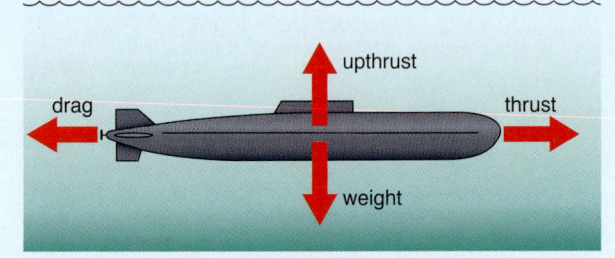

Figure 9.41

3. A cyclist accelerates from rest at 0.9 m/s² for 8 s. Calculate the final velocity of the cyclist.

4. As Sue and Steve walk down the high street, they stop and look in two shop windows.

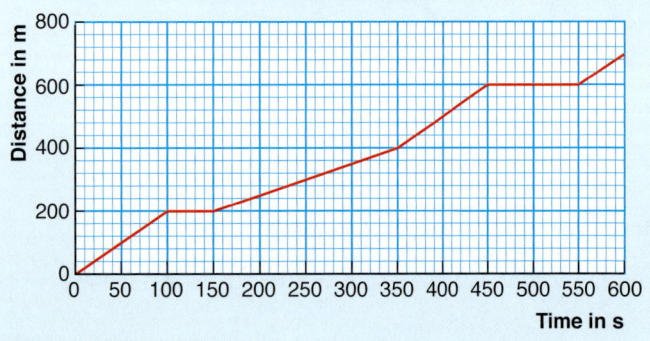

Figure 9.42 The distance–time graph for Sue and Steve's walk
   a How long did they spend looking in the shop windows?
   b How far apart were the two shops they looked in?
   c Between which two times did they walk the slowest?
      Calculate this speed.

5. A train pulls out from Maidenhead station and accelerates at a constant rate, reaching a velocity of 25 m/s after 90 seconds. The train then travels at this speed for 180 seconds, at which point the brakes are applied and the train comes to a stop at Taplow station 60 seconds later.

   Draw a velocity–time graph for the train and use it to calculate:
   a the acceleration of the train
   b the distance between Maidenhead and Taplow stations.

6. Table 9.5 gives the thinking distance and braking distance for a car at different speeds.

Speed in m/s	Thinking distance in m	Braking distance in m
0	0	0
9	6	6
13	9	14
18	12	24
26	18	56
36	24	96
52	36	224

   Table 9.5
   a Plot a graph of thinking distance against speed.
   b What pattern links thinking distance and speed?
   c Plot a graph of braking distance against speed.
   d What pattern links braking distance and speed?
   e Use the graphs to find the stopping distance for a car travelling at 72 m/s. (Hint: Extend the graph lines assuming that they continue to follow the same pattern. Remember, this is called extrapolating the graph.)

7. The velocity–time graph in Figure 9.35 is for a skydiver weighing 600 N.
   a How big is the air resistance force when the skydiver is falling at her terminal velocity? Explain the reason for your answer.
   b Estimate the size of the air resistance force when the skydiver first opens the parachute. Explain the reason for your answer.
   c After landing, the skydiver stands still on the ground. Explain why the resultant force on the skydiver is zero.

208 AQA GCSE Physics

Chapter 9 Exam corner

Exam corner

a The distance a car travels during the time it takes a driver to react is called the 'thinking distance'

Figure 9.43 shows how the thinking distance of a car driver changes with the speed of the car.

(i) Describe the pattern linking thinking distance and speed. *(1 mark)*

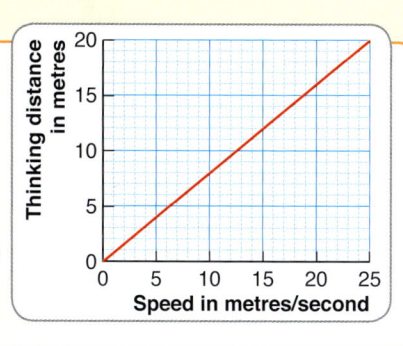

Figure 9.43

Student A
They are directly proportional. ✓

Examiner comment A correct answer.

Student B
If one doubles the other one doubles as well. ✓

Examiner comment Although the term 'proportional' has not been used, this is a clear description of the pattern.

(ii) Some people drive after drinking alcohol.

Draw a new line on the graph to show how thinking distance changes with speed for a driver who has recently drunk alcohol. *(1 mark)*

Student A

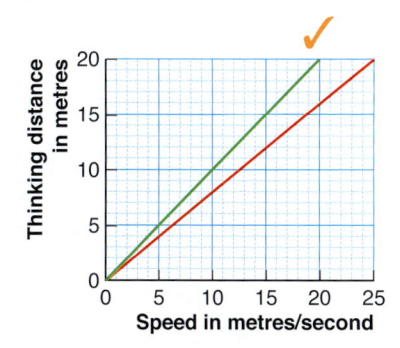

Examiner comment A straight line drawn steeper anywhere above the first line is correct.

Student B

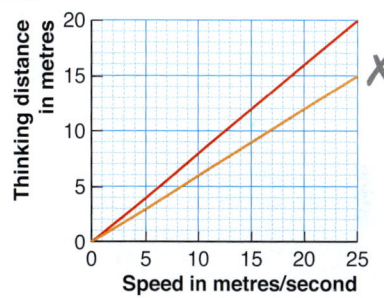

Examiner comment Although some people think that drinking alcohol speeds up their reactions, this is not true. Drinking alcohol will always slow reactions, so the thinking distance for any particular speed will increase, not decrease.

b The 'braking distance' is the distance a car travels once the brakes have been applied. State two factors, other than speed, that would increase the braking distance of a car. *(2 marks)*

Student A
Taking drugs ✗ and listening to music ✗

Examiner comment Make sure that you read the question carefully. It has now changed to asking about 'braking distance'. Both of these factors affect the *driver*, so they affect thinking distance, not braking distance.

Student B
The brakes of the car ✗ and a wet road ✓

Examiner comment You must be specific — the question asks about increasing 'braking distance'. The answer could have been 'badly worn' or 'badly maintained' brakes. Simply saying 'the brakes' is not good enough — if the brakes had just been replaced then they could reduce braking distance.

Physics 2

Physics 2

Chapter 10
Work, energy, power and momentum

Setting the scene
Modern cars have many safety features. In a collision the crumple zones, seat belt and air bag are all designed to increase the time it takes for you to slow down and stop. Why is this so important?

ICT
In this chapter you can learn to use multimedia to show what you know about momentum.

Practical work
In this chapter you can learn to:
- use light gates to measure speed more accurately
- identify the best type of graph for an investigation on crumple zones
- carry out a risk assessment before investigating your power

P2 10.1 Work and energy

Learning outcomes

- Understand that whenever a force makes an object move, work is done and energy is transferred.
- Know that the work done against frictional forces causes energy transfer by heating.

Work is done when a force makes an object move. Work is measured in joules (J).

Worked example

Calculate the work done by a weightlifter when lifting a weight of 750 N to a height of 2.4 m.

$W = F \times d$
$W = 750 \times 2.4$
$W = 1800 J$

Figure 10.2 These cranes can do more work than a human. Why?

Force and work

Ask people what they think 'work' is and you will get lots of different answers — but they usually involve doing something, maybe mowing the lawn or lifting a box, or even writing an essay.

In science, **work** has a specific meaning. Work is done whenever a force makes an object move. The bigger the force, and the further the force makes the object move, the more work is done.

The work done in moving an object can be calculated using the following equation:

Figure 10.1 Curtis wants to remove a tree stump from his garden. He tries to force the stump out of the ground with a steel bar. But no matter how hard he pushes nothing moves. How much work is Curtis doing?

$$W = F \times d$$

work done = force applied × distance moved in the direction of the force
(joules, J)   (newtons, N)   (metres, m)

So, for a force to do work it must make something move. No matter how big the force, if it doesn't make something move it hasn't done any work.

Work and energy

Richard works on a building site. One of his jobs is to move bricks in a wheelbarrow. When he starts pushing the wheelbarrow, it moves, so work is being done. When it moves, the wheelbarrow gains kinetic energy. This energy has been transferred from Richard. Whatever work Richard does to move the wheelbarrow and bricks, the same amount of energy is transferred.

Figure 10.3 Applying a force to move the wheelbarrow means that work is done and energy is transferred. How much work does Richard do if he pushes the wheelbarrow 20 m using a force of 80 N?

Work done and friction

Imagine pushing a heavy crate across the floor — it's hard work! As you push, friction acts with an equal force but in the opposite direction to your push. Because the resultant force on the crate is zero, the crate moves across the floor at a constant velocity. Even though you are doing work, the kinetic energy of the crate is not changing. The work you do is against the frictional forces that are trying to stop the crate moving. Where the crate rubs against the floor, both the crate and the floor get a little hotter.

Physics 2

Chapter 10 Work, energy, power and momentum

pushing force

friction force

Figure 10.4 The work done by the force moving the crate is mainly transferred into thermal energy.

The work done against frictional forces causes energy to be transferred, increasing the temperature of the moving object and the floor.

Worked example

Ahmed uses a force of 225 N to drag a heavy box 5 m across a rough floor at a steady speed. Calculate the work done against friction.

$W = F \times d$
$W = 225 \times 5$
$W = 1125\,J$

Test yourself

1 Copy and complete Table 10.1.

Task	Force applied	Distance moved	Work done
Opening a door	12 N	0.8 m	
Pushing a loaded trolley		80 m	2000 J
Lifting a suitcase	180 N		288 J
Stretching a spring	30 N	5 cm	

Table 10.1

2 Explain why falling over and sliding on an artificial Astroturf pitch can burn you.

P2 10.2   Gravitational potential energy and kinetic energy

Learning outcomes

Know that when work is done to lift an object, the object gains gravitational potential energy.

Know that the kinetic energy of a moving object depends on its mass and its speed.

Work and gravitational potential energy

To lift an object, work is done against gravity and the gravitational potential energy of the object increases.

Mark has a job stacking shelves at the local supermarket (Figure 10.5). To lift a box onto a shelf, Mark must exert an upward force on the box. The greater the weight of the box and the higher the shelf the more work Mark must do.

upward force on the box = weight of the box (downwards)
weight = mass × gravitational field strength (Section 9.8)
work done = force × distance moved in the direction of the force
change in gravitational potential energy (GPE) = work done
change in GPE = mass × gravitational field strength × change in height

In symbols:

$$E_p = m \times g \times h$$

212   AQA GCSE Physics

P2 10.2 Gravitational potential energy and kinetic energy

where E_p is in joules, J
m is in kilograms, kg
g is in newtons per kilogram, N/kg
h is in metres, m

Worked example

Calculate the GPE gained by a 0.5 kg box of cereal that is lifted 1.5 m.
(Gravitational field strength = 10 N/kg)

$E_p = m \times g \times h$
$E_p = 0.5 \times 10 \times 1.5$
$E_p = 7.5 \text{ J}$

Boost your grade ✓

You don't need to be able to derive the equation $E_p = m \times g \times h$, but you do need to be able to use it.

Figure 10.5 Each time Mark lifts a box, he exerts a force on the box and work is done. The box gains gravitational potential energy equal to the work done by the upward force on the box.

Worked example

A jogger of mass 60 kg runs at 3 m/s. What is the kinetic energy of the jogger?

$E_k = \frac{1}{2} \times m \times v^2$
$E_k = \frac{1}{2} \times 60 \times 3^2$
$E_k = 270 \text{ J}$

Boost your grade ✓

In this equation, it is only the speed that is squared — not everything on the right-hand side.

Work and kinetic energy

Moving objects have kinetic energy. The kinetic energy of an object increases when a force does work to make it speed up. Its kinetic energy decreases when the object does work and slows down.

Kinetic energy is not just to do with the speed of an object, it is also to do with the mass of the object. A lorry with a large mass moving at 20 m/s has more kinetic energy than a car with a small mass moving at 20 m/s. So, the kinetic energy of a moving object depends on:

- its mass
- its speed

The kinetic energy of a moving object can be calculated using the following equation:

$$E_k = \frac{1}{2} \times m \times v^2$$
$$\text{kinetic energy} = \frac{1}{2} \times \text{mass} \times \text{speed}^2$$

where   kinetic energy is in joules, J
mass is in kilograms, kg
speed is in metres per second, m/s

Losing gravitational potential energy; gaining kinetic energy

HOW SCIENCE WORKS / PRACTICAL SKILLS

Drop a tennis ball from 20 cm above a light gate.

When the ball passes the light gate it starts and then stops the timer. This gives the time it takes for the ball to fall the same distance as its diameter.

Measure the diameter of the ball and then calculate the speed of the ball. (speed = distance/time)

Measure the mass of the ball.

You now have enough data to calculate the kinetic energy gained by the ball and the gravitational potential energy lost by the ball. Are the two numbers the same? Or almost the same? If they are not the same, why not?

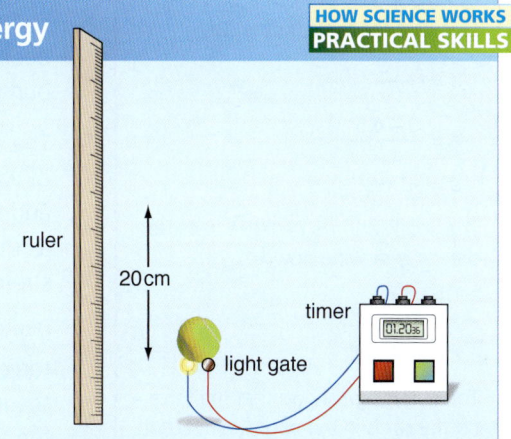

Figure 10.6 Why is the kinetic energy gained by the falling ball slightly less than the gravitational potential energy lost by the falling ball?

Physics 2

Chapter 10   Work, energy, power and momentum

Regenerative braking

Braking and then accelerating a vehicle means that energy stored in the fuel is used to replace the energy lost by braking. So, stopping and starting leads to a big proportion of the energy that could have been used to do work being wasted.

Regenerative braking improves a vehicle's efficiency by usefully using some of the kinetic energy lost when the brakes are applied. Applying the brakes on an electric vehicle makes a generator work to charge the vehicle's battery. So some of the energy that would have been wasted has been stored for future use.

An alternative way of storing the wasted energy is in a heavy, rotating flywheel. This system, first suggested in the 1950s, is still undergoing development.

Car safety

Figure 10.7 An all-electric or a hybrid car uses regenerative braking to improve its overall efficiency.

Figure 10.8 This car has been in a serious collision — but the important part, where the passengers and driver sit, is undamaged.

Modern cars are easily damaged, even in minor accidents — but this is part of the safety design. Cars are designed with 'crumple zones' at the front and back. In a collision, the crumple zones crush. As they crush, the car slows down and gradually stops. The longer it takes the car to stop, the smaller the force exerted by the car on the occupants — and the more likely they are to survive the accident without serious injury.

Look at the worked example. Even at this low speed a considerable force has been exerted on the car. How would the size of the force change if the front bumper had been made from rigid steel?

The advantage of cars having 'crumple zones' can also be explained in terms of work and energy. In a collision, forces act and work is done to slow down and stop a car. The work done to stretch and crumple the metal is transferred as thermal energy. So the car's kinetic energy is spread out as it does work to stretch and deform the 'crumple zone'.

If you are in a car that suddenly slows down, you keep moving forwards until a force acts to slow you down. Without a seat belt, this force may be provided by the impact between your head and the windscreen — often a large and lethal force.

A seat belt is not rigid — in a crash it is designed to stretch slightly. This is important because it increases the time taken for your velocity to be reduced to zero. So both the force on your body and the risk of injury are reduced.

> **Worked example**
>
> A car, of mass 800 kg, crashes into a brick wall at 4.5 m/s. The car slows down and stops in 1.5 seconds. Calculate the force on the car.
>
> $a = \dfrac{v - u}{t}$
>
> $a = \dfrac{0 - 4.5}{1.5}$
>
> $a = -3 \text{ m/s}^2$
>
> The acceleration is negative because the car is slowing down.
>
> $F = m \times a$
> $F = 800 \times (-3)$
> $F = -2400 \text{ N}$
>
> The minus sign shows that the force on the car is in the opposite direction to the velocity.

P2 10.2 Gravitational potential energy and kinetic energy

You can also explain why seat belts are effective in terms of energy and work. The kinetic energy lost by the person wearing the seat belt is equal to the work done in stretching the seat belt. So, the seat belt stretching has the effect of absorbing the kinetic energy.

Investigating the idea of a crumple zone

HOW SCIENCE WORKS — PRACTICAL SKILLS

To investigate the idea of a crumple zone Tracy used the apparatus shown in Figure 10.9. In a collision, Tracy expected the trolley and test material to behave in the same way as a car with a crumple zone.

A constant force was applied to the trolley by a falling weight. When the trolley hit the barrier, it decelerated and stopped. The 500 g mass on top of the trolley slid forwards until it also stopped. Tracy marked the graph paper to show how far the 500 g mass moved. Tracy tested five different types of material for the crumple zone, each of the same area and thickness.

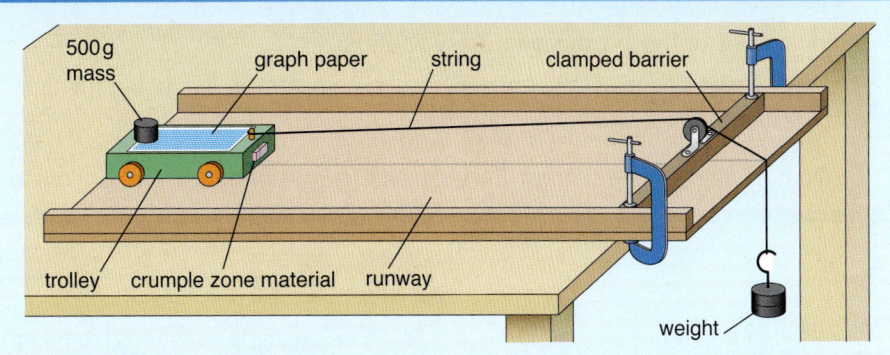

Figure 10.9 Using a trolley and test material to model a car with a crumple zone.

Because a constant force was applied, the trolley always accelerated to hit the barrier at the same speed. This was one factor that made the investigation a fair test.

1 Write down one more way that Tracy made this investigation a fair test.
Reminder: in a 'fair test', only the independent variable affects the dependent variable.

2 Suggest another way that Tracy could have produced a constant force on the trolley.

3 In Tracy's investigation, what was:
  a the independent variable?
  b the dependent variable?

4 What sort of variable was the 'type of crumple zone material' used in Tracy's investigation?

Figure 10.10 shows two sets of data obtained using different accelerating forces.

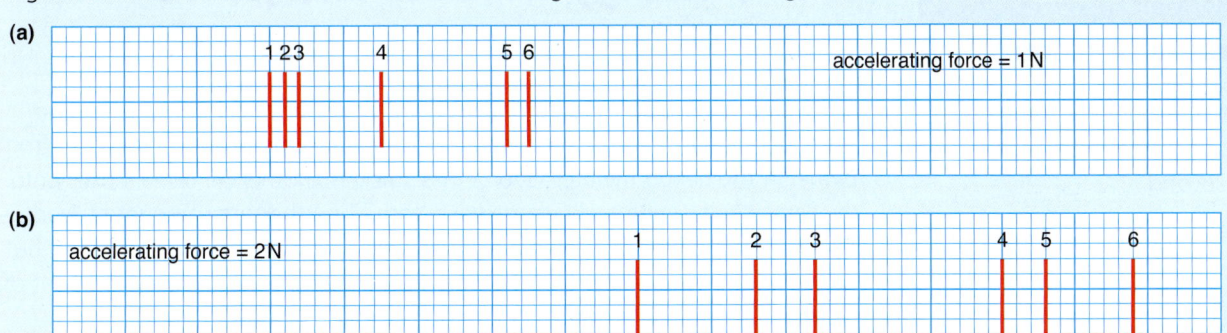

Key
1 = no material; 2 = polystyrene; 3 = foam rubber; 4 = carpet; 5 = fibre wool; 6 = rubber underlay

Figure 10.10 (a) Stopping distance using an accelerating force of 1 N. (b) Stopping distance using an accelerating force of 2 N.

Physics 2 215

Chapter 10 Work, energy, power and momentum

5 Which accelerating force gave the biggest range of results?

Reminder: the 'range' refers to the maximum and minimum values used or recorded.

6 Draw a bar chart for each set of data.

7 Why is it more appropriate to draw a bar chart than a line graph?

In her analysis of the investigation, Tracy wrote:

'Crumple zones do reduce the force on a colliding object by increasing the time taken to stop. Of the five materials that I tested the best was rubber underlay.'

8 Do both sets of data support Tracy's analysis? Explain the reasons for your answer.

Modelling crash barriers
HOW SCIENCE WORKS ACTIVITY

Use the internet to research methods for testing motorway crash barriers. Now plan a laboratory experiment to investigate a model crash barrier. Include a risk assessment as part of your plan.

Test yourself

3 An aeroplane passenger's hand luggage has a mass of 8 kg. It has to be lifted 1.8 m to put it into the overhead locker. Calculate the gain in gravitational potential energy of the luggage.

4 Calculate the kinetic energy of:
  a a motorbike of mass 350 kg travelling at 20 m/s
  b a 14 kg baby crawling at 0.5 m/s
  c a 6 kg cat walking at 60 cm/s
  d a 50 g bird flying at 40 m/s

5 A child of mass 36 kg sits on a swing. Her mum pulls the swing back and the child gains 180 J of gravitational potential energy.
  a How high was the child lifted when her mum pulled the swing back?
  b What was the maximum speed of the child after her mum let go of the swing?

P2 10.3 Work, energy transfer and power

Learning outcomes

Know that power is the work done, or energy transferred, each second.

Know that power is measured in watts.

It takes Lynne 3 minutes to lift and stack 50 boxes of cereal onto the shelf in the supermarket. George, who works at the same supermarket, takes over from Lynne. In the next 5 minutes George has stacked another 50 cereal boxes onto the same shelf. Lynne and George have both done the same amount of work — they have both lifted the same weight of boxes through the same distance. But Lynne did the work faster than George. Because Lynne did the same amount of work in a shorter time, she has more power than George.

The rate at which something or someone does work or transfers energy is called the power.

> **Power** is the rate of transfer of energy or the rate of doing work.
> 1 kilowatt (kW) = 1000 watts.

AQA GCSE Physics

P2 10.3 Work, energy transfer and power

Worked example

A conveyor belt is used to lift materials on a building site. Calculate the power required to lift five 25 kg bags of cement through a height of 4.2 metres in one minute.

gravitational field strength (g) = 10 N/kg

Work is done and the bags of cement gain gravitational potential energy.

$E_p = m \times g \times h$
$E_p = 5 \times 25 \times 10 \times 4.2$
$E_p = 5250 \text{ J}$
$P = \dfrac{E}{t}$
$P = \dfrac{5250}{60}$
$P = 87.5 \text{ W}$

Boost your grade ✓

Remember that in most equations time must be in seconds.

Power can be calculated using this equation:

$$P = \dfrac{E}{t}$$

$$\text{power} = \dfrac{\text{energy transferred}}{\text{time taken}}$$

where power is in watts, W
energy transferred is in joules, J
time taken is in seconds, s

Measuring your own power

HOW SCIENCE WORKS — PRACTICAL SKILLS

The Olympic cyclist Chris Hoy can produce nearly 2 kW or 2000 watts (enough to power a kettle). What is your maximum power output?

You can measure your own power by timing how long it takes you to do some work. So you could lift weights from the floor onto a table, do step-ups onto a box or run up a flight of stairs. Whatever you do, make sure it is safe and that you are unlikely to hurt yourself or anyone else.

Figure 10.11 When you go up a flight of stairs, you are lifting your own body weight and doing work against the force of gravity. The faster you go up the stairs, the quicker you do the work, and the greater the power developed.

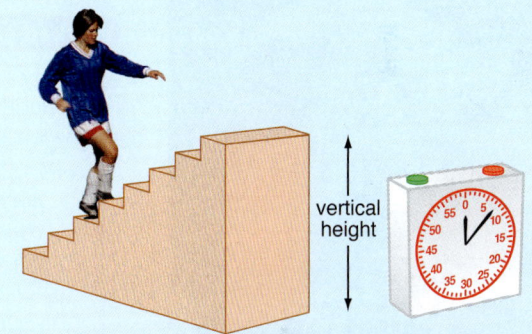

Measure your weight in newtons. If the scales give your mass in kilograms just multiply by 10 (Section 9.8).

Measure the vertical height of the stairs. This is the important distance because you are doing work against the force of gravity and that acts vertically.

Ask someone to time how long it takes you to run up the stairs.

You should now have enough data to calculate the work that you did and the power you developed.

You could now find out if you develop the same power when you lift some weights.

Test yourself

6 What is the connection between energy transferred and power?

7 What is the unit of power?

8 A crane lifts a weight of 10 000 N through a height of 50 m in 200 s. Calculate the output power of the crane.

Nick tells his sister Ros that he is more powerful because he is bigger. Ros disagrees with him saying that she might be smaller but she is quicker. To settle the argument they time each other running up the stairs. Table 10.2 gives the data obtained by Nick and Ros.

	Weight	Fastest time
Nick	720 N	3.6 s
Ros	560 N	2.8 s

Table 10.2 Nick and Ros both lifted their body weight the same distance, 3.2 m, the vertical height of the stairs. Nick has the biggest weight, so he will have done more work than Ros. But who developed the most power? Calculate the power developed by Nick and by Ros. Was Nick right?

Physics 2

Chapter 10 Work, energy, power and momentum

P2 10.4 Momentum

Learning outcomes

- Be able to use the equation momentum = mass × velocity.
- Know that momentum is measured in kilogram metres per second (kg m/s)
- Understand that in a closed system, the total momentum is conserved.

Momentum is the mass of an object multiplied by the velocity of the object.

What is momentum?

Any moving object has momentum. The bigger the mass of the object and the faster it moves, the more momentum it has.

The momentum of an object can be calculated using the following equation:

$$p = m \times v$$
$$\text{momentum} = \text{mass} \times \text{velocity}$$

where momentum is in kilogram metres per second, kg m/s
mass is in kilograms, kg
velocity is in metres per second, m/s

Just like velocity, momentum has a direction. If two objects move in opposite directions, one has a positive momentum and the other has a negative momentum.

Worked example

A lorry and a car are travelling in opposite directions. The lorry has a mass of 9000 kg and a velocity of 14 m/s to the right. The car has a mass of 900 kg and a velocity of 25 m/s to the left. Calculate the momentum of each vehicle.

Lorry:
$p = m \times v$
$p = 9000 \times 14$
$p = 126\,000$ kg m/s to the right

Car:
$p = m \times v$
$p = 900 \times -25$
$p = -22\,500$ kg m/s to the left

The momentum of the lorry moving to the right has been given as positive, so the momentum of the car moving to the left must be given as negative.

Figure 10.12 What is the momentum of a charging rhino of mass 3000 kg, moving at 11 m/s?

Collisions

When two objects collide, each will exert a force on the other. This causes the momentum of each object to change. However, the total momentum of both objects after the collision in a particular direction is the same as the total momentum in that direction before the collision:

total momentum after a collision = total momentum before a collision

The total momentum is unchanged by the collision. It doesn't go up and it doesn't go down. The total momentum is **conserved**.

This is always true — provided that the colliding objects are in a closed system. This means that when the objects collide, no forces outside of the system (external forces) act on the objects.

Look at the white pool ball in Figure 10.13. It is moving with velocity v directly towards a red pool ball. Before they collide, the red ball is not moving — the red ball has zero momentum.

Boost your grade ✓

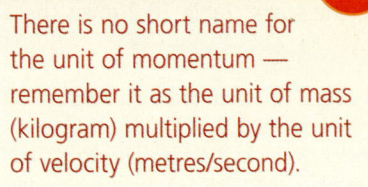

There is no short name for the unit of momentum — remember it as the unit of mass (kilogram) multiplied by the unit of velocity (metres/second).

218 AQA GCSE Physics

P2 10.4 Momentum

Momentum is conserved in a collision. The total momentum after a collision equals the total momentum before the collision.

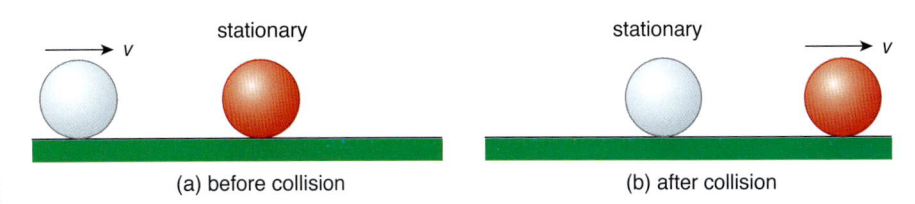

Figure 10.13

During the collision, an equal size force acts on each ball (with the same mass), but in opposite directions. This causes the momentum of each ball to change by the same amount. If, after the collision, the white ball is stationary — the white ball has zero momentum. The red ball will move with a velocity v in the same direction as the white ball was moving.

So, although momentum has been transferred from the white ball to the red ball, the total momentum has been conserved.

Worked example

A van of mass 3200 kg, travelling at 5 m/s, collides into the back of a stationary car of mass 800 kg. The two vehicles join and move off together. What is the velocity, v, of the two vehicles immediately after the collision?

momentum before collision = momentum after collision

$(\text{mass} \times \text{velocity})_{van} + (\text{mass} \times \text{velocity})_{car} = (\text{mass}_{van} + \text{mass}_{car}) \times v$

$(3200 \times 5) + (800 \times 0) = (3200 + 800) \times v$

$16000 + 0 = 4000 \times v$

$16000 = 4000 \times v$

$v = \dfrac{16000}{4000}$

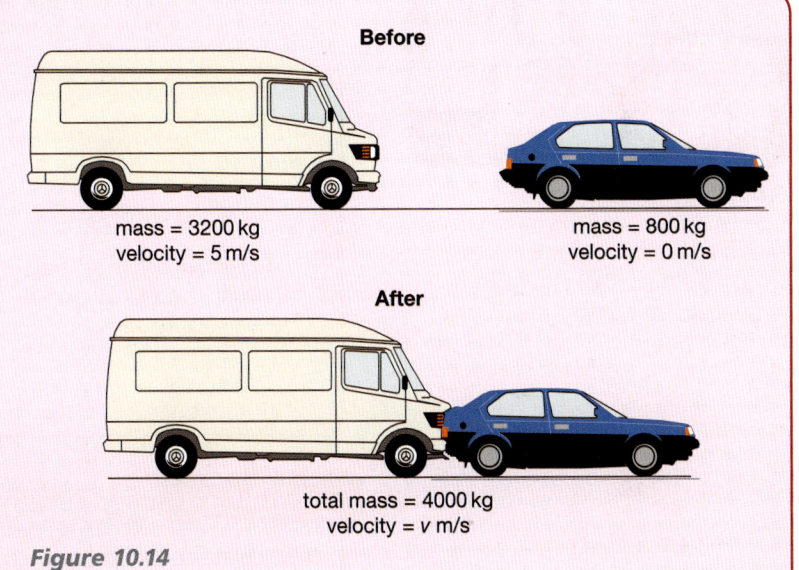

$v = 4$ m/s

Before
mass = 3200 kg
velocity = 5 m/s

mass = 800 kg
velocity = 0 m/s

After
total mass = 4000 kg
velocity = v m/s

Figure 10.14

Investigating collisions

HOW SCIENCE WORKS ACTIVITY

The two trolleys, A and B, have the same mass. Pushing trolley A will make it collide with the stationary trolley B. Trolley A will stop and trolley B will start to move. How can you tell that momentum has been conserved?

By using a pin in trolley A and a cork in trolley B (instead of the rubber bands), the two trolleys can be made to join together. If momentum is conserved, how will the speed of the two trolleys after the collision compare with the speed of trolley A before the collision?

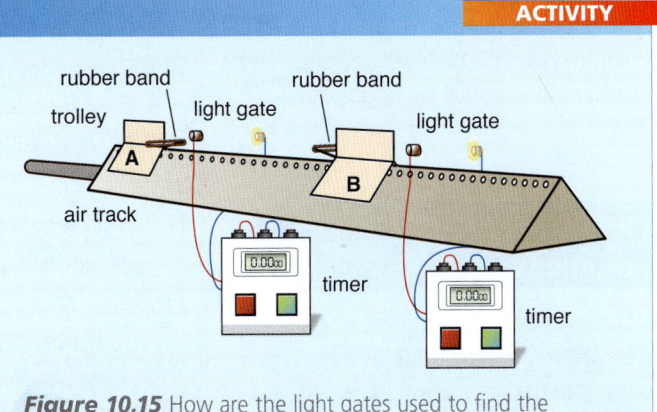

Figure 10.15 How are the light gates used to find the speed of a trolley before and after a collision?

Physics 2

Chapter 10 Work, energy, power and momentum

Momentum is conserved in an explosion. The total momentum before an explosion is the same as the total momentum after the explosion.

Explosions

An explosion is the opposite of a collision — instead of moving together, objects move apart. But, as in a collision, the total momentum of the objects involved in an explosion remains constant. In an explosion momentum is conserved.

Worked example

When Simon jumps off his stationary skateboard, the skateboard rolls backwards. Use the data in Figure 10.16 to calculate the initial speed of the skateboard backwards.

mass of Simon = 60 kg
mass of skateboard = 2.4 kg
velocity = $-v$
velocity = 0.6 m/s

Figure 10.16 The speed of the skateboard backwards is called the recoil speed.

If the initial backwards speed of the skateboard is $-v$ m/s:

momentum before jumping off = momentum after jumping off
$0 = 60 \times (+0.6) + 2.4 \times (-v)$
$0 = 36 - 2.4v$
$v = \dfrac{36}{2.4}$
$v = 15$ m/s

Figure 10.17 The principle of momentum conservation is used in the launch of a space rocket. Before the launch, the total momentum of the rocket and fuel is zero. After lift-off, the total momentum must still be zero. The downward momentum of the hot exhaust gases must be equal to the upward momentum of the rocket.

Boost your grade ✓

Remember: when objects move in opposite directions, one velocity will be positive and the other velocity will be negative.

Principles of momentum   ICT

Using videos from the internet or photographs, create a multimedia resource that shows the principles of momentum.

Test yourself

9 Calculate the momentum of:
  a a 60 kg cyclist pedalling at 8 m/s
  b a football of mass 0.4 kg moving at 7 m/s
  c a bullet of mass 5 g moving at 250 m/s
  d a 500 kg pony galloping at 16 m/s

10 A 2 kg trolley moving at 3 m/s collides with and sticks to a stationary 1 kg trolley. Calculate:
  a the total momentum before the collision
  b the speed of the two trolleys after the collision

11 Explain why a rowing boat moves backwards when you step ashore from the boat.

12 At the 'Circus on Ice' a clown throws a 'custard pie' forwards with as much force as possible. Before throwing the pie, the clown was not moving. Calculate the recoil velocity of the clown.

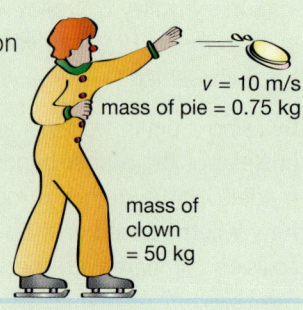

$v = 10$ m/s
mass of pie = 0.75 kg
mass of clown = 50 kg

Figure 10.18

220 AQA GCSE Physics

Chapter 10 Homework questions

Homework questions

1. The Falkirk Wheel (Figure 10.19) is the first ever rotating boat lift. It lifts 600 000 kg of water plus boat through a height of 35 metres.
   a Calculate the weight that the Falkirk Wheel is able to lift.
   b Calculate the work done by the Wheel in one rotation.

   Figure 10.19 The Falkirk Wheel.

2. Figure 10.20 shows Matthew pulling a heavy sack up a ramp. The ramp is a simple machine that makes the job of lifting the sack easier to do. Explain why the sack gains less gravitational potential energy than the work done in pulling the sack up the ramp.

   Figure 10.20

3. Calculate the work done in each of the following:
   a a shopper pushing a trolley with a force of 40 N for 75 m
   b a pole-vaulter lifting her own weight of 520 N a height of 3.6 m
   c a car using a braking force of 700 N to stop in 23 m

4. A ball thrown at 40 m/s has 160 J of kinetic energy. Calculate the mass of the ball.

5. In a car crash, it takes about 0.04 s for an air bag to inflate. A fraction of a second later, the driver's head hits the air bag. The bag, which has tiny holes in it, will start to deflate as the force of impact pushes some of the gas out.
   a What happens to the movement of the driver's head when it hits the air bag?
   b Why is it important for an air bag to deflate?
   c Explain how the air bag reduces the risk of serious injury to the driver's head.

6. Explain the reason for each of the following.
   a Some trainers have gas-filled heels.
   b Ice-hockey goalkeepers wear well-padded kit.

7. John is driving his car at 15 m/s when it is in a collision. John's seat belt slows him down and stops him moving in 0.12 s.
   a Taking John's mass to be 70 kg, calculate:
      (i) John's deceleration
      (ii) the average force exerted by the seat belt on John
   b Explain why seat belts need to be able to stretch.

8. A racing car moving at 85 m/s has a momentum of 42 500 kg m/s. Calculate the mass of the racing car.

9. A 1800 kg elephant has a momentum of 10 800 kg m/s. Calculate the velocity of the elephant.

10. Jackie and Jasmin are on roller blades. They stand facing each other, and then give each other a gentle push. Figure 10.21 shows what happens next.
    a Explain why both Jackie and Jasmin roll backwards.
    b What is their total momentum before they push each other?
    c Calculate Jackie's momentum just after the push.
    d Calculate Jasmin's velocity just after the push.

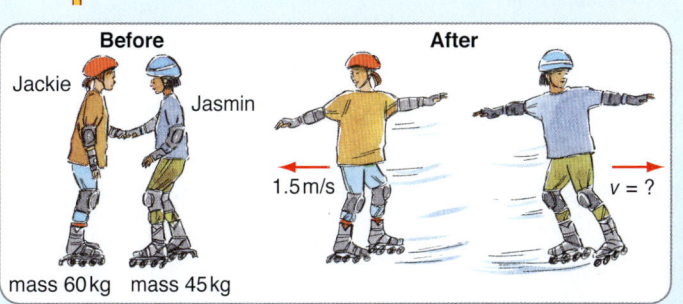

Figure 10.21

Physics 2

Chapter 10 Exam corner

Exam corner

A car, of mass 900 kg, is travelling at 10 m/s when the driver sees a queue of traffic ahead. The driver applies the brakes and the car decelerates to 4 m/s.

a Calculate the kinetic energy lost by the car as it decelerates.
Write down the equation you need to use, then show how you work out your answer. *(3 marks)*

Student A

KE = ½ × mass × speed^2
KE before = ½ × 900 × 10^2 = 45 000 J ✓
KE after = ½ × 900 × 4^2 = 7200 J ✓
KE lost = 45 000 − 7200
= 37 800 J ✓

Examiner comment Each step in the calculation is correct, so full marks.

Student B

E_k = ½ × m × v^2
E_k = ½ × 900 × (10 − 4)^2 ✓
E_k = 450 × 6^2
E_k = 16 200 J

Examiner comment The maths is wrong. You cannot subtract the two speeds and then square the answer. However, the student tries to use the correct equation, so 1 mark is awarded.

b While decelerating, the car travels 40 metres. Calculate the braking force applied. *(3 marks)*

Student A

Work done = loss in KE ✓
F × d = 37 800
F = 37 800 / 40 ✓
F = 945 N ✓

Examiner comment The method and calculation are shown clearly and are correct.

Student B

Work done = KE lost ✓
F × 40 = 16 200 ✓
F = 405 N ✓

Examiner comment The answer is wrong because the wrong answer is used for kinetic energy. However, the method is correct and gains full marks.

c Some electric cars are fitted with a regenerative braking system. Explain the advantage of such a system over friction brakes. *(3 marks)*

Student A

With ordinary brakes, friction causes a lot of energy to be wasted as the car slows down. ✓ With regenerative brakes some of this energy can be stored and used later. ✓

Examiner comment A comparison between the two systems has been made and an advantage of regenerative brakes is given.

Student B

With friction brakes the work done to slow the car results in energy being wasted. ✓ A car with regenerative brakes has a generator that as the kinetic energy decreases charges the car battery. ✓ So some of the energy that would have been wasted is usefully used; this improves the car's efficiency. ✓

Examiner comment This gives just a little more detail than the answer from student A.

AQA GCSE Physics

Physics 2

Chapter 11
Using electricity

Setting the scene

The first Blackpool Illuminations, held in 1879, had just eight electric lamps. Today's illuminations, having one million lamps connected by more than 320 kilometres of cables and wiring, stretch out over a distance of 10 kilometres. If one light bulb were to 'blow', would all the lights go out?

ICT

In this chapter you can learn to:
- use multimedia to provide a clear warning about the danger of overloading sockets
- use multimedia to enhance a presentation on a.c. and d.c.
- create a podcast to explain to a wider audience why you should use RCCBs

Practical work

In this chapter you can learn to:
- take careful measurements of potential difference and current
- draw conclusions from data on current and potential difference for different lengths of wire
- explain why recorded data can vary

Chapter 11 Using electricity

P2 11.1 Static electricity

Learning outcomes

- Know that some insulating materials become electrically charged when rubbed against each other.
- Understand that electrons are transferred from one material to another when materials become charged by friction.
- Know that objects with the same type of charge repel one another and objects with different types of charge attract.
- Understand that an electric current is a flow of charge.

What is static electricity?

People often call static electricity just 'static' for short. 'Static' results in lightning flashes, it makes your hair stand on end and it makes dust stick to television screens. The word 'static' is normally used to mean 'stationary' and **static electricity** results when materials have stationary charges.

Invisible electrostatic forces

Try this simple experiment. Comb your hair briskly with a plastic comb and then use the comb to pick up tiny pieces of paper. (It only works if your hair is dry and not gelled.) When you comb your hair, the comb becomes **charged**, and the charged comb can pick up bits of paper because of electrostatic forces.

You may have noticed some other effects of electric charges.
- When you take off a shirt or top, it tends to cling to your body or T-shirt.
- A balloon can stick to a wall and possibly the ceiling after you have rubbed it against your sweatshirt or fleece.

These effects show that when certain materials are rubbed against each other or move over each other, they become electrically charged. People sometimes say the materials have 'static electricity' on them.

Where do the charges come from?

When certain materials are rubbed against each other, friction causes negatively charged electrons to move off one material onto the other — off hair onto the comb, off your body and onto your shirt or top, off your fleece and onto a balloon.

When you comb your hair briskly, the comb removes electrons from some atoms in your hair. The comb now has more electrons than protons, so it has an overall negative charge. Your hair has lost the electrons. It now has more protons than electrons and so it has an equal positive charge.

When two different insulating materials are rubbed against each other, one gains electrons and become negatively charged. The other material that has lost electrons is left with an equal positive charge.

Have you noticed that your hair sometimes stands up when it is combed? This is more likely to happen if your hair is clean and dry. It happens because each hair becomes positively charged, and objects with the same charges **repel**.

When electrically charged objects are close together, they exert electrostatic forces on each other:
- Objects with the same type of charge **repel** each other.
- Objects with different (opposite) charges **attract** each other.

Insulators and conductors

Materials like plastic, hair, rubber and paper can hold their charge. The charges on them are 'static' (stationary) and electrons do not flow through them. Because of this, these materials are called **insulators**.

Figure 11.1 Plastics like polythene and polyvinylchloride (PVC) are good insulators. They are used to insulate electrical wires and cables.

224 AQA GCSE Physics

P2 11.2 Circuits: starting out

> An **electric current** is a flow of charge. The charge may be carried by electrons or ions.

In contrast to insulators, electrons flow easily through metals. These materials are described as conductors. The flow of negative charges (i.e. electrons) forms an electric current. An electric current can also be carried by ions, which can have either positive or negative charges.

Test yourself

1 When a Perspex rod is rubbed with a cloth, the Perspex rod becomes positively charged. Explain how the rod gets its charge.

2 Look at Figure 11.2. The dome on which the boy is resting his hand has a large negative charge and the boy is standing on a thick rubber mat.
  a What is the charge on the boy's hand?
  b What is the charge on the boy's hair?
  c Why is the boy's hair sticking out?
  d Why is the boy standing on a rubber mat?

Figure 11.2

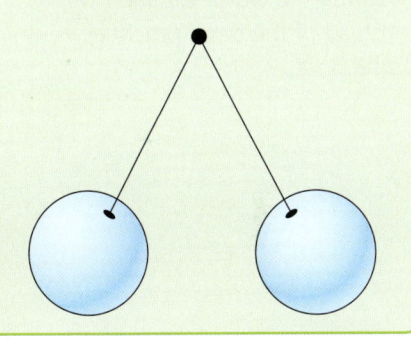

3 Figure 11.3 shows two identical lightweight plastic balls attached to the same hook by nylon threads. The balls are charged.
  a What can you say about the charges on the balls?
  b Copy Figure 11.3 and show the forces acting on each ball.

Figure 11.3

P2 11.2 Circuits: starting out

Learning outcomes

- Know the standard symbol for each component in a circuit.
- Be able to draw circuit diagrams to show how components are joined.
- Know that a potential difference is needed for a current to flow.
- Know that a battery is a number of cells joined together.

For a current to flow through any component in a complete circuit, there must be a potential difference across the component.

Current and charge

Remember that an electric current is a flow of charge (Section 11.1). In a metal wire, an electric current is a flow of negatively charged electrons. A larger current means there is a greater rate of flow of charge.

The **current** in a circuit is measured with an ammeter connected in **series**. The size of an electric current can be calculated using the following equation:

Physics 2

Chapter 11 Using electricity

$$I = \frac{Q}{t}$$

$$\text{current} = \frac{\text{charge}}{\text{time}}$$

where current is in amperes, A
charge is in coulombs, C
time is in seconds, s

Boost your grade ✓
Although current is measured in amperes, it is all right to write it as amps.

1 coulomb is the charge that passes a point in a circuit when a current of 1 ampere flows for 1 second.

Worked example
To boil the water in an electric kettle, 1200 coulombs of charge pass through the heating element in 2 minutes. Calculate the current flowing through the heating element.

$I = \frac{Q}{t}$

$I = \frac{1200}{2 \times 60}$ (Remember that time must be in seconds.)

$I = 10$ amps

Potential difference and charge

Charge flows between two points in a circuit only if there is a potential difference between the two points. The potential difference is equal to the work done (or energy transferred) for each coulomb of charge that passes between the two points. The potential difference can be calculated using the following equation:

$$V = \frac{W}{Q}$$

$$\text{potential difference} = \frac{\text{work done}}{\text{charge}}$$

where potential difference is in volts, V
work done is in joules, J
charge is in coulombs, C

Boost your grade ✓
Potential difference (p.d. for short) and voltage mean the same thing. In an exam, you can use the word 'voltage' instead of the term 'potential difference'.

Potential difference is measured with a voltmeter connected in **parallel** across a component.

Worked example
A voltmeter connected in parallel across the lamp in a circuit gives a reading of 12 volts. Calculate the work done when 300 coulombs of charge pass through the lamp.

$V = \frac{W}{Q}$

Rearranging the equation gives:

$W = V \times Q$
$W = 12 \times 300$
$W = 3600$ joules

Boost your grade ✓
Remember that in a foundation-tier exam paper, you will not be asked any questions that involve rearranging an equation.

Components and circuits

Each component in a circuit has a standard symbol. Rather than drawing a complicated picture, we use symbols to draw circuit diagrams (Figure 11.4).

P2 11.2 Circuits: starting out

Boost your grade ✓

You need to know the symbols and names of the components shown in Figure 11.4.

A **component** is a single part of an electric circuit.

A **series circuit** has only one path that current can take.

A **parallel circuit** has more that one path that current can take. Each path is independent of the others.

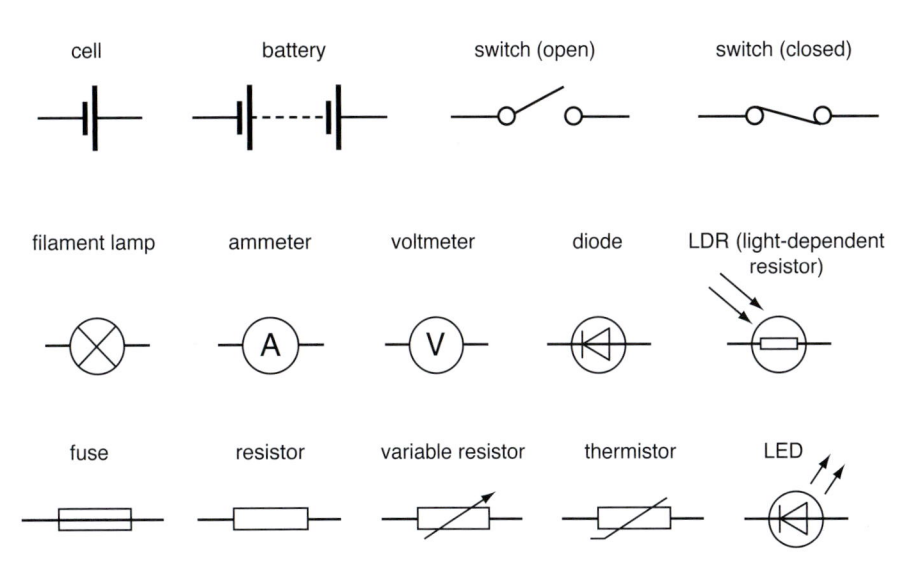

cell | battery | switch (open) | switch (closed)

filament lamp | ammeter | voltmeter | diode | LDR (light-dependent resistor)

fuse | resistor | variable resistor | thermistor | LED

Figure 11.4 Using standard symbols for components makes a circuit easier to understand.

How are series circuits different from parallel circuits?

Components joined in series follow each other in one complete loop. The current has only one path. So the same current must flow through each component and all the wires in the circuit.

The ammeters in Figure 11.5 show the same reading. This is because the current is the same at all points in a series circuit. So, only one ammeter is needed in a series circuit.

A parallel circuit gives more than one pathway for the electric current. If there is a break in one pathway, the other pathways would still work.

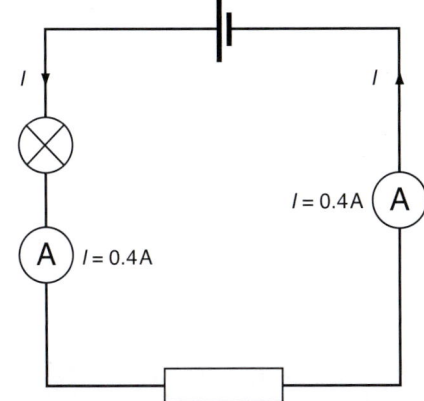

Figure 11.5 The same current flows through components joined in series. The arrow shows the direction of current flow, from the positive to the negative terminal. This is called 'conventional current'. In fact, current (electron flow) goes the other way: from negative to positive.

Cells in series

Cells joined together make a battery. When cells are joined in series, the total potential difference of the battery is worked out by adding the separate potential differences together. This only works if the cells are joined together correctly, positive (+) to negative (−).

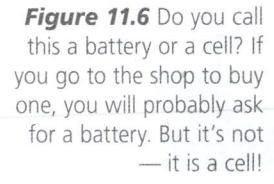

A **battery** is two or more cells connected together.

Figure 11.6 Do you call this a battery or a cell? If you go to the shop to buy one, you will probably ask for a battery. But it's not — it is a cell!

Figure 11.7 A 12 V car battery has six identical cells joined together. What is the p.d. of each cell?

Physics 2   227

Chapter 11 Using electricity

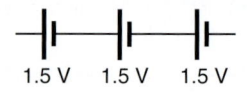

Figure 11.8 The cells are joined correctly — the separate potential differences add to give 4.5V.

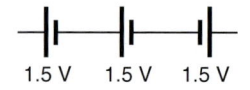

Figure 11.9 One cell is the wrong way round. The p.d. of this cell cancels out the p.d. of one of the other cells — the total p.d. is only 1.5V.

Test yourself

4 Why is only one ammeter needed in a series circuit?

5 Look at the circuit shown in Figure 11.10. Which switches have to be closed to make these lamps light?
 a P and Q only
 b N only
 c P, Q and N

6 Calculate the potential difference of each cell combination shown in Figure 11.11.

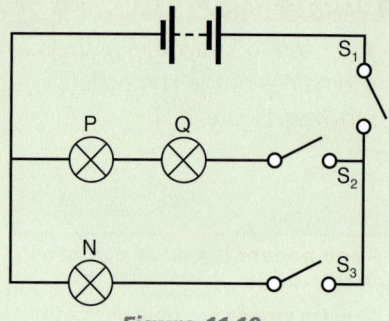

Figure 11.10

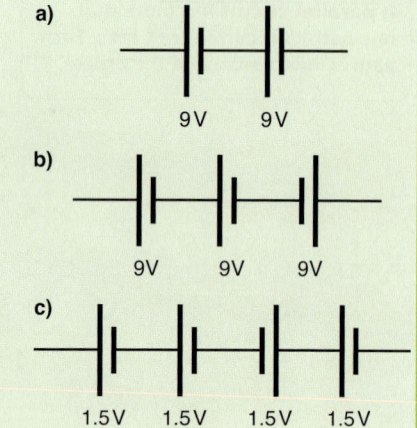

Figure 11.11

7 A horse got an electric shock from touching an electrified fence. Calculate how much charge passed through the horse if a current of 0.020A flowed for 0.1s.

P2 11.3 Resistance

Learning outcomes

- Know that the resistance of a component is found by dividing potential difference by current.
- Know that resistance is measured in ohms (Ω).

All components resist the flow of an electric current. The more easily a current flows, the less resistance the component has. A high resistance component lets only a small current flow.

Resistance is measured in ohms (Ω). The resistance (R) of a component is defined as the potential difference (V) across it divided by the current (I) through it.

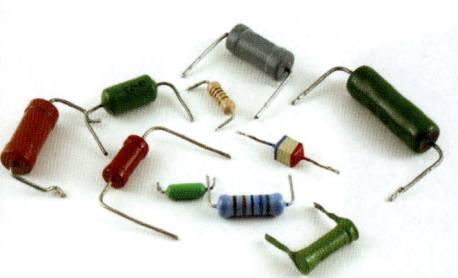

Figure 11.12 These resistors can be used to change the size of the current in a circuit, for example to limit the current through a sensitive electrical component.

Figure 11.13 The resistance of a variable resistor is changed by moving a sliding contact. The variable resistor can be used to vary the current and change the potential difference across a component.

AQA GCSE Physics

P2 11.3 Resistance

Linking potential difference, current and resistance

The potential difference across a component can be calculated using the following equation:

$$V = I \times R$$
potential difference = current × resistance
(volts, V)   (amps, A) (ohms, Ω)

Quantity	Symbol	Unit	Symbol
Current	I	Amp	A
Potential difference	V	Volt	V
Resistance	R	Ohm	Ω

Table 11.1 If you use symbols, make sure they are the right ones.

Boost your grade

You might find the equation easier to use if you remember the triangle shown in Figure 11.14.

V = potential difference
I = current
R = resistance

Figure 11.14

Worked example

What current flows through a 24 Ω resistor which has a potential difference of 12 V across it?

If you are not sure how to get the equation for 'I', use the equation triangle in Figure 11.14. Put your finger over the 'I'. This shows that 'V' is above 'R'. So, the equation is:

$$I = \frac{V}{R}$$

$$I = \frac{12}{24}$$

$$I = 0.5\,A$$

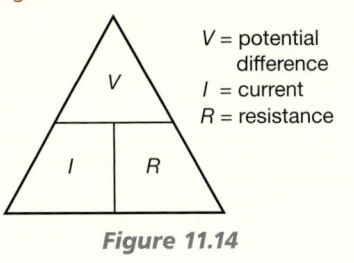

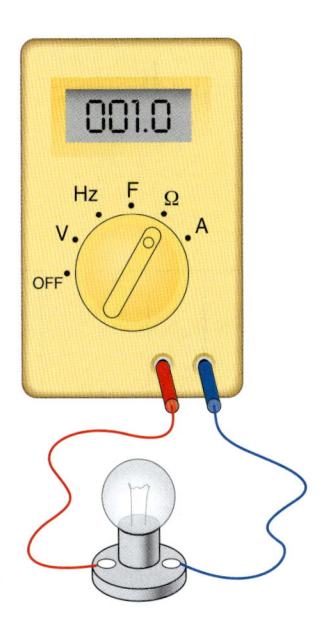

Figure 11.15 The resistance of a lamp, or any other component, can be measured directly using an ohmmeter. The resistance of this lamp is small because the lamp is cold.

Investigating the resistance of different lengths of wire

HOW SCIENCE WORKS — PRACTICAL SKILLS

Set up the circuit shown in Figure 11.16.

Measure the current flowing through the piece of wire (I) and the potential difference (V) across the wire. Use the equation $R = \frac{V}{I}$ to calculate the resistance of the wire.

Use different lengths of wire to obtain the data needed to plot a graph of resistance against length of wire.

1. What was the independent variable in this investigation?
2. Each length of wire needed to be the same thickness and made from the same metal. Why?
3. Provided the wire does not get hot, the graph should give a straight line going through the origin (0, 0). What is the relationship between the resistance of a wire and the length of the wire?

voltmeter reading = 9.5 V
ammeter reading = 1.9 A
resistance = ?

Figure 11.16 Measuring the potential difference and current allows the resistance to be calculated.

Physics 2

Chapter 11 Using electricity

Test yourself

8 Copy and complete Table 11.2

Device	Potential difference across device in volts	Current through device in amps	Resistance of device in ohms
Resistor	10		20
Lamp	12	3	
Kettle	230	10	
Motor		5	83
Iron	230		57.5

Table 11.2

P2 11.4 More about series and parallel circuits

Learning outcomes

Know that for components connected in series:
- the individual resistances add to give the total resistance
- the potential differences add to give the potential difference of the power supply

Know that for components in parallel:
- the potential difference across each component is the same
- the total current is the sum of the currents through the separate components

The **potential differences** across the components in a series circuit add to give the potential difference of the power supply.

Worked example

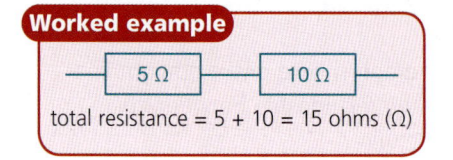

— 5 Ω — 10 Ω —

total resistance = 5 + 10 = 15 ohms (Ω)

Series circuits

The current through components that are joined in series is the same. But the potential difference across the components may be different. The potential differences are the same when the components have the same resistance.

The voltmeters in Figure 11.17 are measuring the potential difference across the lamp and the resistor. The lamp and the resistor cannot have the same resistance as the voltmeters are giving different readings.

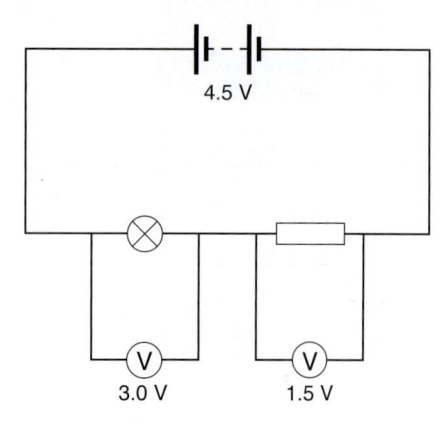

Figure 11.17 The potential difference across a component depends on the resistance of the component. The reading across the lamp is the largest, so the lamp must have the largest resistance.

Adding the voltmeter readings gives the total potential difference across the two components. In this case:

total potential difference = 3.0 + 1.5 = 4.5 V

This is the same as the potential difference of the battery. So the potential difference of the battery has been shared between the two components.

Adding resistors in series

If you have two or more resistors connected in series then each one adds to the total resistance of the circuit (Figure 11.18).

The total resistance of a series circuit is found by adding all the individual resistances together.

AQA GCSE Physics

P2 11.4 More about series and parallel circuits

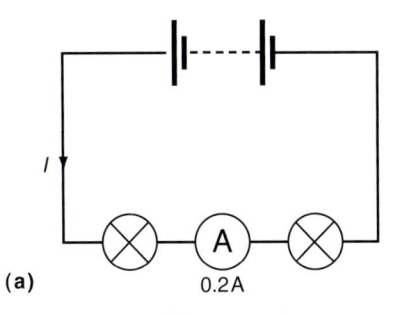

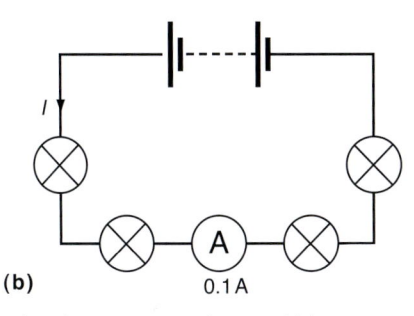

(a) 0.2A   (b) 0.1A

Figure 11.18 Adding more lamps to a circuit makes the current go down, which means that the resistance of the circuit has gone up.

Parallel circuits

When components are joined in parallel, the potential difference across each component is the same. Joining components in parallel gives the electric current different paths. Some current goes down each path. The path with the biggest resistance has the smallest current flowing through it.

In Figure 11.19, half the total current flows through each lamp. So the resistances of the lamps must be the same. The current joins back together where the parallel paths meet.

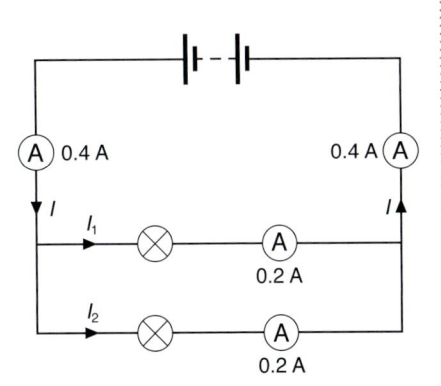

Figure 11.19 The current in a parallel circuit divides between the different paths — the total current is given by $I = I_1 + I_2$.

In a parallel circuit, the total current flowing through the whole circuit is the sum of the currents flowing through the separate components.

Overloaded circuits

An 'adaptor' lets you plug more than one appliance into a single mains socket. The appliances are connected in parallel. So the total current taken from the socket is the sum of the currents flowing through each appliance.

If the computer, stereo and television in Table 11.3 are plugged into an adaptor, then:

total current = 1 + 1 + 4 = 6 A

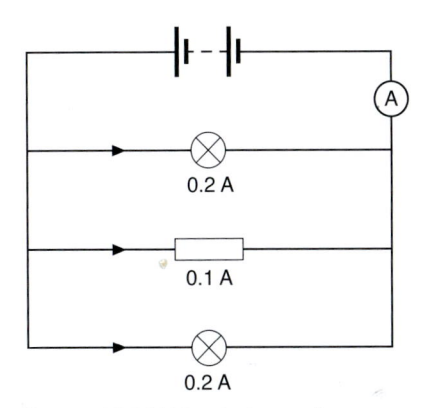

Figure 11.20 What is the reading on the ammeter?

Appliance	Typical working current in amps
Computer	1
Hairdryer	5
Television	4
Stereo	1
Heater	10

Table 11.3 The current drawn by an appliance from the mains supply depends on its power.

This is not a problem. Up to 13 amps can safely flow through the cable connecting the adaptor to the mains electricity supply.

However, plugging the hairdryer, television and heater into an adaptor could cause a big problem:

total current = 5 + 4 + 10 = 19 A

Figure 11.21 An overloaded circuit can cause a fire.

The total current is too large for the connecting wire. The wire could get so hot that it causes a fire.

Physics 2

Chapter 11 Using electricity

Test yourself

9 Calculate the total resistance of each of the arrangements in Figure 11.22.

(a) —[10 Ω]—[10 Ω]—

(b) —[5 Ω]—[15 Ω]—[30 Ω]—

(c) —[50 Ω]—[1 Ω]—

(d) —[4.7 kΩ]—[47 Ω]—

Figure 11.22

10 Calculate the reading on the voltmeter shown in Figure 11.23.

11 A computer, stereo and heater (Table 11.3) are plugged into the same socket using an adaptor. Is this likely to be a problem? Give a reason for your answer.

Figure 7.23

Dangers of overloading sockets — ICT

Create a multimedia resource explaining the danger of overloading sockets, and how you can calculate whether it is safe to use adaptors to plug several appliances into the same socket.

P2 11.5 Current–potential difference graphs

Learning outcomes

- Understand that the current through a resistor kept at a constant temperature is directly proportional to the potential difference across the resistor.
- Be able to explain why the resistance of a filament lamp increases as the temperature of the filament increases.
- Know that a diode lets current flow in only one direction.
- Know that an LED emits light when a current flows through it in the forward direction.

Obtaining the data needed to plot an I–V graph — HOW SCIENCE WORKS / PRACTICAL SKILLS

Set up the circuit shown in Figure 11.24. The component can be a resistor, a filament lamp or a diode.

Adjusting the variable resistor changes the potential difference across the component and the current through the component. This allows a set of potential difference and current readings to be taken for a component. You can use the readings to draw a graph.

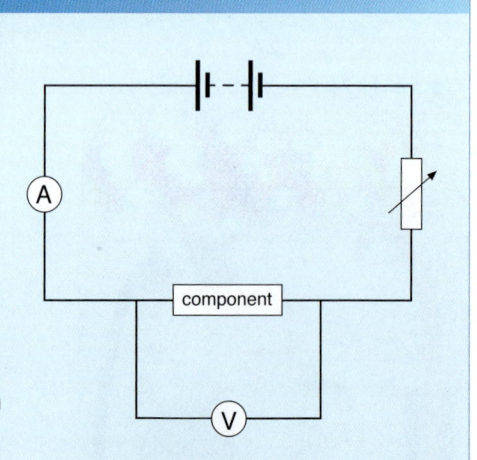

Figure 11.24 This circuit can be used to show how the current through a component depends on the potential difference across the component.

AQA GCSE Physics

P2 11.5 Current–potential difference graphs

When the component is a resistor

As long as the temperature of the resistor (or metal wire) stays the same, the graph is always a straight line going through the origin (0, 0). This shows that the current is directly proportional to the potential difference.
- Reversing the battery does not change the shape of the graph, the line is still straight.
- Changing the direction of the current through the resistor does not change its resistance.

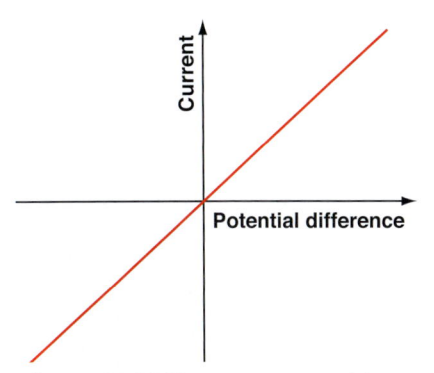

Figure 11.25 The current–potential difference graph for a resistor or metal wire at a constant temperature always gives a straight line passing through the origin.

When the component is a filament lamp

A filament lamp uses a hot wire (filament) to give out light. As the potential difference across the lamp increases, the filament gets hotter and the light gets brighter. As the filament gets hotter its resistance increases (Figure 11.26).

Connecting the battery the other way round makes no difference to the way the resistance of the lamp changes. The resistance always increases when the temperature of the filament increases.

The filament, made from a metal, contains free electrons and ions. As the temperature of the filament increases, the ions vibrate faster and with a bigger amplitude. The free electrons collide more frequently with the ions. If the potential difference stays the same, this has the effect of making the free electrons move more slowly around the circuit. So, the resistance has increased.

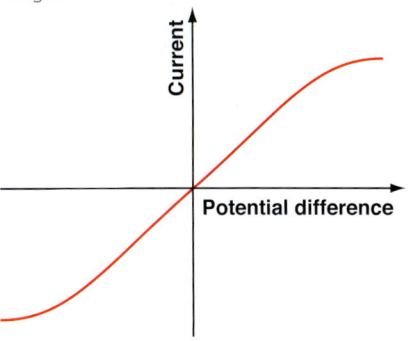

Figure 11.26 As the resistance of the filament increases, the current does not increase in the same proportion as the potential difference.

When the component is a diode

The I–V graph for a diode is different from that for a resistor or a filament lamp (Figure 11.27). The resistance of a diode does depend on which way round it is connected in a circuit.

Diodes let current flow in only one direction (called the forward direction). In the reverse direction diodes have a high resistance, so no current flows.

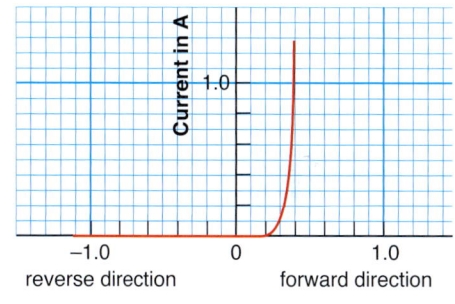

> A **diode** is a device that allows current to flow in only one direction, because the diode has a high resistance in the reverse direction.

Figure 11.27 The resistance of a diode does depend on the direction of the potential difference. If a current flows in one direction, turning the battery round will stop the current flow.

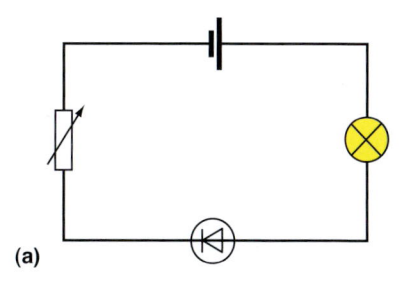

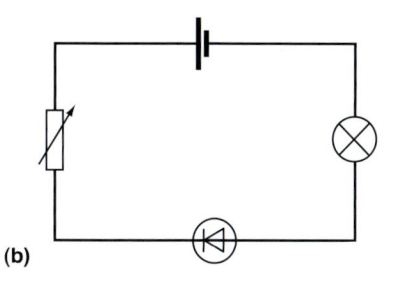

Figure 11.28 (a) The diode is in the 'forward direction' and the lamp is on; (b) the diode is the 'reverse direction', the resistance is high and the lamp is off.

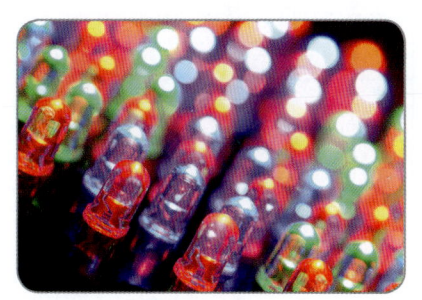

Figure 11.29 Light-emitting diodes (LEDs) are a special sort of diode. When an electric current flows through an LED in the forward direction, it emits light. Lamps that use LEDs are available, but compared with fluorescent lamps they are expensive. However, they use little energy, have a long lifetime and are very reliable.

Physics 2 — 233

Chapter 11 Using electricity

Quantum tunnelling composite

**HOW SCIENCE WORKS
PRACTICAL SKILLS**

Invented in 1997 by David Lussey, quantum tunnelling composite (QTC) is an amazing material. In one form it looks just like a piece of black rubber. But when you squash it, its electrical resistance falls as it changes from an insulator to a conductor.

The range of applications for QTC is huge — from toys to textile switches, robotic sensors to space travel.

Josh designed a simple experiment to check out what he had read about QTC. The apparatus he used is shown in Figure 11.32. Josh used different weights to squeeze the QTC, which was sandwiched between two metal plates. For each new weight, Josh waited 20 seconds before taking the ohmmeter reading. He did this because he had noticed that although the reading often changed rapidly in the first few seconds. Even after 20 seconds the reading still changed, but more slowly. For each weight, Josh recorded a range of resistance values. These are shown in Figure 11.33.

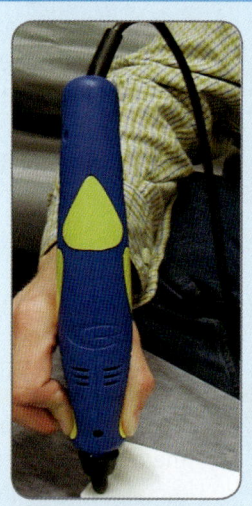

Figure 11.30 The outside casing of this drill contains QTC — squeezing the drill controls its speed.

Figure 11.31 QTC textile switches can be incorporated into 'smart' clothing and bags. Tapping the switches controls your mobile phone and MP3 player. QTC switches are even used in space suits and flexible roll-up computer keyboards.

3. With a weight of 20 N, Josh measured the resistance as 10 Ω. Why has Josh not plotted this value on his results chart?

4. Josh added a different weight to the QTC 'sandwich'. The reading on the ohmmeter was a fairly steady 3 Ω. How many newtons do you think this weight was? Give a reason for your answer.

5. Josh noticed that there is no overlap between the resistance values produced by two different weights. State, with a reason, whether or not this means that QTC could be used as a reliable way to tell one weight from a different one.

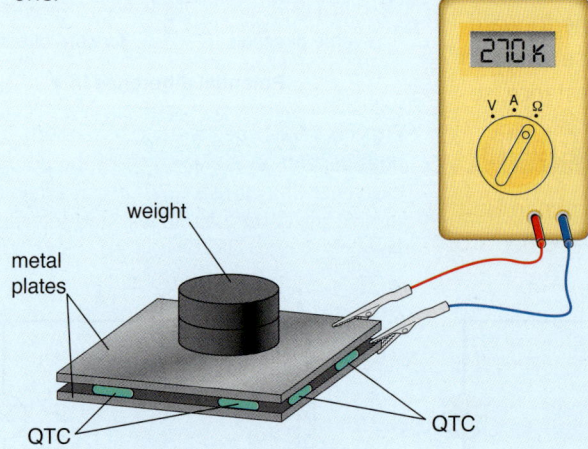

Figure 11.32 Measuring the resistance of QTC with an ohmmeter.

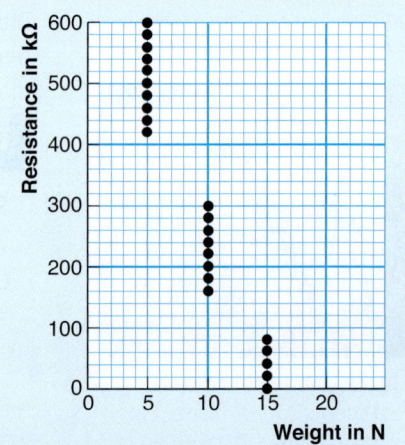

Figure 11.33 For each weight, the QTC has a range of resistance. What is the range of resistance when using a 10 N weight?

AQA GCSE Physics

P2 11.6 Two special types of resistor

> **Test yourself**
>
> 12 Copy Figure 11.25 and label the line 'low resistance'. Draw a second line on the graph for a resistor with a higher resistance. Label the second line 'high resistance'.
>
> 13 What is an ion?
>
> 14 Greg plots the *I–V* graph for a resistor. The line on the graph is curved and not straight. What has happened to the resistor used by Greg?
>
> 15 How does the resistance of a diode in 'reverse direction' compare with its resistance in the 'forward direction'?

P2 11.6 Two special types of resistor

Learning outcomes

- Know that the resistance of a light-dependent resistor (LDR) decreases as the intensity of the light increases.
- Know that the resistance of a thermistor decreases as its temperature increases.

> A **light-dependent resistor (LDR)** is a component in an electric circuit whose resistance decreases as the intensity of light falling on it increases.

Light-dependent resistor (LDR)

The resistance of a **light-dependent resistor (LDR)** changes as the intensity (brightness) of light shining on it changes.

LDRs can be used as sensors in light-operated circuits, such as security lighting. The change in the resistance of LDRs is used in digital cameras to control the total amount of light that enters the camera.

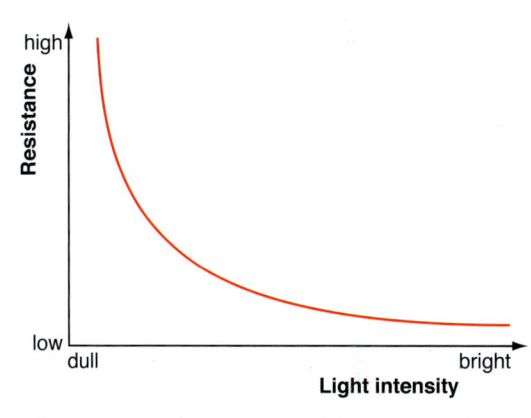

Figure 11.34 The resistance of an LDR goes down as the intensity of light goes up.

Drawing an *I–V* graph for an LDR
HOW SCIENCE WORKS / PRACTICAL SKILLS

Set up the circuit shown in Figure 11.24, using an LDR as the component. Take your own measurements of *I* and *V* and plot a graph.

Now turn off the laboratory lights and, if possible, close the blinds. Obtain a second set of measurements for *I* and *V*. Plot a second graph.

Under which condition does the LDR have the highest resistance — in the light or in the dark?

Thermistor

The resistance of a **thermistor** changes with temperature. As the temperature of the thermistor rises its resistance decreases.

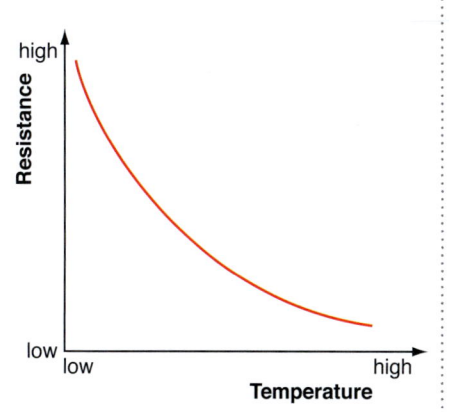

Figure 11.35 The resistance of a thermistor decreases as the temperature of the thermistor increases.

Physics 2 — 235

Chapter 11 Using electricity

A **thermistor** is a component in an electric circuit whose resistance decreases as the temperature increases.

A thermistor can be used as the sensor in a temperature-operated circuit, such as a fire alarm. Some electronic thermometers use a thermistor to detect changes in temperature. The change in the resistance of a thermistor can be used to switch on (or off) other electrical circuits automatically.

Designing an oven thermometer

**HOW SCIENCE WORKS
PRACTICAL SKILLS**

Misti has designed the circuit shown in Figure 11.37 to replace the broken temperature display on her oven. The thermistor, which Misti is using as a temperature sensor, goes inside the oven. The rest of the circuit is outside the oven.

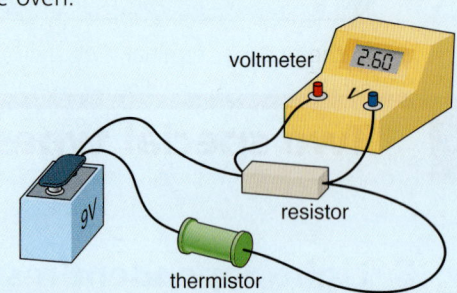

Figure 11.37 Misti's design for an oven thermometer.

1. Draw a circuit diagram of Misti's oven thermometer design.
2. Explain why the voltmeter reading changes as the oven temperature changes.

Before using her oven thermometer, Misti must **calibrate** it. She does this by putting the thermistor in a beaker of water and heating the water using the circuit shown in Figure 11.36.

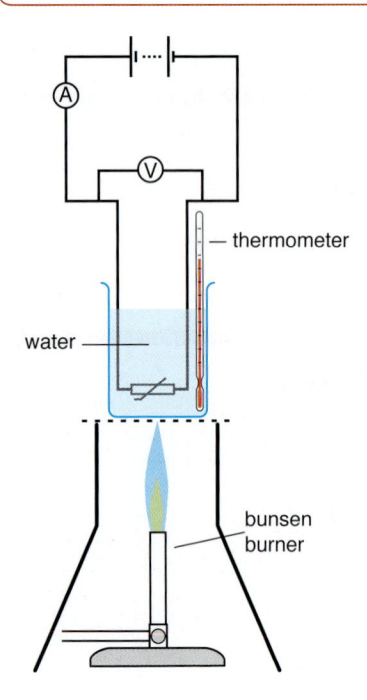

Figure 11.36 By gently heating the water, the resistance of the thermistor can be found at different temperatures. You could use an ohmmeter to measure the resistance directly.

She recorded different water temperatures and the reading on the voltmeter at those temperatures. Misti then drew the calibration graph shown in Figure 11.38.

Calibrate means to make a standard scale between two fixed points so that you can give a value to a measured quantity.

3. Explain how Misti can use the calibration graph to convert voltmeter readings into temperature values.
4. When Misti tried her thermometer in her oven, she obtained voltmeter readings that varied between 11.5 V and 8.0 V. What range of temperatures does this give for the oven?
5. Suggest why the voltmeter reading was not constant.
6. Misti realises there is a problem with her design — it cannot be used to measure high oven temperatures. Why can this design not measure temperatures above 100°C?

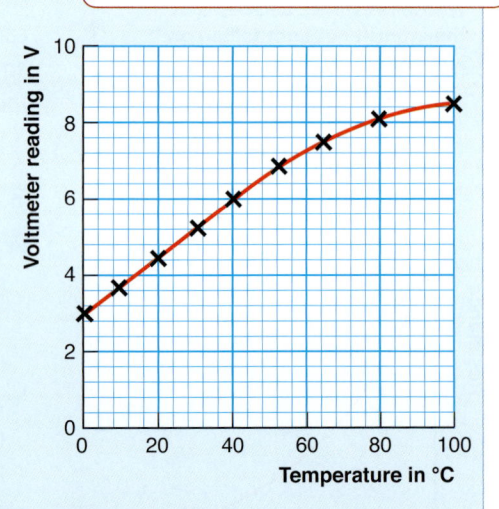

Figure 11.38 Misti's calibration graph.

Test yourself

16 Copy and complete the following sentences.
  a The resistance of an LDR _____ when a bright light shines on it.
  b The resistance of a _____ depends on temperature.
  c An LDR can be used in a _____ sensing circuit.

236 AQA GCSE Physics

P2 11.7 Direct current and alternating current

Learning outcomes

- Know that direct current (d.c.) flows in one direction.
- Know that alternating current (a.c.) changes direction.
- Know that in the UK the mains supply is at 230 V a.c. with a frequency of 50 Hz.

Direct current (d.c.) is an electric current that flows in the same direction all the time.

A **cathode ray oscilloscope (CRO)** is a device that displays how a potential difference varies over time.

Alternating current (a.c.) is an electric current that continually changes its direction.

The **frequency** of an alternating current is the number of cycles per second.

A **direct current (d.c.)** always flows in the same direction around a circuit. It is shown in circuit diagrams by an arrow, usually labelled *I*, going from the positive terminal to the negative terminal of the power supply. This is called 'conventional' current. In terms of electrons, the current flows in the opposite direction — from the negative terminal of the power supply to the positive terminal. Cells and batteries produce a constant direct current.

You can use a **cathode ray oscilloscope (CRO)** to show that the potential difference of a d.c. supply is constant. The display or 'trace' on the CRO screen shows the potential difference in volts across the supply. In Figure 11.39, connecting a cell to the input terminals has made the line on the CRO jump up above the red zero line.

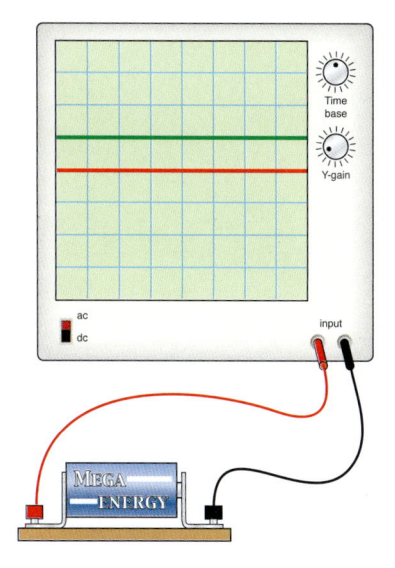

Figure 11.39 The horizontal line (trace) on the CRO shows that the cell produces a constant potential difference. The zero value is shown as a red line.

Connecting the cell the other way around would make the line jump down below the red zero line. The bigger the potential difference of the cell, the further the line jumps. Therefore, the CRO gives us a way of comparing the potential differences of different d.c. supplies.

Alternating current (a.c.) is like having a cell in a circuit, but you keep turning the cell around. First the current goes one way, and then it goes back again. The current flows backwards and forwards as the terminals of the power supply change from positive to negative.

An alternating current is one that is constantly reversing its direction. The number of times the current changes direction in one second is the **frequency** of the supply.

Comparing CRO traces for two different a.c. supplies

Figure 11.40 shows the CRO traces for two a.c. supplies. The controls on the oscilloscope were the same for both supplies. Every time the trace changes from above to below the red zero line (or from below to above), the current in the circuit changes direction.

Trace A is $1\frac{1}{2}$ times taller than trace B. So the peak potential difference (V_0) of supply A is $1\frac{1}{2}$ times bigger than that of supply B.

The CRO trace for supply B shows twice as many waves in the same time as supply A. This means the potential difference (and current) of supply B changes direction twice as often as the potential difference (and current) of supply A. So the frequency of supply B is twice the frequency of supply A.

Physics 2

Chapter 11 Using electricity

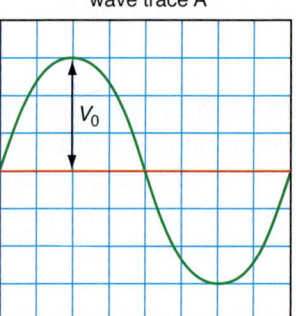

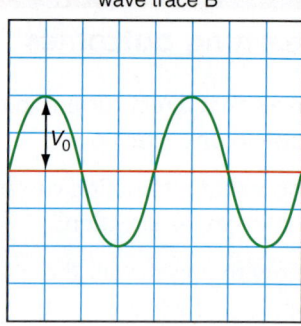

Figure 11.40 The CRO traces show that the two a.c. supplies have different frequencies and peak potential differences (V_0).

Calculating the period and frequency of an a.c. supply

The screen of a CRO is like a graph. The vertical and horizontal divisions each have a scale. Each vertical division represents a certain number of volts; each horizontal division represents a certain length of time. When you know the scale you can work out three important quantities:
- peak (maximum) potential difference (V_0)
- time to complete one cycle (called the **time period**)
- frequency of the supply

> The **time period** is the time taken for an alternating current to complete one cycle.
> $$\text{frequency} = \frac{1}{\text{time period}}$$

Worked example

Look at trace A in Figure 11.40. If each vertical division on the CRO represents 2 volts and each horizontal division represents 0.005 seconds, then:

peak potential difference = 3 divisions × 2 volts per division = 6 V
time period = 8 divisions × 0.005 s per division
time period = 0.04 s
$$\text{frequency} = \frac{1}{\text{time period}} = \frac{1}{0.04} = 25\,\text{Hz}$$

The mains electricity supply

The UK mains electricity is an a.c. supply at about 230 V. It has a frequency of 50 Hz — this means the current flows one way, then back again, 50 times in each second.

Explaining a.c. and d.c. — ICT

Create an animation or presentation that shows and explains all of these: a.c., d.c., frequency, time period and peak potential difference.

Test yourself

17 Explain the difference between an alternating current and a direct current.

18 Figure 11.41 shows the CRO traces for different a.c. supplies. Which of the supplies has the:
 a largest peak voltage?
 b highest frequency?
 c lowest frequency?
 d smallest peak voltage?

19 Each horizontal division on the CRO screen in Figure 11.41 represents 0.002 s. Calculate the frequency of each trace.

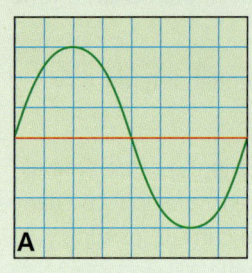

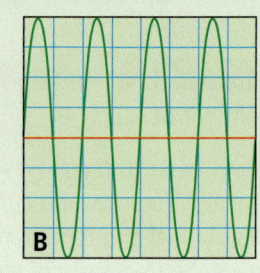

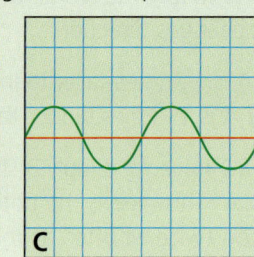

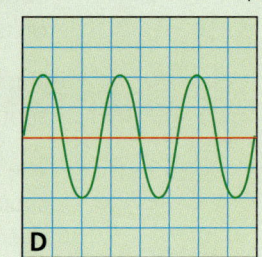

Figure 11.41

238 AQA GCSE Physics

P2 11.8 Cables and plugs

Learning outcomes

- Know that two-core or three-core cables are used to connect appliances to the mains.
- Know how colour-coded insulation is used to identify the live, neutral and earth wires inside a cable.
- Know that the outside of a plug is made from an insulating material.

Many electrical appliances are designed to work from the mains electricity supply. Most of these appliances are connected to the mains using a cable and a three-pin plug. Both the cable and plug are designed to make the appliance safe to use.

Cables

Inside the cables are either two or three copper wires (the inner cores). Copper is used because it is a good electrical conductor. A layer of plastic covers each wire, and another layer forms the outside of the cable. Plastic is used because it is a good insulator and is flexible.

Three-core cable must be used with appliances that have any outside metal parts — for example an electric iron. Two-core cable does not have an earth wire. It is used with appliances that have an outside plastic case.

Choosing the right cable

The power of an electric kettle is usually about 2 kW, but the power of an electric shower can be as high as 10 kW. This means that the shower draws a much larger current from the mains supply than the kettle. The larger the current, the thicker the wires inside the connecting cable need to be. If the wrong cable is used with an appliance, the cable could overheat, melt and cause a fire.

Wall sockets in most homes are connected to the mains supply using 2.5 mm² cables. But to connect high-power appliances, such as cookers and showers, thicker 6 mm² or 10 mm² cables must be used.

Cable size	Maximum current
1.0 mm²	5 A
2.5 mm²	18 A
4.0 mm²	25 A
6.0 mm²	32 A
10.0 mm²	45 A

Table 11.4 The 'cable size' refers to the cross-sectional area of one of the wires inside the cable.

Plugs

The **outside case** of the plug is made of plastic or rubber. These materials are used because they are good electrical insulators.

Pushing the plug into a socket joins the **three brass pins** of the plug to the terminals of the socket. Brass is used for the pins because it is a good electrical conductor. The earth pin is the longest so that it makes contact inside the socket first.

Inside the plug there is a **cable grip** and a **fuse**. The cable grip is there to hold the cable firmly in place. If the cable is pulled, the cable grip stops the copper wires inside the plug being pulled loose.

Connecting an appliance to a three-pin plug

If you ever fit a plug, you must connect the correct wires from the appliance to the correct pins in the plug. The wires are colour coded to help you connect them in the right place (Figure 11.42).

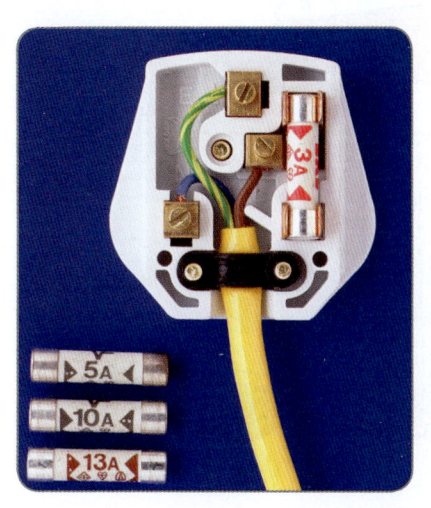

Figure 11.42 A correctly wired three-pin plug:
- The brown wire (live wire) is joined to the fuse, which is joined to the live pin.
- The blue wire (neutral wire) is joined to the neutral pin.
- The green/yellow wire (earth wire) is joined to the earth pin.

Boost your grade ✓

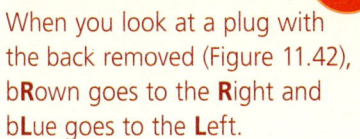

When you look at a plug with the back removed (Figure 11.42), b**R**own goes to the **R**ight and b**L**ue goes to the **L**eft.

Physics 2

Chapter 11 Using electricity

Test yourself

20 What do appliances connected to the mains with two-core cable have in common?

21 Say why each of the plugs shown in Figure 11.43 is not safe to use.

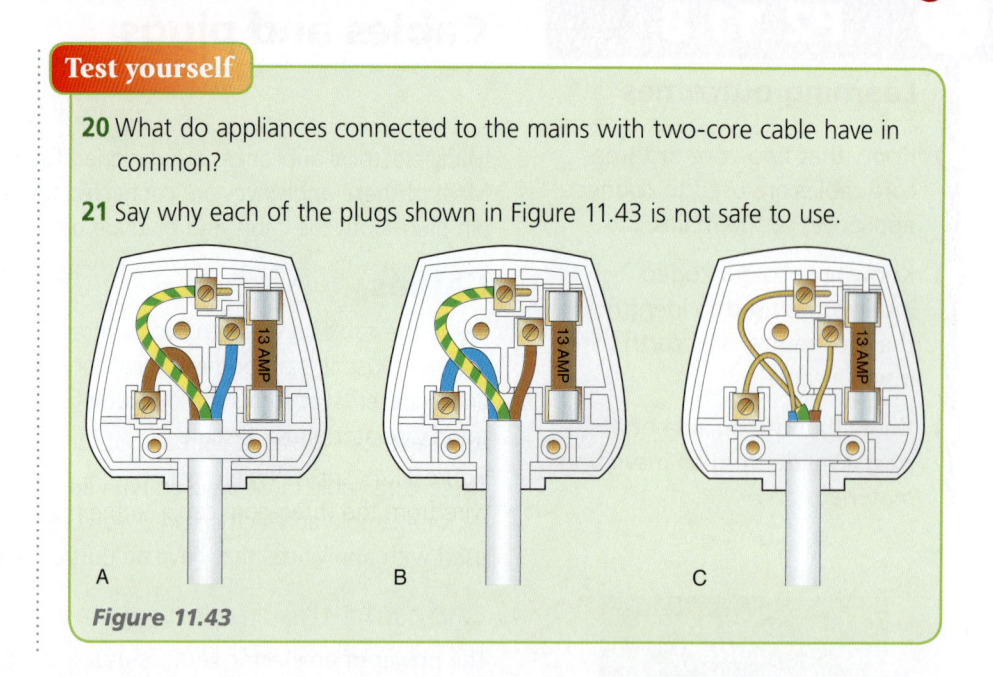

A B C

Figure 11.43

P2 11.9 Using the mains electricity supply safely

Learning outcomes

- Understand that a fault can cause a fuse or circuit breaker to disconnect a circuit.
- Explain how a residual current circuit breaker protects a circuit.
- Know that appliances with metal cases should be earthed.

Danger, electricity being used badly!

Each situation in Figure 11.44 shows mains electricity being used in a dangerous way. What are the problems? (See answers on page 242).

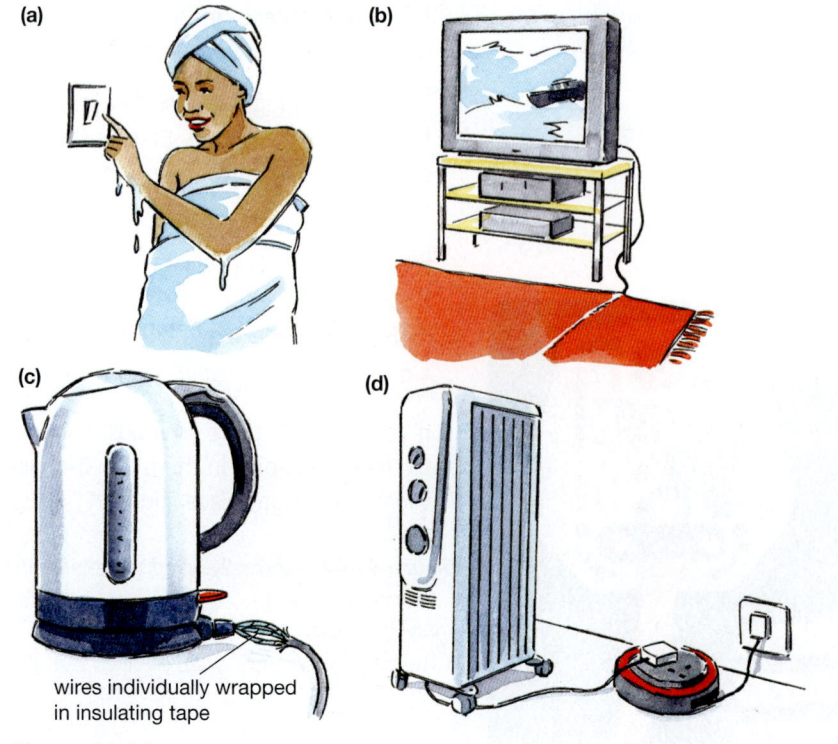

(a) (b)

(c) (d)

wires individually wrapped in insulating tape

Figure 11.44

240 AQA GCSE Physics

P2 11.9 Using the mains electricity supply safely

> A **fuse** is a thin wire that is designed to melt and break the circuit when the current through it exceeds the rating on the fuse.

Fuses

A **fuse** is a thin piece of wire that allows a current of up to a certain value flow through it. Above this value, the fuse overheats and melts. We often say that such a fuse has 'blown'.

The fuse joins the live pin to the live wire inside the plug. If a fault causes the fuse to melt, the live wire is disconnected. This protects the appliance from taking a current that might damage it or cause its wiring to overheat and start a fire.

Earthing

Any electrical appliance with an outside metal part is usually earthed. Figure 11.45 shows an electric toaster. It has been earthed by joining the earth wire from the three-core cable to the metal casing.

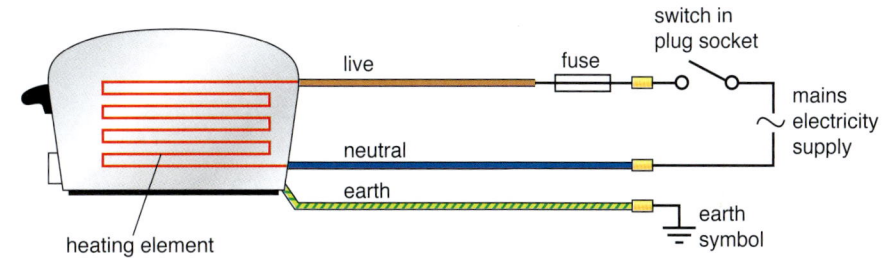

Figure 11.45 The earth wire is an important safety feature of appliances with outer metal parts.

If a fault causes the live wire to touch the metal case, and the toaster is switched on, a large current flows from the live wire to the earth wire. This large current causes the fuse to melt, which switches the toaster off. Without the earth wire the toaster would become live. Anyone who touched the metal case would get an electric shock as the current flows through them to earth.

The earth wire, together with the fuse, also protects the cables used to wire the circuit. In the event of a fault, the fuse will melt rather than the cable getting so hot that it overheats and starts a fire.

Figure 11.46 If you see this symbol on an appliance, the appliance is 'double insulated'. This means that there are two layers of insulating material around any live parts — so these appliances do not need an earth wire.

Circuit breakers

The mains supply to a house starts at a consumer unit. This used to be called the 'fuse box'. However, most consumer units now contain **circuit breakers** rather than fuses. Each circuit breaker leads to a different circuit supplying electricity to a different part of the house.

Figure 11.47 This consumer unit contains several circuit breakers and RCCBs.

Physics 2 241

Chapter 11 Using electricity

Figure 11.48 A plug-in RCCB is used in a socket not already connected to an RCCB in the consumer unit.

The circuit breaker is a fast-acting switch that automatically turns off ('trips') when the current through it goes above a set value. Having mended the fault you press a button to reset the circuit breaker.

A second type of circuit breaker, a **residual current circuit breaker (RCCB)**, works by detecting any difference between the currents in the live and neutral wires of the supply cable. If everything is working properly, the two currents are the same. If something happens to make the current in the wires different, the RCCB automatically switches the circuit off.

For safety, Vince plugs his lawn mower into an RCCB (Figure 11.49). Just as well, because Vince has been careless and has cut into the cable insulation. This causes a current to flow through Vince to earth. This means that the currents in the live and neutral wires are now different. The RCCB detects this difference and rapidly switches the circuit off saving Vince from a nasty electric shock.

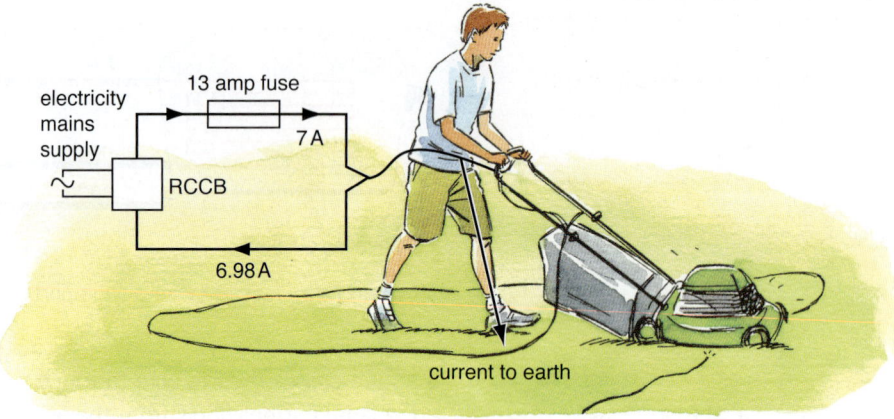

Figure 11.49 Using an RCCB saves Vince from a shocking experience. Before the RCCB acts, what current flows through Vince to earth?

One big advantage of any type of circuit breaker over a fuse is that they operate much faster. So, in the event of a fault, the circuit is disconnected much quicker.

Danger, electricity being used badly! (page 240)

Answers

a   Damp skin has a much lower resistance than dry skin. If you were to get a shock, the current through you could be large enough to kill.
b   A cable under a rug can become worn, and if it starts sparking then it can cause a fire. The cable may also overheat causing the possibility of a fire.
c   Wires joined by twisting together cause 'hot spots'. This may cause the tape to melt and sparking to happen.
d   Leaving most of the extension cable inside the plastic case means that the cable can heat up excessively and catch fire.

Test yourself

22 An electric coffee-maker takes a current of 2 A from the mains supply. Explain why the plug of the coffee-maker should not be fitted with a 13 A fuse.

23 Give two ways in which a circuit breaker is a better safety device than a fuse.

Advertising circuit breakers   ICT

You work for a company that makes plug-in RCCBs for electrical appliances such as electric lawnmowers, hedge trimmers and power tools. Create an advert or podcast explaining why people need to use an RCCB with such appliances.

P2 11.10

What links current, charge, energy and power?

Learning outcomes

- Know that when charge flows through a resistor, the resistor gets hot.
- Know that the rate at which energy is transferred is called power.

Energy and charge

When charge flows through a resistor, the resistor gets hot. This is because the electrons collide with the atoms of the resistor as they pass through it. The atoms gain kinetic energy and vibrate faster, making the resistor become hotter. So, the energy stored in the battery goes down and the temperature of the resistor goes up.

The amount of energy transferred by a resistor in a circuit depends on two factors:
- the potential difference across the resistor
- how much charge flows through the resistor

The energy transferred can be calculated using the following equation:

$$E = V \times Q$$
energy transferred = potential difference × charge
(joules, J)   (volts, V)   (coulombs, C)

This equation also applies to the electrical energy transferred from a power supply to an appliance.

Worked example

Each time Sue uses her 230 V hair straighteners, 110 coulombs of charge flow around the circuit. Calculate the energy transferred each time the straighteners are used.

$E = V \times Q$
$E = 230 \times 110$
$E = 25\,300\,J$

Power and energy

Sometimes, knowing how quickly energy is transferred is more important than knowing how much energy is transferred. For example, an electric kettle increases the temperature of the water inside the kettle. But if the increase takes a long time, the water warms up only slowly. To know how quickly the water warms up, you need to know how much energy the kettle transfers every second. What you need to know is the **power** of the kettle, which can be calculated using the following equation:

$$P = \frac{E}{t}$$
$$\text{power} = \frac{\text{energy}}{\text{time}}$$

where   power is in watts, W
            energy is in joules, J
            time is in seconds, s

Remember that not all of the energy transferred by the kettle may be used to raise the temperature of the water. Some energy is wasted.

Worked example

An electric kettle transfers 36 000 J of energy when it is switched on for 3 minutes. Calculate the power of the kettle.

$P = \frac{E}{t}$
$P = \frac{360\,000}{180}$
$P = 2000\,W\,(2\,kW)$

Figure 11.50 When you buy a new appliance, you should consider how efficiently the appliance transfers energy. Both of these kettles boil the same volume of water in the same time, but they are not equally efficient? Which kettle do you think is more efficient and why?

Chapter 11 Using electricity

Power is the rate at which a device transfers energy.

If you put the two equations in this section together with the equation $I = \frac{Q}{t}$ (Section 11.2), you end up with a fourth equation. This is the equation for electrical power:

$$P = I \times V$$
power = current × potential difference
(watts, W) (amps, A) (volts, V)

Most appliances have an information plate. This gives the power of the appliance and the potential difference of the supply. From this you can work out the current that the appliance uses and the fuse it needs.

Boost your grade ✓

Remembering this equation in the form of a triangle (Figure 11.51) will help you to rearrange the equation if you have to.

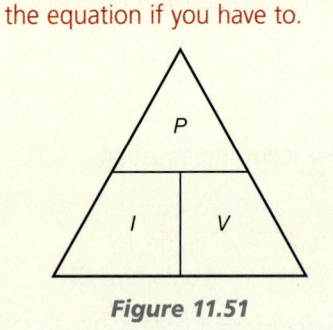

Figure 11.51

Worked example

Figure 11.52 shows the information plate from an electric grill. Calculate the current draw from the mains and suggest which fuse, 3 A or 13 A, should be used in the plug.

$P = I \times V$

Rearranging the equation gives:

$I = \frac{P}{V}$

$I = \frac{750}{230}$

$I = 3.26\,A$

Grilling Machine
230 V   750 W
~   50 Hz
Do not immerse in water

Figure 11.52

Although only just over 3 A is drawn from the mains, if the only choice is a 3 A fuse or a 13 A fuse then the 13 A must be used.

What would happen if the 3 A fuse were fitted to the plug? Why is the warning 'Do not immerse in water' included on the information plate?

This equation can also be used to calculate the current that will flow through a cable and then chose the correct cable thickness (Table 11.4).

Worked example

An electrician is about to install an 8.2 kW electric shower in Izzy's bathroom. What size cable should be used to connect the shower to the 230 V mains supply?

(Remember: power must be in watts)

Using the equation triangle:

$I = \frac{P}{V}$

$I = \frac{8200}{230}$

$I = 35.7\,A$

Looking at Table 11.4 the electrician needs to use 10 mm² cable.

Test yourself

24 Calculate the power rating in watts of:
   a a car starter motor operating from a 12 V supply, taking a 100 A current
   b a torch bulb operating from a 4.5 V supply, taking a 0.3 A current
   c a pocket calculator operating from a 3 V supply, taking a 0.0001 A current.

25 A 12 V battery drives a motor taking a current of 2 A for 20 seconds. Calculate the electrical energy transferred to the motor.

26 A 3 A or 13 A fuse is usually fitted to the plug of a 230 V household appliance.
   a Calculate the current each of the appliances in Table 11.5 draws from the mains supply.
   b State the correct fuse to use with each appliance.

Appliance	Power
Microwave	750 W
Oven	2.5 kW
Toaster	350 W
Radio	15 W

Table 11.5

AQA GCSE Physics

Chapter 11 Homework questions

Homework questions

1. Figure 11.53 shows a simplified picture of the inside of a small fan heater. The electrical wiring is not shown.

   Draw, using circuit symbols, a diagram to show how the heating elements, fan and switches would be connected together. It is essential that:
   - when connected to the mains and switched on the fan comes on
   - both heating elements can be switched on independently
   - on hot days the heater can be used as a fan

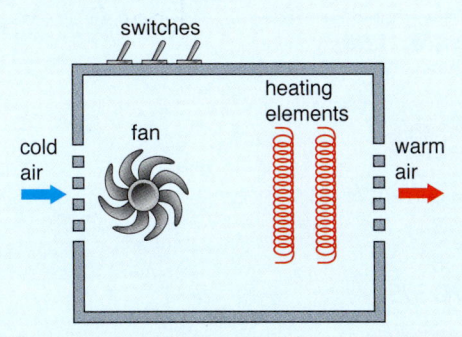

   Figure 11.53

2. Figure 11.54 shows a simplified circuit diagram for the front lights of a car. The metal car body acts like a wire in the circuit.
   a. Which lights would not work if fuse A melted?
   b. Which switch operates the headlights?
   c. Can the sidelights be switched on without the headlights? Give a reason for your answer.
   d. If one headlight breaks, will the other one still work? Give a reason for your answer.
   e. How would this circuit need to be changed if the car had a plastic body?

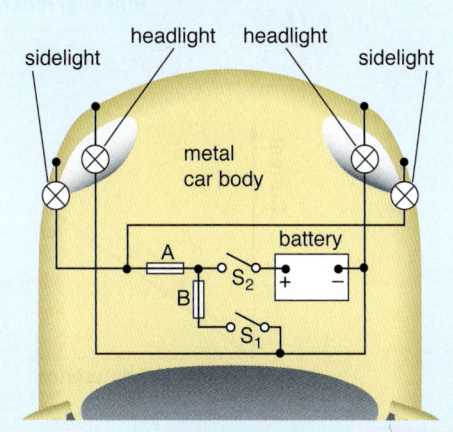

   Figure 11.54

3. Explain why the resistance of a filament lamp increases with temperature.

4. Figure 11.55 shows a thermistor in series with a high-power lamp and positioned close to it.
   a. Explain why the lamp only shines dimly when the switch is closed.
   b. Explain why the lamp gets slowly brighter.

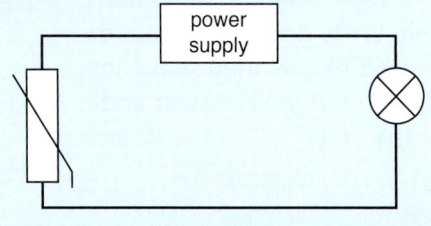

   Figure 11.55

5. Ramesh uses the ammeter in Figure 11.56 as a simple light intensity meter.
   a. Explain why the reading on the ammeter changes when the light intensity changes.
   b. Sketch a graph to show how the ammeter reading changes with light intensity.

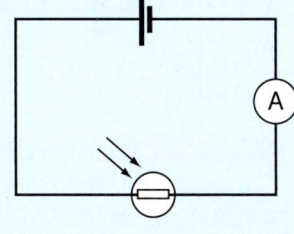

   Figure 11.56

6. Zafran leans against his washing machine. The live wire inside the machine has become loose and is touching the metal casing. Explain how the earth wire, which is correctly connected, protects Zafran from a serious electric shock.

7. Table 11.6 shows how long it takes three different electric kettles, X, Y and Z, to boil some water. The starting temperature of the water in each kettle is the same.

   Assuming the kettles are equally efficient, which one has the highest power? Explain your choice.

   Table 11.6

Kettle	Mass of water in the kettle	Time to boil the water
X	1.0 kg	4 minutes
Y	500 g	2½ minutes
Z	1.2 kg	4 minutes

Physics 2

Chapter 11 Exam corner

Exam corner

Figure 11.57 shows the circuit for a simple light meter.

a Figure 11.58 shows how the reading on the ammeter changes with light intensity.
Calculate the resistance of the LDR when the light intensity is equal to the value marked 'X' on the graph. Write down the equation you need to use, then show how you work out your answer and give the unit. *(3 marks)*

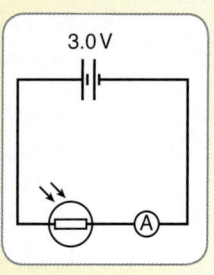

Figure 11.57

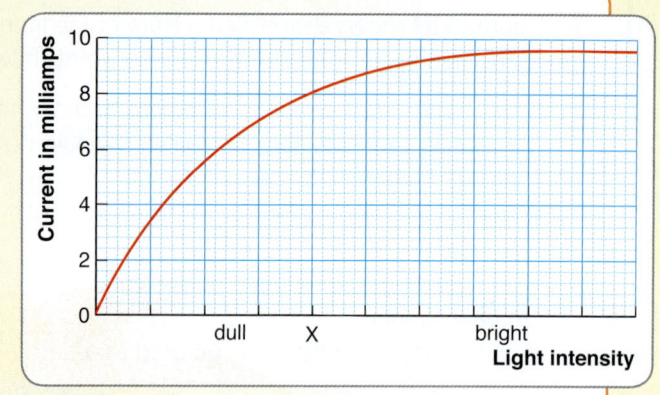

Figure 11.58

Student A

$V = I \times R$

$R = \dfrac{3.0}{0.008}$ ✓

$R = 375 \, \Omega$ ✓✓

Examiner comment The equation has been rearranged correctly. The current in milliamps has been changed to amps and the unit is correct. Remember: be careful if you use symbols; they must be fully correct.

Student B

voltage = current × resistance

current = $\dfrac{3}{8}$

current = 0.375 ✓

Examiner comment The answer is wrong but really only one mistake has been made — the current has not been changed to amps. It is always worth trying to give a unit; just ignoring it loses a mark automatically.

b Figure 11.59 shows the current–potential difference graph for an LDR in a darkened room.

Copy the graph and draw a second line to show how the current would change with potential difference if the LDR was in bright sunlight. *(1 mark)*

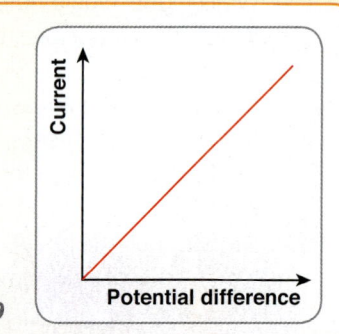

Figure 11.59

Student A

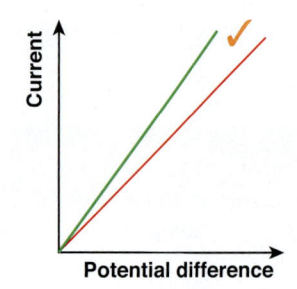

Examiner comment Correct — the line is steeper because the resistance is less in bright sunlight.

Student B

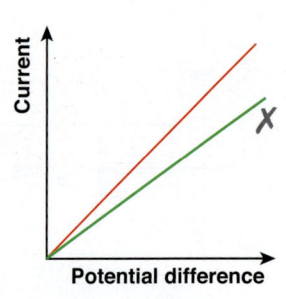

Examiner comment This shows an increase in resistance, which is wrong.

246 AQA GCSE Physics

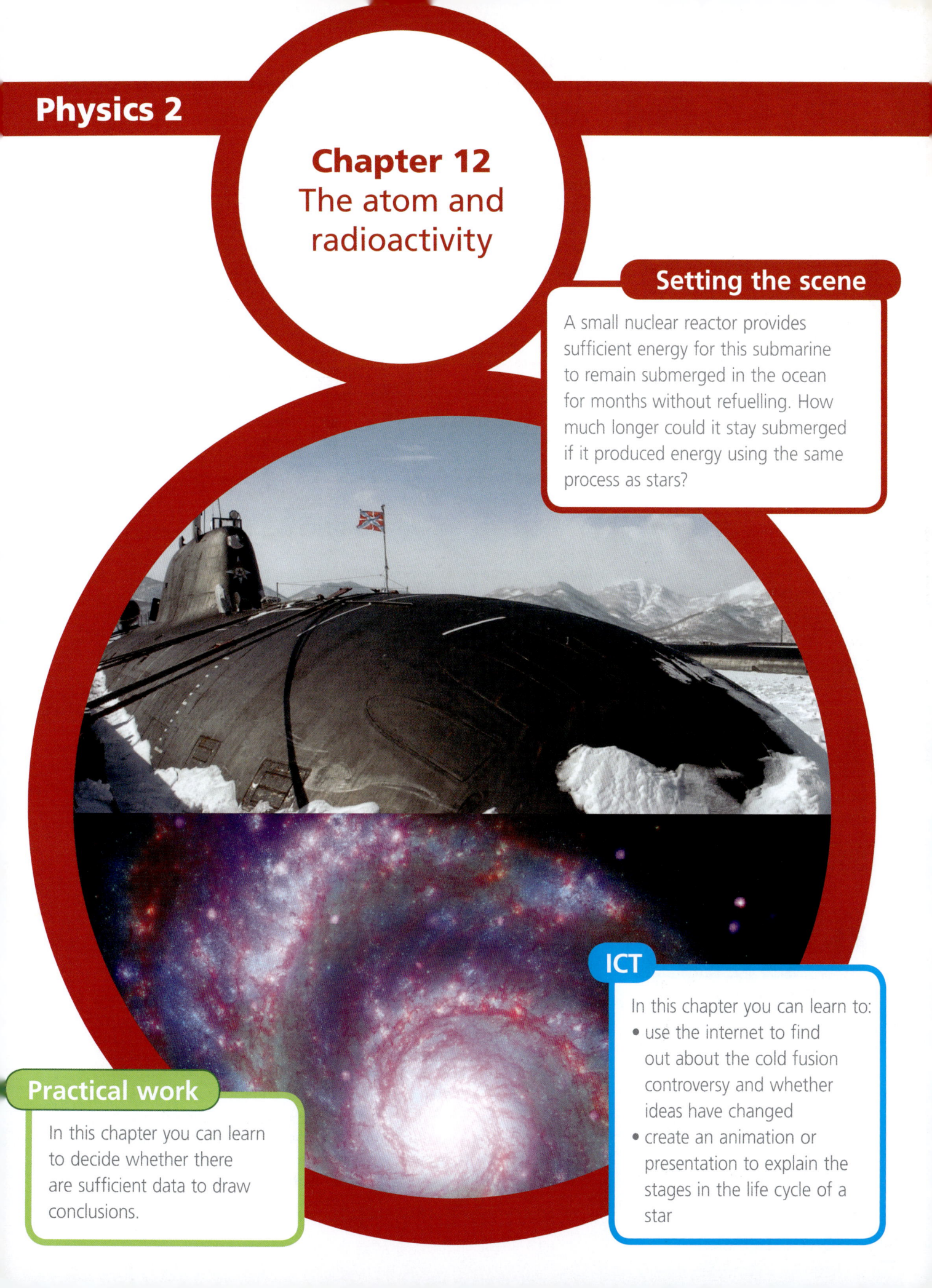

Physics 2

Chapter 12
The atom and radioactivity

Setting the scene

A small nuclear reactor provides sufficient energy for this submarine to remain submerged in the ocean for months without refuelling. How much longer could it stay submerged if it produced energy using the same process as stars?

Practical work

In this chapter you can learn to decide whether there are sufficient data to draw conclusions.

ICT

In this chapter you can learn to:
- use the internet to find out about the cold fusion controversy and whether ideas have changed
- create an animation or presentation to explain the stages in the life cycle of a star

Chapter 12 The atom and radioactivity

P2 12.1 Atomic structure

Learning outcomes

- Know that an atom consists of protons, neutrons and electrons.
- Know that all atoms of the same element have the same number of protons.
- Know that the number of protons in an atom is called the atomic number.
- Know that the number of protons + the number of neutrons = the mass number.

The **nucleus**, at the centre of an atom, contains positively charged **protons** and neutral **neutrons**. Negatively charged **electrons** orbit the nucleus.

Protons, **neutrons** and **electrons** are the building blocks for all atoms. Hydrogen atoms are the simplest, with just one proton and one electron (Figure 12.1a). The next simplest atoms are those of helium with two protons, two electrons and two neutrons. Next comes lithium with three protons, three electrons and four neutrons (Figure 12.1b).

(a) hydrogen atom (b) lithium atom

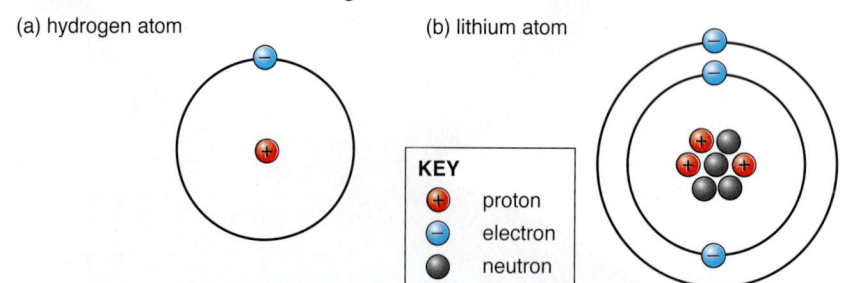

KEY
⊕ proton
⊖ electron
● neutron

Figure 12.1 Protons, neutrons and electrons in a hydrogen atom and a lithium atom. Draw a labelled diagram like these to represent a helium atom.

The relative masses and relative charges of a proton, a neutron and an electron are given in Table 12.1.

Particle	Relative mass	Relative charge
Proton	1	+1
Neutron	1	0
Electron	$\frac{1}{2000}$	−1

Table 12.1 The mass of an electron is negligible compared with the mass of a proton or a neutron.

At the centre of an atom, in the **nucleus**, are the protons and neutrons. The electrons whiz rapidly around the nucleus.

In any atom there are always equal numbers of protons and electrons. Effectively, the positive charges on the protons cancel the negative charges on the electrons and the atom has no overall charge.

Atomic number and mass number

Atoms of different elements have different numbers of protons. Hydrogen atoms are the only atoms with one proton. Helium atoms are the only atoms with two protons, and so on. This means that the number of protons in an atom tells you which element it is. Scientists call this number the **atomic number**. So, hydrogen has an atomic number 1 and helium has an atomic number 2.

The mass of an atom depends on the total number of protons and neutrons in its nucleus. Scientists call this number the **mass number** of an atom.

Figure 12.2 If the nucleus of an atom was enlarged to the size of a pea and put on the top of Nelson's Column, the electrons would be on the pavement below.

The **atomic number** is the number of protons in an atom. All the atoms of an element have the same atomic number.
The **mass number** is the number of protons plus the number of neutrons in an atom.

Atoms of an element that have different numbers of neutrons, and so different mass numbers, are called **isotopes**.

All the atoms of a particular element have the same number of protons, but they don't all have the same number of neutrons. Atoms of the same element with different numbers of neutrons are called **isotopes**.

AQA GCSE Physics

P2 12.2 Models of the atom

mass number = 63
atomic number = 29
Cu

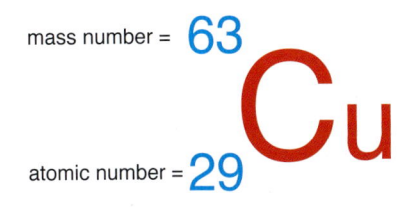

Figure 12.3 The symbol $^{63}_{29}$Cu shows the mass number and atomic number for a copper atom. How many neutrons are there in an atom of copper-63?

Isotopes have the same	Isotopes have different
• number of protons • number of electrons • atomic number	• number of neutrons • mass numbers

Table 12.2 Comparing isotopes of the same element.

From atoms to ions

An atom that loses or gains one or more electrons forms a charged particle called an **ion**. An atom that loses electrons forms a positively charged ion. An atom that gains electrons forms a negatively charged ion.

> **Test yourself**
>
> 1 Copy and complete the following sentences.
>
>   Uranium has two isotopes. Each isotope has 92 protons and _____ electrons and therefore an _____ number of 92. But one of these isotopes has 143 neutrons and the other has 146 neutrons. Their mass numbers are therefore _____ and _____.
>
> 2 Copy and complete Table 12.3 for the isotopes of carbon.
>
> | | Carbon-12, $^{12}_{6}$C | Carbon-14, $^{14}_{6}$C |
> |---|---|---|
> | Number of protons | | |
> | Number of electrons | | |
> | Number of neutrons | | |
> | Mass number | | |
> | Atomic number | | |
>
> *Table 12.3*

P2 12.2 Models of the atom

Learning outcomes

- Understand that the plum-pudding model represents the atom as a positively charged sphere with electrons embedded in it.
- Understand that the nuclear model represents the atom as a small positively charged nucleus surrounded by electrons.
- Be able to explain how new evidence can cause a model to change

For most of the nineteenth century, scientists thought that atoms were hard, solid particles, like tiny marbles. Then, in 1897, J. J. Thomson discovered that atoms contained negative particles that were about 2000 times lighter than hydrogen atoms. He called these particles electrons. Thomson knew that atoms had no overall electrical charge. So the rest of the atom must have a positive charge balancing the negative charge of the electrons.

In 1904, Thomson suggested a model for the structure of atoms. This became known as the **'plum-pudding' model**. Thomson said that atoms were tiny balls of positive material with electrons embedded in them, like currants scattered throughout a cake.

> In Thomson's **'plum-pudding' model**, each atom was thought to be a positively charged sphere with electrons embedded in it.

Physics 2

Chapter 12 The atom and radioactivity

Discovery of the nucleus

In 1909, three scientists called Rutherford, Marsden and Geiger investigated the 'plum-pudding' model. The experiment designed by Rutherford and carried out by Marsden and Geiger involved bombarding a thin gold foil with a beam of alpha particles (Section 12.3). Although the scientists did not know exactly what alpha particles were, they did know that the particles were small and had a positive charge.

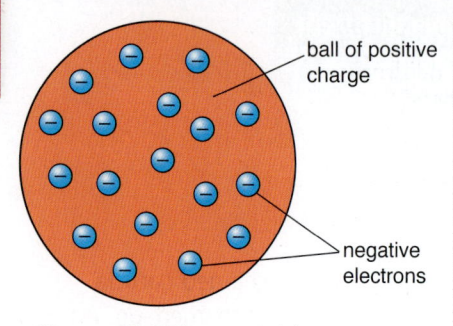

Figure 12.4 Thomson's 'plum-pudding' model for the structure of atoms.

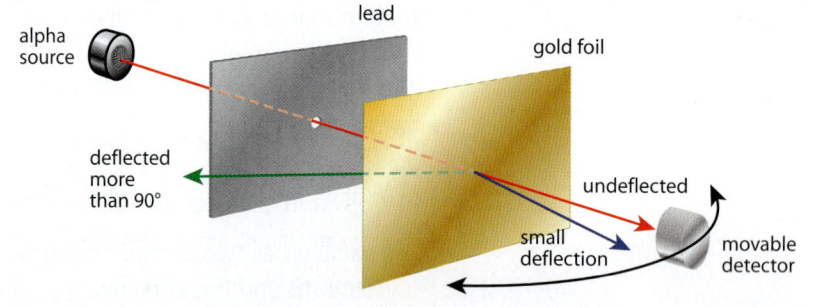

Figure 12.5 How Marsden and Geiger set up the alpha-scattering experiment.

Assuming the 'plum-pudding' model to be correct, the scientists expected that the alpha particles would pass straight, or almost straight, through the foil. However, the results collected over many months and involving over 100 000 measurements were not what was expected and could not be explained using the 'plum-pudding' model.

The results showed that:
- Most of the alpha particles went straight through the foil.
- Some of the alpha particles were scattered or deflected by the foil.
- A few alpha particles seemed to rebound from the foil.

Figure 12.6 The paths taken by some of the alpha particles were surprising. Although most alpha particles passed straight through the foil, about 1 in 10 000 were deflected by more than 90°.

Rutherford's nuclear model

By considering the results from the experiment, Rutherford deduced that at the centre of every atom is a nucleus. This nucleus must be:
- small compared with the atom itself, because most alpha particles had passed straight through undeflected
- positively charged, because some of the positive alpha particles had been repelled and deflected
- dense, because a small number of alpha particles were repelled back the way they came

Rutherford's model of the atom, in which most of the mass of the atom is concentrated in a positively charged nucleus surrounded by an empty space in which electrons orbit, was called the **nuclear model**.

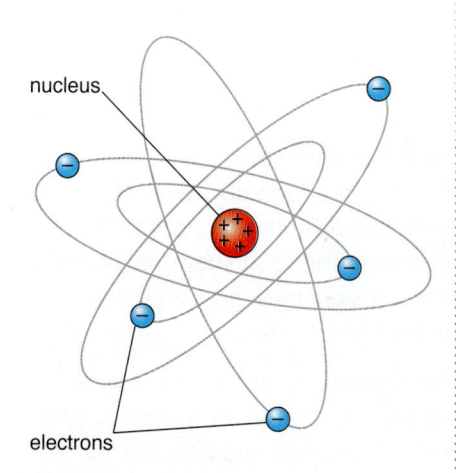

Figure 12.7 Rutherford pictured atoms to be like miniature solar systems with electrons orbiting the nucleus, like planets around the Sun.

AQA GCSE Physics

P2 12.3 Atoms and radiation

> In Rutherford's **nuclear model**, each atom was thought to have a small, positively charged nucleus surrounded by orbiting electrons.

Using his model, Rutherford was able to work out an equation to predict the proportion of alpha particles that would be deflected through various angles. The results obtained by Marsden and Geiger agreed with the predictions made by Rutherford using his equation.

The thorough way in which the experiment had been carried out, along with the way in which the predictions were supported by the experimental evidence, quickly convinced other scientists that Rutherford's idea of a nuclear atom was correct.

Test yourself

3 How is the mass of an atom distributed in the 'plum-pudding' model?

4 Where is most of the mass of an atom found in the nuclear model?

5 Why did most alpha particles pass through the gold foil undeflected?

6 Suggest why it was important that Marsden and Geiger took thousands of measurements.

P2 12.3 Atoms and radiation

Learning outcomes

- Know that radioactive substances emit radiation from the nuclei of their atoms.
- Know that the nuclei of the atoms of a radioactive substance are unstable.
- Understand that radioactive decay is random.
- Know that background radiation is the radiation around us all of the time.

Henri Becquerel, and his assistant Marie Curie, carried out the first investigations into radioactivity in 1896. They discovered that all uranium compounds emitted radiation. This radiation could pass through paper and could affect photographic film, just like light. Becquerel called the uranium compounds **radioactive** substances, and he described the process by which they emit radiation as **radioactivity**.

In later investigations, Marie Curie discovered that the rate at which radiation was emitted from a sample of uranium depended only on the amount of uranium. Nothing she did to the uranium — such as heating it — made any difference. 'This', she said, 'shows that radioactivity is an atomic property'.

Why are some substances radioactive?

Most atoms have a nucleus that does not change — a stable nucleus. But the nuclei of radioactive atoms are unstable. An unstable nucleus tries to become stable by emitting an **alpha particle**, a **beta particle** or a **gamma ray**. This is the process called **radioactive decay**. The decay of an unstable nucleus is a random process. This means that you cannot tell when an unstable nucleus will decay — it just happens.

Background radiation

Nuclear radiation is all around us. This is because there are naturally occurring radioactive elements in many materials. There is also nuclear radiation, called cosmic rays, reaching us from space. This means that our bodies are exposed to radiation all of the time. We call this **background radiation**.

> A **radioactive** atom is one that emits radiation from its nucleus.
> **Radioactive decay** is the random emission of radiation from the unstable nuclei of radioactive atoms as alpha particles, beta particles and gamma rays.

Physics 2

Chapter 12 The atom and radioactivity

Fortunately, for most people, the level of background radiation is low and the dose of radiation our bodies receive is not a health risk. However, certain people, because of their jobs or where they live, may be exposed to a higher dose of background radiation.

> **Background radiation** is the nuclear radiation that is present all around us. It comes mainly from natural sources.

- Various rocks in the Earth, particularly granite, contain small percentages of radioactive uranium. As the uranium decays it emits radioactive radon gas.
- Rocks and soils, as well as building materials, contain small amounts of radioactive elements that give out radiation as they decay.
- Our food contains traces of radioactive materials.
- Cosmic rays from the Sun and other stars are a major source of radiation in the air.

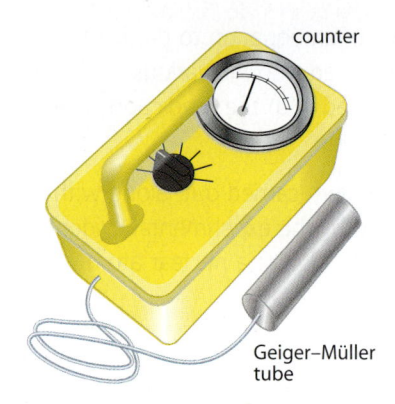

Figure 12.8 Radiation is detected using a Geiger–Müller tube. The counter gives a reading even when there are no obvious sources of radiation nearby. The Geiger–Müller tube is detecting background radiation. The higher the count rate, the higher the level of background radiation.

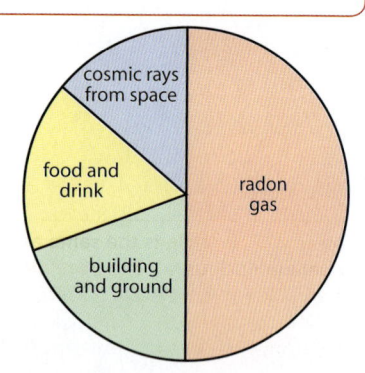

Figure 12.9 The proportions of background radiation from natural sources for people living in the UK.

Other sources including the fallout from nuclear weapons testing, nuclear power stations and nuclear accidents account for less than 1% of the total background radiation level.

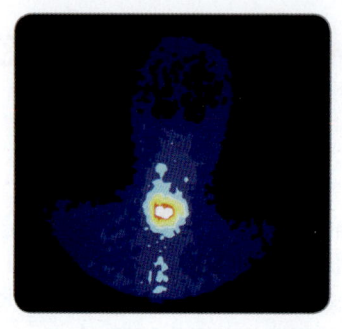

Figure 12.11 Your annual radiation dose would be increased if a radioactive source were used to help in diagnosing a medical problem. Here the radioactive isotope iodine-131 has been injected into a patient to investigate the thyroid gland.

Figure 12.10 Cosmic rays are absorbed by the atmosphere. The higher you go into the atmosphere, the thinner it gets. How does flying or living at high altitude change the radiation dose received from cosmic rays?

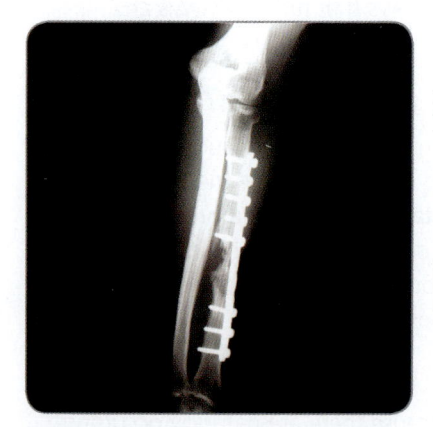

Figure 12.12 X-rays are not nuclear radiation, but they are an ionising radiation. Having an X-ray increases your annual radiation dose.

> **Test yourself**
>
> 7 On average, what percentage of background radiation comes from radon gas?
> 8 What is meant by radioactive decay being a 'random process'?
> 9 Where do cosmic rays come from?
> 10 Who is most likely to receive the highest annual radiation dose — a gardener, an office worker or an airline pilot? Give a reason for your answer.

AQA GCSE Physics

P2 12.4

Learning outcomes

- Know that alpha particles are the same as a helium nucleus: 2 protons and 2 neutrons.
- Know that beta particles are electrons.
- Know that gamma rays are electromagnetic radiation.

> An **alpha particle** is the same as a helium nucleus, a **beta particle** is an electron emitted from the nucleus, and **gamma rays** are electromagnetic radiation.

P2 12.4 Identifying alpha, beta and gamma radiation

Identifying alpha, beta and gamma radiation

The nuclei of radioactive atoms emit three main kinds of radiation — **alpha particles** (α particles), **beta particles** (β particles) and **gamma rays** (γ rays).

Range in air

Figure 12.13 (a) Measuring background radiation; (b) measuring the count rate from the source and background; (c) the count rate has dropped to the background value. The distance between the source and the Geiger–Müller tube gives the range in air.

Gamma radiation has a huge range in air, but the radiation gets weaker as it travels from the source and spreads out.

Penetrating power

Different types of radiation have different penetrating powers:
- Alpha particles cannot pass through a sheet of paper.
- Beta particles pass through a sheet of paper, but are absorbed by 5 mm thick aluminium.
- Gamma rays pass through a sheet of paper and 5 mm thick aluminium, but are absorbed by thick lead.

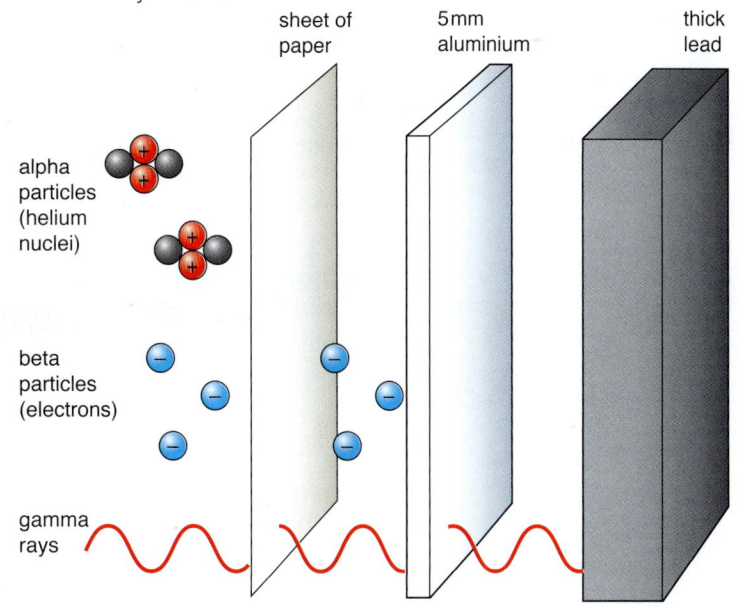

Figure 12.14 Gamma rays are much more penetrating than beta particles, which are much more penetrating than alpha particles.

Ionising power

Alpha, beta and gamma radiations can remove electrons from atoms. So these radiations (along with X-rays) are ionising radiations. Although alpha particles

Physics 2

Chapter 12 The atom and radioactivity

are the least penetrating, they have the strongest ionising power. On the other hand, gamma rays are the most penetrating but have the weakest ionising power.

Effect of magnetic fields and electric fields

A magnet creates a magnetic field. Positively charged or negatively charged particles that pass through a magnetic field are deflected.

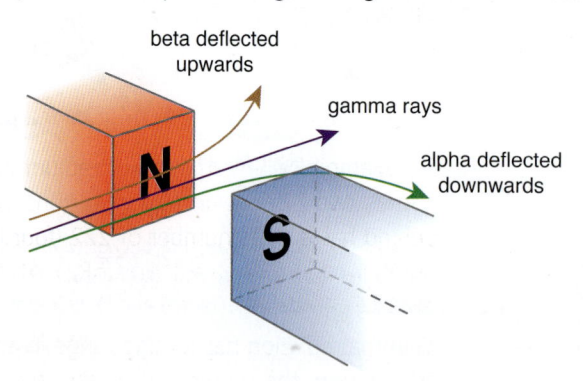

Figure 12.15 Alpha particles and beta particles are both deflected by the magnetic field, but in opposite directions. This tells us that both types of particles are charged, but one is positive and the other is negative. Gamma rays are not deflected — what does this tell us about gamma rays?

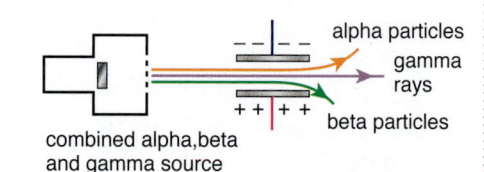

Figure 12.16 Alpha particles are deflected towards the negatively charged metal plate — so, alpha particles have a positive charge. (Remember: a negative charge attracts a positive charge.) Beta particles are deflected towards the positively charged metal plate — what does this tell us about beta particles? Again gamma rays are not deflected — why not?

An electric field is created in the gap between two electrically charged metal plates. Electric fields affect charged particles in a similar way to magnetic fields.

Beta particles are much easier to deflect than alpha particles. This is because beta particles have a lot less mass than alpha particles.

Radiation	Range in air	Stopped by	Ionising power	Effect of magnetic and electric fields	Nature of radiation
Alpha particles	About 5 cm	Paper or thin card	Strong	Small deflection	Two protons and two neutrons — the same as a helium nucleus
Beta particles	About 1 m	5 mm of aluminium	Moderate	Large deflection	An electron
Gamma rays	At least 1 km	About 10 cm of lead	Very weak	No effect	Electromagnetic radiation

Table 12.4 Properties and nature of the three main kinds of nuclear radiation.

> **Test yourself**
>
> 11 What type of nuclear radiation is uncharged?
>
> 12 Why are alpha particles deflected less by a magnetic field than beta particles?
>
> 13 How is an alpha particle similar to a helium atom?
>
> 14 Which types of radiation are stopped by a thin sheet of lead?

AQA GCSE Physics

P2 12.5 Radioactive decay and half-life

Learning outcomes

- Be able to explain why the mass number of a nucleus decreases by four and its atomic number decreases by two during alpha decay.
- Be able to explain why the mass number of a nucleus is unchanged and its atomic number increases by one during beta decay.
- Understand that a half-life is the time it takes for the number of nuclei in a sample to halve.

When atoms emit alpha particles or beta particles, they become different atoms with different numbers of protons and neutrons in the nucleus.

Alpha decay

Unstable isotopes with atomic numbers above 83 often decay (break up) by emitting an alpha particle — this is called alpha decay. These isotopes tend to decay because their nuclei are just too heavy. Their nuclei become more stable by losing mass in the form of an alpha particle.

For example, when an atom of radium-226 ($^{226}_{88}Ra$) emits an alpha particle, it loses two protons and two neutrons from its nucleus. The atom left behind has a mass number of 222 (four less than $^{226}_{88}Ra$) and an atomic number of 86 (two less than $^{226}_{88}Ra$). All atoms of atomic number 86 are those of radon (Rn).

The decay of radium-226 is summarised by this equation:

$$^{226}_{88}Ra \rightarrow {}^{222}_{86}Rn + {}^{4}_{2}\alpha$$

Boost your grade ✓

In all decay equations, the numbers to the right of the arrow (top and bottom) must add to equal the numbers to the left of the arrow (top and bottom).

Beta decay

During beta decay, a neutron in the nucleus splits up into a proton and an electron. The proton stays in the nucleus, but the electron is ejected as a beta particle. So the mass number of the remaining nucleus stays the same, but its atomic number increases by one. For example, the nucleus produced when carbon-14 ($^{14}_{6}C$) emits a beta particle ($^{0}_{-1}\beta$) has an atomic number one more than $^{14}_{6}C$. This is $^{14}_{7}N$.

The decay of carbon-14 is summarised by this equation:

$$^{14}_{6}C \rightarrow {}^{14}_{7}N + {}^{0}_{-1}\beta$$

Boost your grade ✓

Remember that beta decay is nothing to do with the electrons that orbit a nucleus.

	Alpha decay	Beta decay
Particles emitted	2 protons and 2 neutrons	1 electron
Change in mass number	−4	0
Change in atomic number	−2	+1

Table 12.5 In both alpha decay and beta decay, the atomic number changes (number of protons) so one element is changed into another element.

Gamma emission

When atoms emit gamma rays, there is no change in the number of protons or neutrons in the nucleus. So, gamma rays do not cause one element to change into another.

Physics 2

Chapter 12 The atom and radioactivity

Half-life

The decay of unstable radioactive nuclei is **random**. You cannot tell when an individual unstable nucleus will decay. But if there are large numbers of unstable nuclei, an average rate of decay will occur. So if a sample contains 10 million radioactive atoms, it should decay at twice the rate of a sample containing 5 million atoms. As the unstable nuclei decay, the rate of decay will fall.

The shape of the decay curve for a sample of iodine-131 (Figure 12.17) is similar in shape to those for other radioactive materials, but the scales on the axes can vary a great deal.

The decay curve shows that:
- the count rate falls from 1000 to 500 counts/minute in the first 8 days
- the count rate falls from 500 to 250 counts/minute in the next 8 days

The time taken for the count rate of iodine-131 to fall by half is always the same: 8 days. This is also the time taken for the number of iodine-131 nuclei in the sample to halve. A similar constant pattern, in halving the count rate and the number of radioactive nuclei, occurs with all radioactive substances.

The time it takes for the count rate or for the number of nuclei in a radioactive isotope to halve is called the **half-life**.

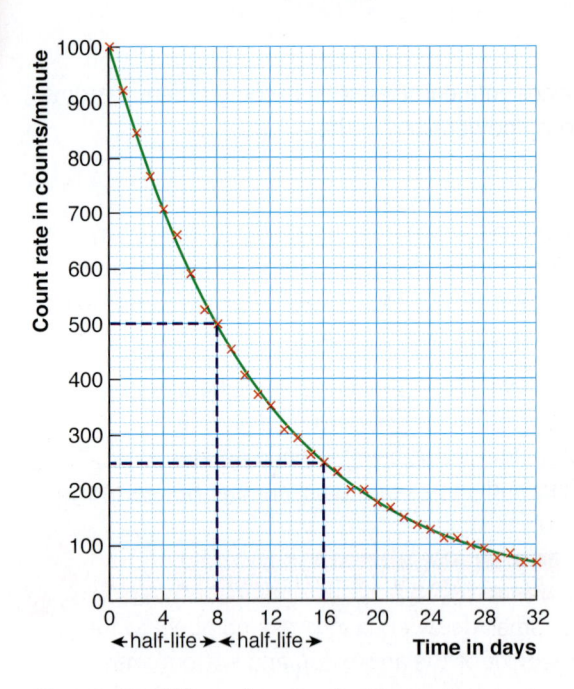

Figure 12.17 The radioactive decay curve for iodine-131. Not all of the points are on the line because the decay is random. How long would it take the count rate to fall from 600 to 300 counts/minute?

> The **half-life** of a radioactive isotope is the time taken for the number of nuclei in a sample or the count rate from that sample to halve.

Half-lives can vary from a few milliseconds to millions of years. The shorter the half-life, the faster the isotope decays and the more unstable it is. The longer the half-life, the slower the decay process and the more stable the isotope. Polonium-234 with a half-life of only 0.000 15 seconds is very unstable. Uranium-238 with a half-life of 4500 million years is 'almost' stable.

Boost your grade ✓

You need to be able to find the half-life from a decay curve. Make sure you look at the units on the axes — time could be in seconds, minutes, days or even years.

Test yourself

15 How can you tell from a decay equation that the atom produced by the decay is not the same as the atom that decayed?

16 Uranium-238 ($^{238}_{92}U$) decays by emitting an alpha particle to form thorium (Th).
   a What is the mass number of the thorium produced?
   b What is the atomic number of the thorium?
   c Write an equation to show the decay process.

17 A hospital gamma ray unit contains 10 grams of cobalt-60. This has a half-life of 5 years.
   a How much cobalt-60 will be left after 5 years?
   b How much cobalt-60 will be left after 15 years?

AQA GCSE Physics

P2 12.6 Using nuclear radiation

Learning outcomes

- Know that radioactive isotopes have important medical and non-medical uses.
- Be able to choose the most appropriate isotope for a particular use by considering the radiation emitted and the half-life.

Radioactive isotopes are used widely in medicine and in industry. Their uses depend mainly on their penetrating power and half-life.

Medical uses

Radiation can damage human cells, so it can be used to treat cancers. Cancer cells are killed more easily by radiation than healthy cells are.

Figure 12.18 shows how radiotherapy is used to treat a cancerous growth. The process requires penetrating gamma rays, which can pass through flesh and kill cancerous cells inside the body. It also requires an isotope with a fairly long half-life so that the dose of gamma radiation can be adjusted to the same level over a few weeks of treatment. Cobalt-60 ($^{60}_{27}$Co), which emits gamma rays and has a half-life of 9.3 years, is used widely.

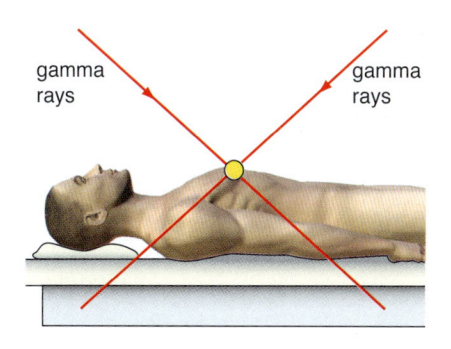

Figure 12.18 The gamma rays emitted from a cobalt-60 source are being used to treat lung cancer. Each session, the gamma rays hit the cancer from a different direction. This means that the cancer gets a higher dose than the surrounding lung tissue.

Skin cancer can be treated with less penetrating beta particles. This is done by strapping a plastic sheet containing phosphorus-32 ($^{32}_{15}$P) with a half-life of 14 days on the affected area.

Radioactive isotopes can be used to find out what happens to chemicals injected into our bodies (Figure 12.19). The isotopes, often called tracers, must emit gamma rays, which can penetrate flesh and bone to be detected outside the body. Ideally, they should also have a short half-life so that the radiation in the body soon falls to a safe level.

Most medical equipment, such as dressings and syringes, must be sterilised before being used. This is done by sealing the equipment in plastic bags and exposing it to gamma radiation from cobalt-60. The gamma radiation passes through the bag to kill any bacteria inside.

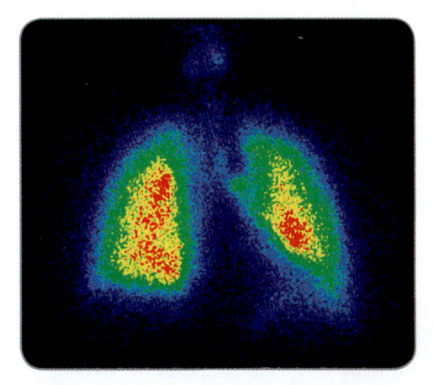

Figure 12.19 After injecting radioactive technetium-99 into a patient's blood, a doctor can study the blood flow through the heart and lungs. Technetium-99 ($^{99}_{43}$Tc) has a half-life of 6 hours.

Non-medical uses

Figure 12.20 Certain types of fresh food may be treated with gamma radiation. The radiation kills the bacteria that make the food go rotten and so gives the food a longer 'shelf-life'. The strawberries on the left were exposed to gamma radiation after they had been picked.

Physics 2

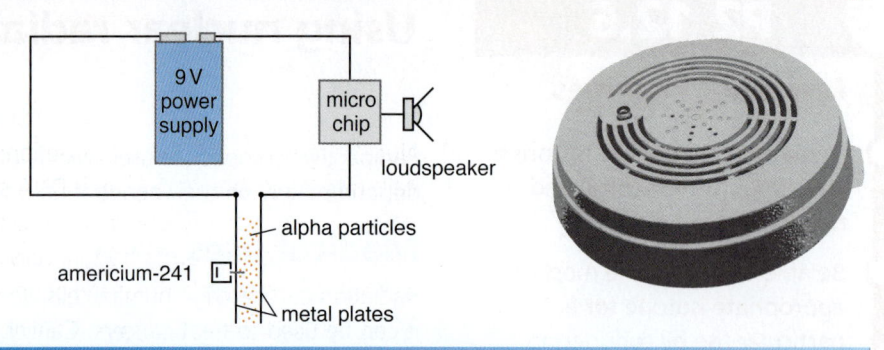

Chapter 12 The atom and radioactivity

Figure 12.21 Smoke alarms contain a small source of americium-241, which is an alpha emitter with a long half-life. The alpha particles ionise the air and create a small electric current between two metal plates. This current is detected by a microchip connected into the alarm circuit. Smoke getting into the alarm absorbs the alpha particles causing the current to fall — this triggers the alarm.

Choosing radioactive isotopes for different industrial uses

ACTIVITY

Radioactive isotopes have a large number of industrial uses. These include their use in detecting leaks (Figure 12.22) and in thickness gauges (Figure 12.23). The choice of isotope is based on the type of radiation needed and a suitable half-life.

1 Look at Figure 12.22.
  a What type of radiation (alpha, beta or gamma) does this use require?
  b Should the radiation used have a long or a short half-life? (Do you want the radiation to remain in the pipe for a long or a short time?)
  c Why is the radioactive source described as a tracer when it is used in this way?

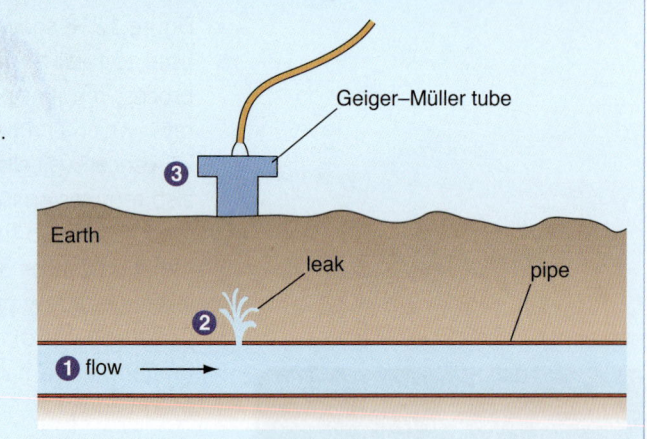

Figure 12.22 Using a radioactive source to detect a leak in an underground pipe. A small amount of radioactive isotope is fed into the pipe; the radioactive isotope leaks into the soil and a Geiger–Müller tube detects the radiation — and so the position of the leak.

① Small amount of radioactive isotope is fed into pipe.
② Radioactive isotope leaks into soil.
③ Geiger–Müller tube detects radiation and position of leak.

2 Look at Figure 12.23.
  a What type of radiation (alpha, beta or gamma) does this use require?
  b Should the radiation used have a long or a short half-life? (Do you want the level of radiation to remain at a constant level for a long time or to disappear fairly quickly?)
  c What happens to the reading on the counter if the thickness of the aluminium increases? How would this affect the rollers?
  d What type of source should be used to monitor the production of a 1 cm thick sheet of steel?

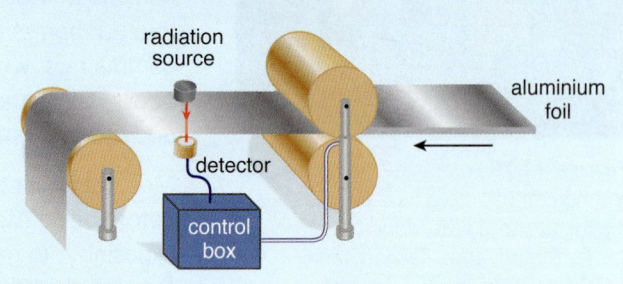

Figure 12.23 A source emits radiation towards the aluminium foil. A Geiger–Müller tube detects the radiation penetrating the foil. If the foil is too thick, less radiation passes through. A signal from the control box causes the pressure on the rollers to be increased, so the thickness of the foil decreases.

Test yourself

18 Why are isotopes that emit alpha radiation not used as medical tracers?
19 Why is gamma radiation suitable for killing the bacteria on food packaged in wooden crates?
20 Give a reason why the radiation from the alpha source inside the plastic case of a smoke detector does not pose any risk to people's health.
21 A small quantity of technetium-99 is injected into a patient. What fraction of the technetium-99 will have decayed after 24 hours?

258 AQA GCSE Physics

P2 12.7 Hazards of using nuclear radiation

Learning outcomes

- Understand that each type of nuclear radiation presents a possible hazard to health.
- Know that exposure to nuclear radiation should be kept to a minimum.

Nuclear radiation and X-rays cause ionisation. If ionisation occurs in our bodies, cells may be damaged and their DNA structure altered. This may stop the cells working properly, and in some cases they become a threat to the living organism, affecting other normal cells and causing cancers. Scientists know that a moderate to large dose of radiation increases the risk of developing a cancer. Most scientists think that even a small dose presents a low level of risk, but the evidence is not conclusive. High doses of radiation kill body cells.

Figure 12.24 Scientists and technicians who work with low-level radioactive materials wear special badges containing photographic film sensitive to radiation. The film is developed at regular intervals, and then replaced, to monitor the total radiation dose the worker has been exposed to.

The effects of radiation depend on its penetration, its ionising power and its half-life, as well as the length of exposure.

In general, outside the body, gamma rays with their greater penetrating power are more harmful than alpha and beta particles. So, people who work with dangerous isotopes emitting gamma rays must take extra safety precautions. These include:

- shields of lead, concrete or thick glass to absorb the radiation
- lead aprons, worn by radiographers who work continually with X-rays and gamma rays
- remote-control handling of dangerous isotopes from a safe distance
- reducing exposure to radiation to the shortest possible time

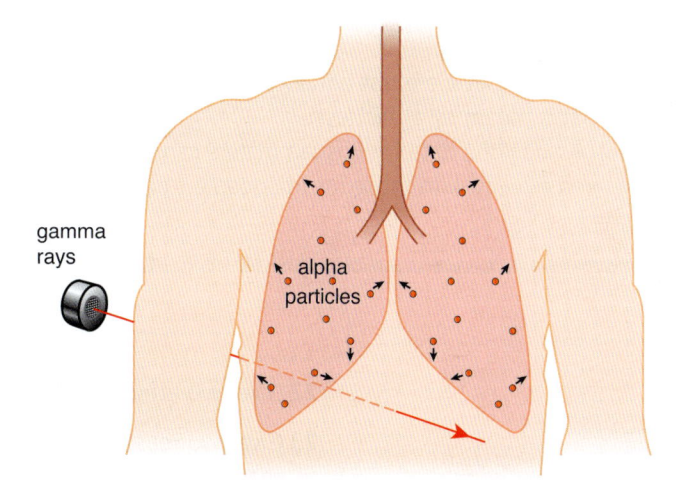

Figure 12.25 Outside the body, the most dangerous sources are those emitting gamma radiation. Having any source inside your body would be dangerous, but the most dangerous are those emitting alpha radiation.

Physics 2

Chapter 12 The atom and radioactivity

What are the long-term effects of radiation?

HOW SCIENCE WORKS ACTIVITY

The data in Table 12.6 show the effects of radiation on uranium miners in Russia and on a group of people from Hiroshima after the atomic bomb in 1945.

Source of radiation	Type of radiation	Number of people studied	Extra deaths due to cancer caused by the radiation
Radon gas from decay of uranium	Alpha particles	3400 uranium miners in Russia	60
Radioactive materials from the Hiroshima bomb	Gamma rays	15 000 inhabitants of Hiroshima	100

Table 12.6

1 There were 100 extra deaths due to cancer in the 15 000 people studied that had been living in Hiroshima after the atomic bomb had been dropped. This works out at one extra death per 150 people.
  a How many people relate to one extra death among the uranium miners?
  b Which group of people, those living in Hiroshima or the uranium miners in Russia, suffered the highest proportion of deaths?
  c Do the data in Table 12.6 suggest that alpha radiation is more or less dangerous than gamma radiation?

2 a What was the source of radiation that affected the uranium miners?
  b What form of cancer do you think the miners suffered from? Explain your answer.

3 It would be wrong to jump to conclusions about the data in Table 12.6 because they were not obtained by a fair test.
  a Explain what is meant by a 'fair test'.
  b State reasons why the data for Russian miners cannot be compared directly with those for the people from Hiroshima.

Figure 12.26 In August 1945, atomic bombs were dropped on the Japanese cities of Hiroshima and Nagasaki. Thousands of people died instantly or within hours of the bombs being dropped. It is estimated that 15–20% of these deaths were due to the effects of radiation. The survivors and subsequent generations have given scientists a unique opportunity to study the long-term effects of radiation on the human body.

Test yourself

22 Give two effects of ionising radiation on body cells.

23 Give two reasons why, inside the body, alpha radiation is more dangerous than gamma radiation.

24 Sometimes when taking an X-ray, a radiographer needs to be close to the patient. Why does the radiographer then wear a lead-lined apron?

P2 12.8 Nuclear fission and nuclear fusion

Learning outcomes

- Know that during nuclear fission the nucleus of an atom splits into smaller parts.
- Know that during nuclear fusion, two nuclei join to form a larger nucleus.
- Know that nuclear fusion is the process by which energy is released in stars.

Nuclear reactions involve changes to the nuclei of atoms. During nuclear reactions, one element is converted to another element by:
- radioactive decay
- nuclear fission or
- nuclear fusion

Nuclear fission

In 1938, Otto Hahn discovered that enormous amounts of energy could be obtained from **nuclear fission**. His discovery happened almost by chance — Hahn and his colleagues were trying to make a new element by bombarding uranium with neutrons. Instead of producing just one new element with a slightly different atomic number, the neutrons caused the uranium nuclei to break up violently, forming two smaller nuclei. More neutrons were also released plus energy (Figure 12.27).

Nuclear fission is the splitting of an atomic nucleus, resulting in the release of energy.

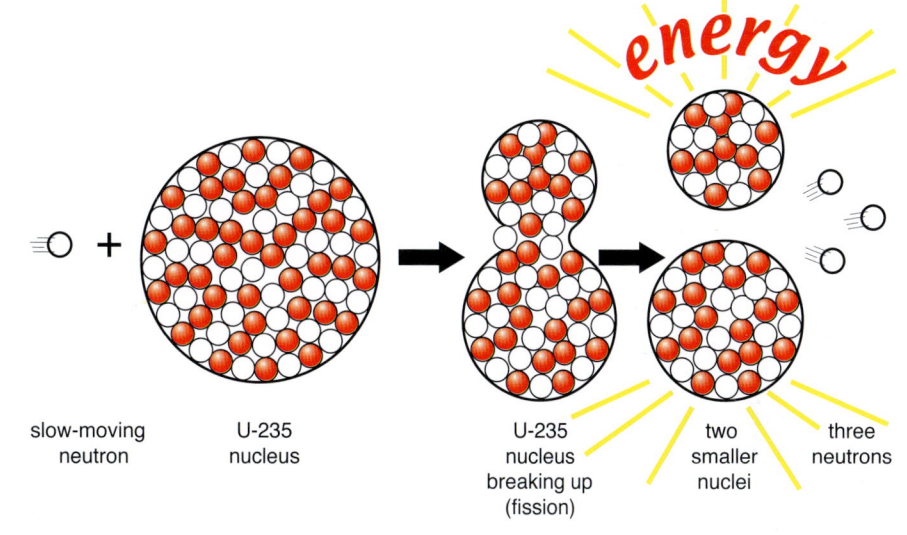

Figure 12.27 Absorbing a slow-moving neutron causes the fission of a uranium-235 nucleus.

Natural uranium contains two isotopes, $^{235}_{92}U$ and $^{238}_{92}U$. Only 0.7% is uranium-235. Experiments show that only uranium-235 takes part in nuclear fission. This has led scientists to realise that if natural uranium is enriched with more uranium-235, the neutrons released during fission can go on to split more uranium nuclei and start a **chain reaction**.

In an atomic bomb, the fission of enriched uranium-235 atoms releases three neutrons. These neutrons collide with three more uranium-235 atoms, releasing nine neutrons. These nine then release 27, and so on. Each time, more and more energy is produced in an uncontrolled reaction resulting in vast amounts of energy.

In a nuclear reactor, a lower concentration of uranium-235 is used to produce a controlled chain reaction. In this case, only one of the neutrons released at each fission goes on to cause further fission and a steady chain reaction occurs. The energy released is used to generate electricity.

Figure 12.28 An uncontrolled chain reaction occurs when all the neutrons released by splitting one nucleus go on to split more nuclei.

Physics 2

Chapter 12 The atom and radioactivity

Nuclear podcasting
HOW SCIENCE WORKS / ICT

Imagine that you work as a safety officer for a nuclear power station and that there has been a leak of radioactive material into a local river. Create a podcast to be broadcast on radio explaining the possible dangers.

Boost your grade ✓
Make sure you spell 'fission' with two s's and 'fusion' with one.

Reactors have also been developed that use plutonium-239 as the fuel. Plutonium-239 can use fast neutrons, unlike uranium-235, which undergoes fission more readily with slow neutrons.

Nuclear fusion

The energy that is released by the Sun and other stars comes from **nuclear fusion**. During nuclear fusion, two nuclei are joined together to form one larger nucleus and energy is released.

> **Nuclear fusion** is the joining together of two nuclei, resulting in the release of energy.

Temperatures reach about 15 000 000°C at the centre of the Sun. At these temperatures, the core forms a 'plasma' of atoms stripped of their electrons. The nuclei collide at high speed with each other and with individual protons and neutrons. Some of the collisions result in fusion. One of the major fusion processes in stars involves hydrogen nuclei (protons) fusing to form helium (Figure 12.29).

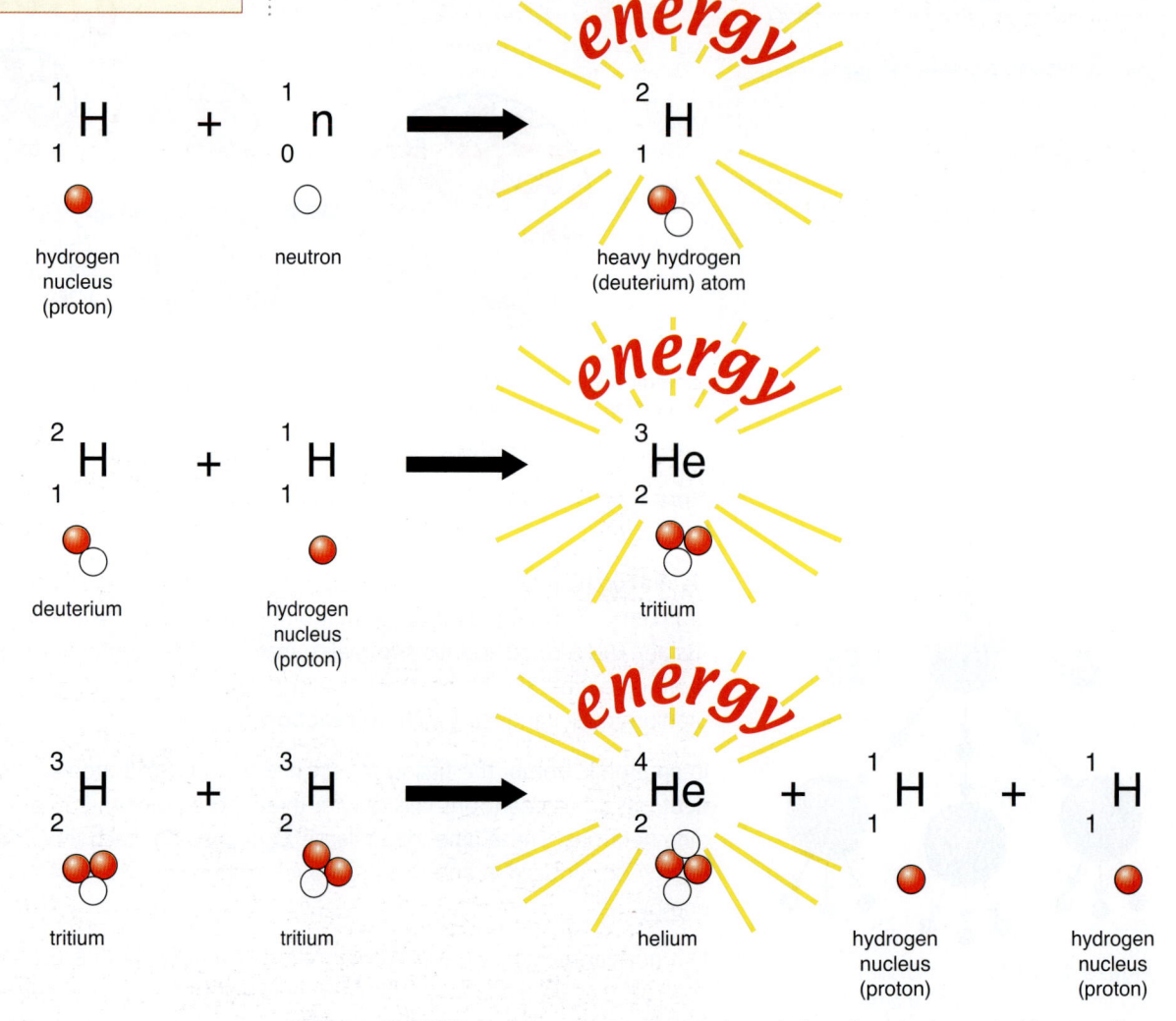

$$^{1}_{1}H + ^{1}_{0}n \rightarrow ^{2}_{1}H + \text{energy}$$

hydrogen nucleus (proton) + neutron → heavy hydrogen (deuterium) atom

$$^{2}_{1}H + ^{1}_{1}H \rightarrow ^{3}_{2}He + \text{energy}$$

deuterium + hydrogen nucleus (proton) → tritium

$$^{3}_{2}H + ^{3}_{2}H \rightarrow ^{4}_{2}He + ^{1}_{1}H + ^{1}_{1}H + \text{energy}$$

tritium + tritium → helium + hydrogen nucleus (proton) + hydrogen nucleus (proton)

Figure 12.29 The fusion processes that produces helium from hydrogen inside a star like the Sun — energy is released at each stage.

AQA GCSE Physics

P2 12.9 The life cycle of a star

Nuclear fusion reactors

The energy released in a fusion reaction is far greater than the energy released in a fission reaction. If this energy could be harnessed and used to generate electricity, it would meet the increasing global demand for electricity for the foreseeable future. But at the moment, fusion reactors are only in the experimental stage and it is likely to be many years before a commercially viable reactor is built.

Two advantages of fusion reactors are:
- There is a plentiful supply of fuel. It is easily derived from sea water and from the element lithium.
- It produces less radioactive waste than fission reactors and the waste is safer to handle.

> **Cold fusion?** HOW SCIENCE WORKS / ICT
>
> In 1989, two physicists claimed to have been able to produce nuclear fusion at room temperature. Use the internet to research why was this potentially such a big deal and what happened next.

Test yourself

25 Name the two fissionable materials commonly used in nuclear reactors.

26 Draw a labelled diagram to show how a controlled chain reaction occurs.

27 Why would slow-moving nuclei be unable to fuse? (Hint: what type of charge does a nucleus have?)

P2 12.9 The life cycle of a star

Learning outcomes

- Know that the life cycle of a star is determined by the size of the star.
- Know that elements are formed inside stars and in supernovas.

> A **nebula** is a huge cloud of dust and gas from which stars are formed.

Life cycle of a star

Like all stars, the Sun was formed from a **nebula** — a huge cloud of gas and dust. The gas was mainly hydrogen.

A huge cloud of dust and gas accumulates in space as a nebula.

⬇

Gravitational forces pull the particles together, causing the temperature to rise.

⬇

The temperature reaches a point at which nuclear fusion reactions start. Energy is produced and a star is formed.

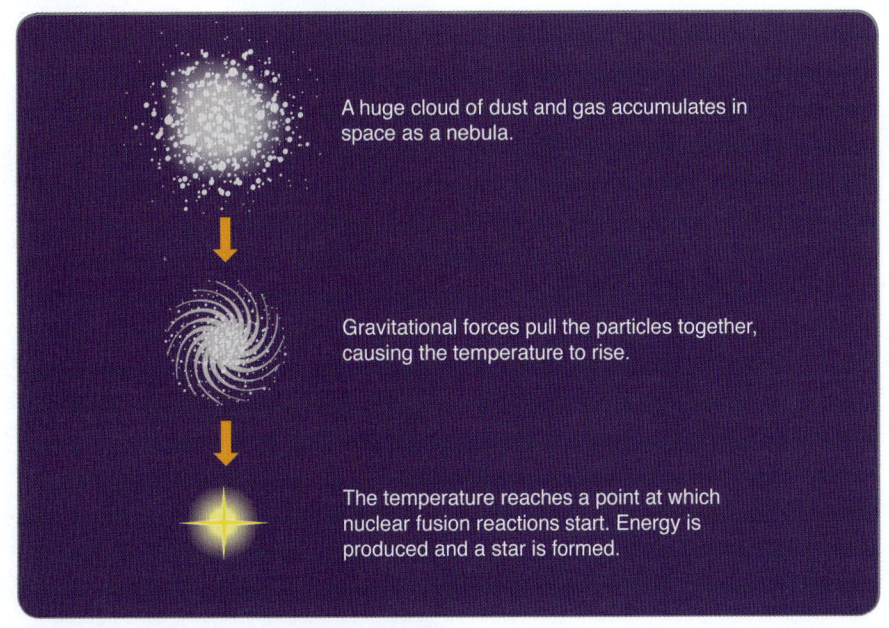

Figure 12.30 The formation of a star from a nebula.

Physics 2 263

Chapter 12   The atom and radioactivity

Over millions of years, the force of gravity pulls clumps of dust and gas particles together, squeezing them into a smaller and smaller volume. Before this mass collapses enough for nuclear fusion reactions to begin, it is called a **protostar**. As more particles are pulled into the mass, the temperature increases until eventually the centre reaches about 15 000 000°C, which is high enough for nuclear fusion reactions to start. The compressed gas ball starts to emit immense amounts of radiation and is now called a **main sequence star**.

Scientists think that at the same time as stars were formed, smaller masses also formed from the same nebula. These masses were too small for the temperature to rise enough for fusion reactions to start, so they did not emit their own radiation. Instead of becoming stars, they joined with other small masses to form planets.

Stable period of a star

During the **main period** of its life, a star remains in a stable state — the star does not shrink or expand. This period of stability may last for billions of years.

Nearing the end of life

As the star's supply of hydrogen runs out, the outward forces produced by the fusion reactions decrease and the gravitational forces cause the star's core to contract. This increases the temperature of the core, making it hotter than it was during the stable period. Eventually, the core temperature becomes high enough for further, more complex, nuclear fusion reactions to occur.

The new fusion reactions produce larger outward forces than before. At this point the inward and outward forces are no longer balanced and the star expands. It expands so much that the outer layers cool to about 3000°C. Because the surface is cooler, it looks red — so this type of star is called a red giant.

Death of a star

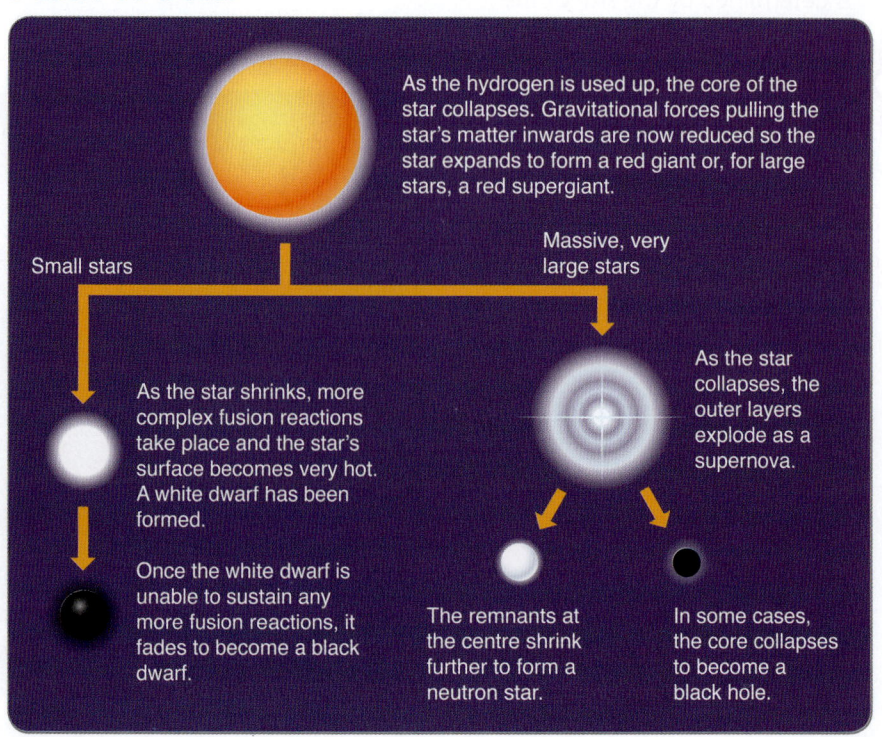

Figure 12.31 During a star's stable period, the forces within the star are balanced.

Figure 12.32 Stars die in different ways, depending on their size.

As the hydrogen is used up, the core of the star collapses. Gravitational forces pulling the star's matter inwards are now reduced so the star expands to form a red giant or, for large stars, a red supergiant.

Small stars: As the star shrinks, more complex fusion reactions take place and the star's surface becomes very hot. A white dwarf has been formed.

Once the white dwarf is unable to sustain any more fusion reactions, it fades to become a black dwarf.

Massive, very large stars: As the star collapses, the outer layers explode as a supernova.

The remnants at the centre shrink further to form a neutron star.

In some cases, the core collapses to become a black hole.

264   AQA GCSE Physics

P2 12.9 The life cycle of a star

What happens next depends on the size of the star. Small stars, like our Sun, contract and their surface becomes hot. This causes the star to glow white-hot, becoming a type of star called a white dwarf. Eventually, the light from the white dwarf fades and the star ends its life as a black dwarf.

Stars many times the size of our Sun first expand to form red supergiants. Gravitational forces cause red supergiants to collapse and huge amounts of energy are released, causing the outer layers to blow up in a spectacular explosion called a supernova.

Afterwards, the remnants at the centre of the original star contract even further to form an incredibly dense object called a neutron star. If the original star had enough mass (at least 10 times the mass of the Sun), the core contracts even further. The gravitational forces become so strong that nothing, not even light, can escape from its surface. It is then a black hole.

Nuclear fusion and the elements

The universe has not always been as it is now. Early on, the universe consisted mainly of hydrogen, but it now contains a variety of elements. These elements came from nuclear reactions within stars.

The fusion reactions in a star begin with hydrogen creating helium (Section 12.8). When the star runs out of hydrogen it collapses, the core temperature rises rapidly and more fusion reactions start to happen. First, helium nuclei are converted to carbon. The carbon nuclei then fuse with other nuclei to form new elements of even higher mass, up to and including iron.

Formation of the heaviest elements

Once we get to iron, the process changes. For iron to fuse with other elements, large amounts of energy must be absorbed, not emitted. This energy can only come from the huge amounts of energy released in a supernova. So, the creation of the heaviest elements occurs only in a supernova.

A supernova not only causes new, heavier elements to be formed but also scatters them throughout the universe. When gravity starts to pull the remnants of the explosion together into a nebula, the whole life cycle of a star begins again. However, the future star starts life with a richer supply of heavier elements than earlier generations of stars.

Nuclei of the heaviest elements are present in the Sun's outer layers, and atoms of these elements are present in the inner planets of our solar system. This is the evidence that scientists have for thinking that our Sun and the rest of our solar system were formed from the remains of a supernova — a case of recycling on an astronomical scale!

Life cycle of a star — ICT

Create an animation or multimedia presentation showing the life cycle of a star.

Test yourself

28 What is a 'nebula'?

29 Why does a protostar not emit radiation?

30 What process provides the energy needed for a star to produce its own light?

31 Why are the heaviest elements formed only in a supernova?

32 What evidence do scientists have for thinking that our solar system was formed from 'recycled' materials?

Chapter 12 Homework questions

Homework questions

1 Lithium atoms have three protons and four neutrons.
   a What is the atomic number of lithium?
   b What is the mass number of lithium?
   c How many electrons are there in a lithium atom?

2 Why did the work carried out by Marsden and Geiger convince many scientists that the 'plum-pudding' model of the atom needed to be replaced?

3 In Geiger and Marsden's experiment, about 1 in every 10 000 alpha particles appeared to bounce back from the gold foil.

Explain how you think the number of alpha particles bouncing back would change if:
   a thicker gold foil was used
   b aluminium foil of the same thickness was used

4 What numbers should replace the letters A–E in the following equation for the possible fission of uranium-235?

$$^1_A n + ^B_{92}U \rightarrow ^C_{56}Ba + ^{88}_D Kr + E^1_0 n$$

5 Write nuclear equations for the beta decay of:
   a $^{43}_{19}K$ to calcium
   b $^{24}_{11}Na$ to magnesium

6 How do atoms of one particular element always differ in atomic structure from those of all other elements?

7 Zak's teacher carried out an experiment to measure the half-life of protactinium-234. His results are shown in Table 12.7

Time in seconds	Count rate in counts/second
0	66
40	44
80	30
120	20
160	13
200	9
240	6
280	4
320	4

Table 12.7

   a In the experiment, what was:
      (i) the independent variable?
      (ii) the dependent variable?
   b What instrument was used to measure the count rate?
   c What is the background count during the experiment?
   d Rewrite the table, subtracting the background count from each count rate value.
   e Plot a graph of the new count rate (vertically) against time (horizontally).
   f How long does it take for the count rate to fall from:
      (i) 60 to 30?
      (ii) 40 to 20?
      (iii) 30 to 15?
   g Calculate an average value for the half-life of protactinium-234.

8 In 1938, German scientists discovered that bombarding uranium-235 with neutrons could produce enormous amounts of energy. This process split uranium atoms in two and was called atomic fission.

What were the consequences of the discovery of atomic fission, or 'splitting the atom'?

9 Plutonium-238 decays by emitting alpha particles.
   a What is an alpha particle?
   b Plutonium-238 would be highly dangerous if inside the body. Explain why.

10 Doctors may inject a radioactive tracer into a patient's body. Why is it important that the isotope used has a *short* half-life of a few hours and not a longer half-life of several days or weeks?

Chapter 12 Exam corner

Exam corner

For most of its lifespan, a star is stable.
a Explain, in terms of the forces acting, why a star is stable. *(2 marks)*

Student A
The forces are balanced. ✓

Examiner comment There is no detail about the directions of these forces. However, the idea of the forces being balanced is worth 1 mark.

Student B
Forces due to gravity act inwards. The high core temperature causes radiation forces which act outwards. ✓ While the star is stable these forces are balanced. ✓

Examiner comment This answer is better because it also gives the directions in which the forces act. Note that there are no marks given for any additional detail about the cause of the forces.

b What happens to end the stable period of a star's life cycle? *(1 mark)*

Student A
It runs out of fuel ✓

Examiner comment This is about the minimum acceptable answer.

Student B
Its supply of hydrogen runs out. ✓ The star then expands into a red giant or super red-giant. What happens next depends on the size of the star.

Examiner comment The first part of the answer is enough for this mark. The question is not asking what happens at the end of the stable period — only what causes it to end.

c Describe how a star like the Sun changes after it has passed through the stable period. *(3 marks)*

Student A
It expands to a red giant, then explodes as a supernova, finally becoming a black hole. ✗

Examiner comment Although the first part is correct, no mark is given because the answer goes on to describe what happens to a massive star and not a star like the Sun.

Student B
The Sun will expand becoming a red giant. ✓ It then shrinks ✓ and changes into a bright, small star called a white dwarf. ✓ After it stops producing energy it is a black dwarf.

Examiner comment Had the question been worth 4 marks a further mark could have been given for 'black dwarf'.

Physics 2 267

Physics 2

Exam questions

Chapter 9

1 Bill and Anthea took part in a sponsored run. The distance–time graph for Bill's run is shown in Figure 9A. Anthea did not start the run until 500 seconds after Bill. She completed the whole run at a constant speed in 3000 seconds.

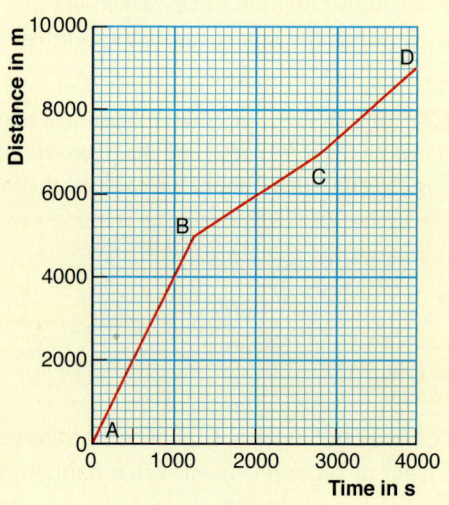

Figure 9A

a During which part of the race was Bill running the fastest? *(1 mark)*
b Copy the distance–time graph for Bill's run. On the same axes draw a distance–time graph for Anthea's run. *(2 marks)*
c How far had Bill run when Anthea overtook him? *(1 mark)*

2 a Figure 9B shows the forces acting on a ball-bearing falling through a tube of glycerine.

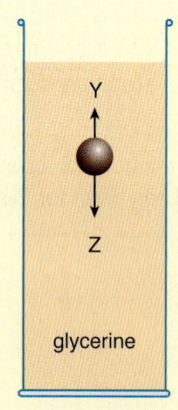

Figure 9B

(i) What causes the force labelled Y? *(1 mark)*
(ii) What causes the force labelled Z? *(1 mark)*

b Figure 9C shows how the velocity of the ball-bearing changes as it falls through the glycerine.

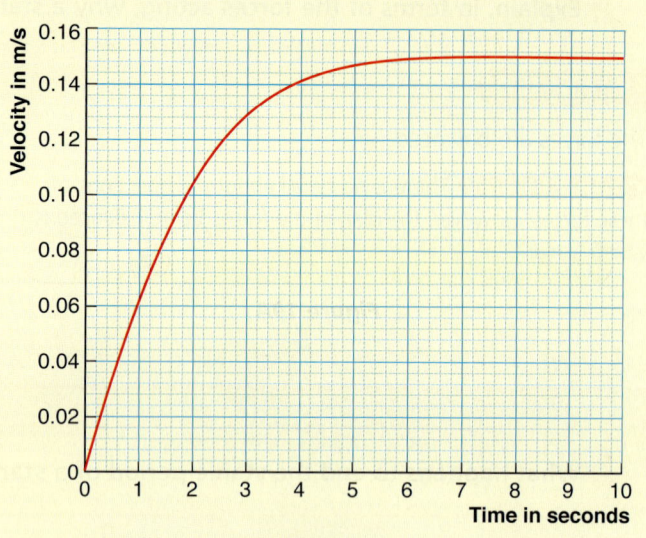

Figure 9C

(i) After 7 seconds the ball-bearing falls at a constant speed.
What name is given to this constant speed? *(1 mark)*
(ii) Explain in terms of the forces, Y and Z, why the ball-bearing accelerates at first but then reaches a constant speed. *(3 marks)*

c Calculate how far the ball-bearing travels in the final 3 seconds of its fall. *(2 marks)*

3 a A cyclist is travelling along a straight road. The resultant force on the cyclist is zero.
(i) What does the term 'resultant force' mean? *(1 mark)*
(ii) Describe the motion of the cyclist when the resultant force is zero. *(1 mark)*

b The cyclist now pedals harder and initially accelerates at 1.5 m/s^2. The mass of the cyclist and bicycle is 90 kg.
Calculate the resultant force needed to produce the initial acceleration.
Write down the equation you need to use, then show how you work out your answer and give the unit. *(3 marks)*

c Acceleration does not always produce a change in speed. What else can acceleration produce? *(1 mark)*

268 AQA GCSE Physics

Physics 2

Chapter 10

1 Figure 10A shows three racing cars.

Figure 10A

Car A has a mass of 500 kg and is moving at 85 m/s along a straight part of the track.

Car B is moving at constant speed around a curved part of the track.

Car C is in the pits and is not moving.

a Which, if any, of the cars has zero momentum?
(1 mark)

b Calculate the momentum of car A.
Write down the equation you need to use, then show how you work out your answer and give the unit. *(3 marks)*

c The momentum of one of the cars is changing. Which one? Give a reason for your answer.
(2 marks)

2 A bridge is used for bungee jumping. The bungee jumpers fall 125 m before the bungee rope starts to stretch.

a A person of mass 60 kg performs the bungee jump.
  (i) Calculate the gravitational potential energy lost by the person in falling 125 m.
  (Gravitational field strength = 10 N/kg)
  Write down the equation you need to use, then show how you work out your answer.
  (2 marks)

  (ii) Calculate the velocity of the person at the bottom of the 125 m fall. Ignore the effect of air resistance.
  Write down the equation you need to use, then show how you work out your answer.
  (2 marks)

b Explain why it is important that the bungee rope is able to stretch. *(3 marks)*

3 A student measures his power by lifting as many 1 kg masses as he can from the floor onto a table in one minute. The top of the table is 0.85 m up from the floor.

(a) (i) Calculate, in joules, the gravitational potential energy gained by a 1 kg mass when it is lifted from the floor to the table.
(Gravitational field strength = 10 N/kg)
Write down the equation you need to use, then show how you work out your answer.
(2 marks)

  (ii) How much work does the student do each time he lifts a mass from the floor to the table? *(1 mark)*

b The student calculated his power as 3.4 W. Calculate the number of 1 kg masses that he lifted in the one minute.
Write down the equation you need to use, then show how you work out your answer. *(2 marks)*

c The student then measured his power by running up a flight of stairs. This time he calculated his power as 740 W.
Suggest why the two values for power are different. *(2 marks)*

4 Figure 10B shows a pellet having just been fired from an air rifle.

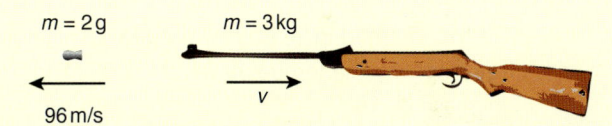

$m = 2\,g$ $m = 3\,kg$
96 m/s v

Figure 10B

a Why is firing a pellet from an air rifle an explosion and not a collision? *(1 mark)*

b In an explosion, the total momentum of the exploding parts is conserved.
What is meant by the term 'momentum is conserved'? *(1 mark)*

c Calculate, in m/s, the recoil velocity of the air rifle.
Write down the equation you need to use, then show how you work out your answer. *(3 marks)*

Physics 2

Chapter 11

1. Figure 11A shows a student using a 12 V heater to warm a block of metal. He is using a joulemeter to measure the energy transferred to the heater.

   Figure 11A

   Before starting the experiment, the student reset the joulemeter to zero. After switching on and leaving for 5 minutes the reading was 7200.
   a (i) Calculate the energy transferred each second to the heater. *(1 mark)*
   (ii) State the power of the heater. *(1 mark)*
   b The student correctly calculated that the temperature of the metal block should go up by 18°C. But, when he measured the temperature he found it had increased by only 15°C. Give two reasons why the temperature of the block did not increase by 18°C. *(2 marks)*
   c Calculate the charge that flowed through the heater in 5 minutes.
   Write down the equation you need to use, then show how you work out your answer and give the unit. *(3 marks)*

2. The ammeters in Figure 11B are identical.

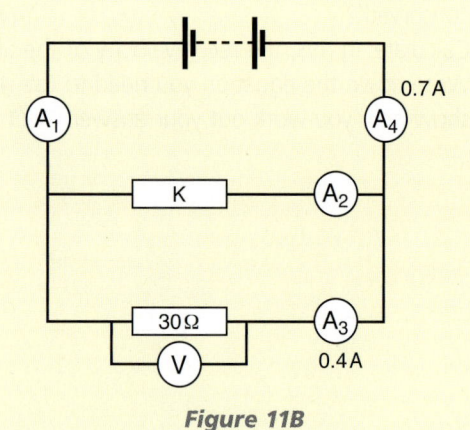

   Figure 11B

   a (i) What are the readings on the ammeters A_1 and A_2? *(2 marks)*
   (ii) Is the resistance of 'K', larger, smaller or equal to 30 Ω? Give a reason for your choice of answer. *(2 marks)*
   b (i) Calculate the reading on the voltmeter. *(2 marks)*
   (ii) State the potential difference of the power supply. Give a reason for your choice of value. *(2 marks)*

3. Figure 11C shows the circuit diagram for a 'flame effect' electric fire.

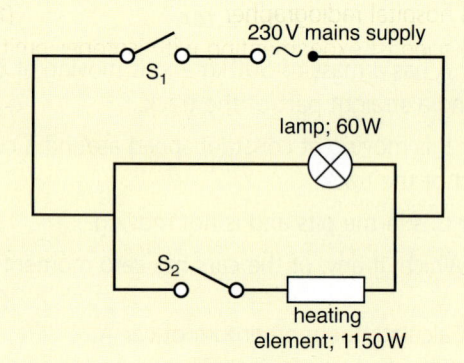

   Figure 11C

   a Explain why you cannot switch only the heating element on. *(1 mark)*
   b With both switches closed, calculate:
   (i) the current taken by the heating element *(2 marks)*
   (ii) the current taken by the lamp (give your answer to the nearest whole number) *(2 marks)*
   (iii) the total current taken from the supply *(1 mark)*
   (iv) the resistance of the heating element *(2 marks)*
   c What size fuse should be fitted to the fire, 3 A or 13 A? Give a reason for your choice. *(1 mark)*

AQA GCSE Physics

Physics 2

Chapter 12

1 Several people who work in hospitals and in industry are exposed to different types of radiation. Three ways to check or reduce their exposure to radiation are:
  A wear a badge containing photographic film
  B wear a lead apron
  C work behind a thick glass screen with remote-handling equipment

Which method (A, B or C) should be used by the following people?
  a a hospital radiographer *(1 mark)*
  b a scientist experimenting with isotopes emitting gamma rays *(1 mark)*
  c a nuclear power station worker *(1 mark)*

2 Some patients who suffer from cancer are given an injection of boron, which is absorbed by cancer cells. The cancerous tissue is then irradiated with neutrons. After this, the reaction that occurs is:

$$^{11}_{5}B \rightarrow ^{7}_{x}Li + ^{y}_{z}He$$

  a Copy the equation and write numbers in place of *x*, *y* and *z*. *(3 marks)*
  b Why do the lithium and helium nuclei repel each other? *(1 mark)*
  c Why can this treatment kill cancer cells? *(1 mark)*

3 Uranium has two natural isotopes, uranium-235 and uranium-238.
  a How is the structure of a uranium-235 nucleus different from the structure of a uranium-238 nucleus? *(1 mark)*
  b Explain how the fission of uranium-235 can lead to an uncontrolled chain reaction. You may want to draw a diagram to help your answer. *(3 marks)*

4 Table 12A gives information about five radioactive isotopes.

Table 12A

Isotope	Radiation emitted	Half-life
A	Alpha	4 minutes
B	Gamma	5 years
C	Beta	12 years
D	Beta	28 years
E	Gamma	6 hours

  a What is a 'beta particle'? *(1 mark)*
  b What does the term 'half-life' mean? *(1 mark)*
  c Which of the isotopes could be used as a medical tracer? Explain the reason for your answer. *(3 marks)*
  d Radioactive waste needs to be stored. One suggestion is to seal it in steel drums and bury these in deep underground caverns. Suggest why people may be worried about having such a storage site close to where they live. *(3 marks)*

5 Stars go through a life cycle of change.
  a Explain how a star is formed. *(3 marks)*
  b Why does the amount of hydrogen in a star decrease as time goes on? *(1 mark)*
  c Describe what happens to a very large star from the time it runs out of hydrogen. *(4 marks)*

6 Our solar system contains atoms of the heaviest elements.
  a Where were these elements formed? *(1 mark)*
  b Explain how these elements are formed. *(2 marks)*
  c What does this tell you about the age of our solar system compared with the age of most of the stars in the universe? *(1 mark)*

Index

Note: A **bold** page number indicates where a term's definition is to be found.

A

abiotic factors **34–35**
absorption spectroscopy 142, 143
acceleration **196–97**
    of a falling object 206
    force and mass 198–99
    velocity–time graphs 200–01
acids **169–70**
    acidity and pH scale 171
    neutralisation reaction 170–71
    salts made with 172–75
activation energy **41**, **152**
actual yield **138**
aerobic respiration 50, **52**
air resistance 192, 193, 204–05
alkali metals
    reaction with halogens 93–94
    reaction with other non-metals 94–95
alkalis (soluble bases) **169–70**
    alkalinity and pH scale 171
    electrolysis 181–82
    neutralisation reaction 170–71
    salts made with 172–73
alleles **66–68**
    and family trees 68–70
    and new species 78
allotropes **104**
alloys 109–11
alpha particles **253**
    alpha decay 255
    and discovery of the nucleus 250–51
    hazards to health 259–60
    properties of 253–54
    in smoke alarms 258
alternating current (a.c.) **237**
    CRO traces for 237–38
    period and frequency of 238
aluminium metal, electrolysis 181

amino acids **7**, 25, 39
    in baby food 48
    and enzymes 40, 44–45
    and genes 61–62
    protein synthesis 50
ammonia 173
amperes/amps (A) 226
amylase **45**
anaerobic respiration **52**, 55
analytical chemists 141–42
animal cells 5–7
    diffusion in 13
animal sampling 32–33
anode (positive electrode) **178**
antibodies 39
appliances
    cables and plugs for 239–40
    circuit breakers 241–42
    energy transferred 243
    fuses and earthing 241
    and overloaded circuits 231–32
    power rating 243–44
atomic bombs 260, 261
atomic number **124**, **248**
    and radioactive decay 255
atomic structure 248–49
atomic weight *see* relative atomic masses
atoms 85
    in metals 107
    and radiation 251–52
    structure of 248–49
    theoretical models of 249–51
    valency of 96, 100
attraction (electrical charge) 224

B

baby foods, use of enzymes 48
background radiation **251–52**
bacterial cells 8–9
bacterium/bacteria **8**
    killed by gamma radiation 257
balanced forces 192
banded snails 78

Index

bases   **169–70**, **174**

bases in genes   61–62

batteries   **227**

Becquerel, Henri   251

belt transects, sampling method   31

beta particles (electrons)   **253**
- beta decay   255
- properties of   253–54
- skin cancer treatment   257

bile   **45**

black holes   265

boiling points   99, 100–01, 120

braking
- braking distance   202–03
- and kinetic energy   204
- regenerative   214

braking distance   **202–03**

Brown, Robert   5

buckyballs   **104**

C

cables   239
- dangerous use of   242

calibration   **236**
- of mass spectrometers   126

cancer
- and embryo screening   73
- and long-term radiation exposure   259–60
- treatment with radioactive isotopes   257

carbohydrase enzymes   **47**

carbon-12 isotope   128

carbon dioxide
- 'dry ice'   99
- from respiration   50, 52, 55
- in greenhouses   27–29
- and photosynthesis   16, 20–21
- and rate of photosynthesis   22–23

carrier (of a disease)   **69**

cars
- braking distance   202–03
- safety features   214–16
- stopping safely   202–03

catalysts   **156–57**
- cost-effective amount of   158–60
- enzymes as biological   39–40
- industrial use of   157–58
- nanoparticles as   117–18

cathode (negative electrode)   **178**

cathode ray oscilloscope (CRO)   **237–38**

cell membrane   **6**

cells   5
- animal cells   5–7
- and diffusion   13–16
- division of   60–64
- enzyme reactions in   50
- membrane of   6
- plant cells   7–9
- specialisation of   9–10
- stem cells   71–74
- and tissues   10

cells (batteries)   227–28

chain reactions   261

charge   *see also* electric current
- electrostatic   224

charged particles *see* ions

chemical analysis, instrumental methods   141–45

chemical bonding theory   92

chemical bonds   **85**

chemical reactions
- collision theory   151–52, 154–55
- efficiency and outcome of   137–38
- finding masses of reactants and products   133–34
- increasing rate of   152, 156–60
- measuring rate of   149–50
- predicting changes in rates of   153–54
- reversible reactions   136–37
- yield of   138–40

chlor–alkali process   181–82

chloroplasts   **7–8**, 20

chromatography   143–45

chromosomes   61, **62**
- and cell division   6, 62–64
- inheritance of characteristics   64–66

Index

circuit breakers   241–42
circuits *see* electric circuits
cold packs   161, 162
collisions
    between particles   151–52
    and nuclear fusion   262
    and total momentum   218–19
collision theory   **151–52**
    visual model for   154–55
community (of animals/plants)   **29**
components   **226–27**
concentration   **14**, 15
concentration gradient   13, 16
conductors   108, 225
conservation of total
    momentum   218–20
cosmic rays   252
cost-effectiveness
    chemical reactions   137–38
    in use of catalysts   158–60
coulombs (C)   226
covalent bonding   89–92
    giant covalent structures   102–06
    simple molecular
        compounds   98–101
covalent bonds   **89**
crop production in greenhouses   28–29
crumple zones, cars   214, 215–16
crystallisation   169
crystals   **169**, 173
Curie, Marie   251
current *see* electric current
cystic fibrosis   **69**
cytoplasm   **6**, 8

D

deceleration   193, 197, 202, 203
decomposition   *see also* electrolysis
    hydrated copper sulfate   163
    of hydrogen peroxide   156–57
'delocalised' electrons   103, 104, 107, 108
denatured enzymes   41, **42**

diamonds   102–03
differentiation   9, 60, **71**
diffusion   **13–16**
digestive system   11
    enzymes of   44–46
diodes   **233**
    symbol for   227
direct current (d.c.)   **237**
diseases, inheritance of   69–70
displacement reactions   **174**
distance–time graphs   199–200
distance travelled, calculation of   201
distraction, effect on reaction time   204
DNA (deoxyribonucleic acid)   **62**
DNA fingerprinting   **61**
dominant alleles   66, **67**
drag force   192, 193, 204–05

E

earthing   241
effective collisions   **152**
elastic potential energy   **194–95**
electric circuits   225–26
    batteries   227–28
    circuit breakers   241–42
    overloaded   231–32
    parallel   227, 231
    series   227, 230–31
    symbols for components   226–27
electric current   **225**
    and charge   225–26
    direct and alternating   237–38
    energy transferred   243
    and overloaded circuits   231–32
    and potential difference   232–33
    and resistance   228–29
    in series and parallel circuits   227, 230–31
electric fields, effect on charged
    particles   254
electricity
    mains electricity supply   238
    safe use of   240–42
    static   224

Index

electric plugs   239–40
electrodes   177, **178**
electrolysis   **177–78**
   half-equations   178
   as oxidation and
     reduction   178–79
   of solutions   179
   uses of   181–83
electrolytes   **178**
electromagnetic radiation *see* gamma rays
electronic theory of chemical bonding   **92**
electrons   **248**
   beta particles   253–54
   and chemical bonds   85
   'delocalised'   103, 104, 107, 108
   in metals   107
electroplating   **182**
electrostatic forces   224
   ionic compounds   95
elements   85
   chemical analysis   141–45
   formation of heaviest   265
embryonic stem cells   71–72
embryo screening   72–73
emission spectroscope   142
empirical formulae   **131**
endothermic reactions   **161**, 163
energy
   activation   41, 152
   change in reactions   160–64
   and charge   243
   elastic potential   194–95
   from nuclear fission   261
   gravitational potential   212–13
   kinetic   215
   and power   243–44
   released by fusion   262–63
   released by respiration   50
   and work   211
environment
   abiotic (physical)   34–35
   animal habitats   29–33

geographical isolation of populations   78
   and rate of photosynthesis   26–29
environmental issues, washing powders   48–49
enzymes   6, **39–40**
   conditions affecting reaction rate   41–42
   controlling aerobic respiration   50
   in the digestive system   44–45
   in the food industry   47–48
   in laundry detergents   48–49
   'lock and key' model 40–41, 43
   production from microorganisms   46
epithelium   **10**
equations
   acceleration   197, 198
   aerobic respiration   50, 52
   anaerobic respiration   52
   current in a circuit   226
   energy transferred   243
   force on a spring   195
   frequency   238
   gravitational potential energy   212–13
   half-equations (electrolysis)   178
   kinetic energy   213
   momentum   218
   neutralisation   170
   for photosynthesis   20
   potential difference   226, 229
   power   217, 243, 244
   resistance   229
   speed   205
   using masses in   133–36
   weight   207
   work done   211
   yield of chemical reaction   138–40
ethical issues, stem cell research   71–74
evolutionary theory   75–77
exercise   51–52
   effects of vigorous   54–55
   measuring effects of   53–54
   response of body to   52–53
exothermic reactions   **160–61**, 163
explosions   220

Index

extension of a spring   194–95
extinction, causes of   76

F

family trees   68–70
fermenters   46, **47**
fertilisation in human   64
filament lamps
    current–potential difference graph   233
    symbol for   227
finches, natural selection   78
Fischer, Emil   43
food
    products made from enzymes   47–48
    treated with gamma radiation   257
forces   *see also* work
    acceleration and deceleration   196–97
    applied to elastic objects   194–95
    distance–time graphs   199–200
    effects on motion   191–93
    mass and acceleration   198–99
    speed and velocity   196
    stopping safely   202–04
    and terminal velocity   204–07
    velocity–time graphs   200–02
formulae   *see also* relative formula masses
    empirical   131
    of ionic compounds   96
    simple molecular substances   98, 100
    structural   90
fossils
    evidence from   75–77
    formation of   74–75
free electrons   233
frequency   **237–38**
frictional forces   192, 193
    acting against gravity   204–05
    and braking distance   203
    and static electricity   224
    work done against   211–12

fructose syrup   47
fullerenes   **104**, 105
fungi   46
fuses   239, **241**
    finding required   244
    symbol for   227

G

Galapagos Islands, adaptation of finches   78
Galileo's theory of gravity   206
gametes (sex cells)   **63**, 64–5
gamma rays   **253**
    for cancer treatment   257
    emission of   255
    health hazards of   259, 260
    killing bacteria   257
    properties of   253–54
gas–liquid chromatography (GLC)   142, 144
Geiger, Hans   250, 251
Geiger–Müller tube   252, 253, 258
gene pool   **78**
genes   **61–62**
    genetic inheritance   64–70
    mitosis and meiosis 62–4
genetic diagrams   66, 67-8
genotypes   67
geographical isolation   78
giant covalent structures   91, **102**, 120
giant ionic lattices   **95–96**, 169
giant metallic structures   **107**, 120
    see also metals
glands   10, **45**
glucose
    aerobic and anaerobic respiration   52
    stored as glycogen   53
    used in plants   25–26
glucose syrup   47
glycogen   **53**
gradient   **200, 201**
    of a distance–time graph   199–200

276   AQA GCSE Additional Science

Index

of a velocity–time graph   201
concentration gradient   13, 16
graphite   102, 103–04
graphs
    current–potential
        difference   232–33
    distance–time   199–200
    rates of reaction   150, 153–54
    velocity–time   200–02, 204–05
gravitational field strength (g)   207, 212–13
gravitational potential energy (GPE)   212–13
gravity
    and formation of a star   264
    Galileo's theory   206
    mass and weight   207
    and terminal velocity   204–05
    working against   212–13
greenhouses
    controlling environment of   27–28
    crop production in   28–29
growth hormone   39

H

Hahn, Otto   261
half-equations   **178**
half-life   **256**
halogens
    electrolysis of solutions   179
    reaction with alkali metals   93–94
HD (high-density) polythene   113–14
heat packs   162
heterozygous   67
HFEA (Human Fertilisation and Embryology Authority)   72, 73
high-density (HD) polyethene   113–14
homozygous   67
Hooke, Robert   5
hormones   39
hydrogen peroxide, decomposition of   156–57
hypothesis   **206**

I

indicators, using to prepare a salt   172–73
industrial reactions and catalysts   158
industrial use of enzymes   46–49
infrared spectroscopy   142
inheritance
    of characteristics   64–66
    chromosomes and genes   61–64
    controlled by a single gene   66–68
    of diseases   69–70
    family trees   68–69
    Mendel's work   59–60
insoluble salts   175–76
instrumental methods of chemical analysis   141–45
insulators   224–25
intermolecular forces   98, **99**
iodine-131, decay curve for   256
ionic bonds/ionic bonding   **86–88**
ionic compounds   93
    alkali metal halides   93–94
    alkali metals and other non-metals   94–95
    comparing   97
    crystallisation   169
    formulae of   96
    melting and dissolving   168
    structure and properties   95–96
ionising radiation   253–54
ions   249
    effects of melting and dissolving   168
    electrolysis   177–83
    and ionic bonding   86–87
    neutralisation process   170–71
    salts   172–77
    solutions of   179
    'spectator' ions   170–71
isomerase enzymes   **47**
isotopes   **127–28**, **248–49**
    radioactive   255–60

Index

J
joules (J) 211

K
keratin 39
kick-sampling 32
kinetic energy
    and braking 204
    and work 213–15

L
lactic acid **54–55**
laundry detergents, enzymes in 48–49
LD (low-density) polyethene 113–14
life cycle of a star 263–65
life, origins of 74–77
light capture by plant cells 20
light-dependent resistors (LDRs) **235**
    symbol for 227
light-emitting diodes (LEDs) 233
light intensity **23**
limiting factor **21**
limit of proportionality 195
lipase **45**
lipids 25, 45
'lock and key' model, enzymes **40–41**, 43
low-density (LD) polyethene 113–14
lungs
    cancer treatment 257
    and cystic fibrosis 69
    effects of exercise 52–53
    oxygen diffusion in 14

M
macromolecular structures 120
magnetic fields
    effect on charged particles 254
    in mass spectrometers 125, 142
mains electricity supply 238
    cables and plugs 239–40
    overloading 231
    safe use of 240–42
malleable **108**

Marsden, Ernest 151, 250
mass
    force and acceleration 198–99
    and kinetic energy 213
    and momentum 218
    and weight 207
masses
    relative atomic masses 125–33
    relative formula masses 130, 133–34, 138–39
mass number **124**, **248**
    and radioactive decay 255
mass spectrometers **125–26**, 127
    identifying elements/compounds 142
    measuring masses of atoms 129
    modelling 128
mean **32**
median **32**
meiosis **63–64**
melting ionic compounds 168
melting points 94, 99, 100–01, 108, 111–12, 120
membranes
    of cells 6
    partially permeable 14
memory metals **109–11**
Mendel, Gregor 59–60, 68
metals
    alkali metals 93–95
    alloys 109–11
    atomic models for 113
    giant structure of 107
    properties of 108, 110, 111
microorganisms, enzyme production 46
mitochondria 6, **7**
mitosis 60, **62–63**, 64
mode **32**
models **41**, 105
    of the atom 249–51
    collision theory 154–55
    giant metallic structures 113
    making molecular 105–06
molecular forces 98–99

Index

molecular substances 98–101

moles **126**

momentum **218**
    conserved in collisions 218–19
    conserved in explosions 220

monomers **113–14**

motion, effects of forces on 191–93

muscle fatigue **54**

muscles
    effect of exercise on 54–55
    protein for 48
    and respiration 51–52
    tissue of 10

mutations 69

N

nanomaterials 117–18

nanoparticles **117**

nanotechnology **117**

nanotubes 104, 105

natural selection 78

nebula(e) **263**

neon, isotopes of 127–28

neutralisation reaction 170–71, 172–73

neutrons **248**

newtons (N) 195

nitinol, smart memory metal 110

non-metals
    covalent bonding 89–92
    ionic bonding 86–88
    ionic compounds 93–97
    simple covalent molecular structures 98–101

nuclear fission **261–62**

nuclear fusion **262**
    and origin of elements 265

nuclear fusion reactors 263

nuclear magnetic resonance (NMR) spectrometer 142

nuclear model of atomic structure **250–51**

nuclear radiation
    hazards of using 259–60
    medical uses of 257
    non-medical uses 257–58

nuclear reactors 261–62

nucleus of a cell **5–6**

nucleus of an atom **248**

O

ohmmeter 229

ohms 228

organs 10–12

oven thermometer, designing 236

overload of electrical circuits 231–32

oxidation (electron loss) **178–79**

oxygen debt **54–55**

oxygen diffusion in the lungs 14

P

paper chromatography 143–45

parallel circuits **227**, 231

Parkinson's disease 73

partially permeable membrane **14**

penetrating power of radiation 253

percentage yield **139**

permanent vacuole 8

phenotypes 67

pH levels
    and enzyme activity 41–2, 44, 45
    scale for measuring 171

photosynthesis **20–21**
    factors affecting rate of 21–24
    factors limiting rate of 26–29
    glucose use by plants 25–26
    role of plant enzymes 50

plants
    diffusion in 13, 16
    plant cells 7–9
    sampling 29–32
    tissues and organs 12

plastics *see* polymers

plugs 239–40

'plum-pudding' model **249–50**

polydactyly **70**

Index

polymers **113–14**
    nanomaterials 117–18
    slime 115–16
    thermosetting and thermosoftening 114–15
polythene (polyethene) 113–14
pooters, collecting insects with 32
population **34**
    geographical isolation of 78
potential difference **230**
    and charge 226
    CRO 'traces' 237–38
    and current 225, 229
    current–potential difference graphs 232–33
    and energy transferred 243
    parallel circuits 231
    and power of appliances 244
    and resistance 229
    series circuits 230
power **216–17**
    and energy 243–4
precipitate **175**
precipitation reactions **175–77**
properties of substances, determining 119–20
protease 40, 44, **45**, **48**
proteins **7**, 39 *see also* amino acids; enzymes
protein synthesis **50**
proton number *see* atomic number
protons **248**
protostars 264
Punnett square 66, **67**, 68

Q

quadrats, sampling with 30–31
quantitative data, collecting 29–31
quantum tunnelling composite (QTC) 234

R

radiation *see* radioactive isotopes
radioactive atoms **251**
radioactive decay **251**, 255–56

radioactive isotopes 253–54
    background radiation 251–52
    dangers of 259–60
    decay of 255
    half-life 256
    industrial uses of 258
    medical uses of 257
randomness of radioactive decay 256
random sampling with quadrats 30
rate of photosynthesis **21**
reactions *see* chemical reactions
reactivity series 175
recessive alleles 66, **67**
red copper oxide 132–33
reduction (electron gain) **179**
regenerative braking 214
relative atomic masses **125–26**
    of common elements 128
    and empirical formulae 131–33
    and isotopes 127–28
    measurement of 129
    and percentage composition 130
    and relative formula masses 130
relative formula masses **130**
    calculations 133–34
reproducible results **32**, **133**
repulsion (electrical charge) 224
residual current circuit breaker (RCCB) 242
resistance 228
    of different lengths of wire 229
    diodes 233
    filament lamps 233
    light-dependent resistors (LDRs) 235
    potential difference and current 229
    quantum tunnelling composite (QTC) 234
    series circuits 230–31
    thermistors 235–36
resistors 228
    adding to a series circuit 230–31
    current–potential difference graph 233

Index

    energy transferred by   243
    light-dependent (LDRs)   235
    symbol for   227
    thermistors   235–36
respiration   6, **51–2**
    aerobic   50, 52
    anaerobic   52
    glucose needed for   25
resultant force   **191–93**
    and acceleration   198
    and air resistance   204–05
reversible reactions   **136–37**
    energy changes in   163–64
    factors affecting efficiency and outcome   137–38
    and yield   140
ribosomes   **7**
Rutherford, Ernest   250–51

S

safety issues
    cars   202–04, 214–16
    electricity   240–42
    nanomaterials   118
salts   169
    electrolysis   179, 181–82
    insoluble   175–76
    making   172–75
sampling methods
    animals   32–33
    plants   29–32
'saviour siblings'   72–73
Schleiden, Matthias Jakob   5
Schwann, Theodor   5
seat belts   214–15
seeds, nutrient stores in   25
self-warming cans   160
series circuits   **227**, 230–31
series dilution   22, **23**
sex cells (gametes)   63
sex chromosomes   **65–66**
sex determination   65–66
shape memory alloys   109–11
simple molecular substances   **99**

    formulae of   98, 100
    melting points   100–01
    properties of   99, 120
    structure of   98
slime   115–16
smart alloys (memory metals)   109–11
smoke alarms   258
snails, natural selection   78
sodium chloride solution, electrolysis   181–82
soft-centred chocolates   48
solid structures   119–20
soluble salts, making   174
solutions
    acids and alkalis   169–74
    electrolysis   177–83
    ionic substances   168–69
    precipitation   175–76
specialised cells   9–10
    and stem cells   71–74
species, formation of new   78
'spectator' ions   170–71
spectroscopy   142–45
speed   196, 205
springs
    elastic potential energy   194–95
    shape-memory   111
stain removal   48–49
starch, testing a leaf for   25
stars, life cycle of   263–65
static electricity   224–25
steel, 'slip' property   109
stem cells   9–10, 71
    and embryo screening   72–73
    research in the news   74
    social and ethical issues   71–73
stents   110
stone-washed jeans   49
stopping distance   **202–03**
structural formulae   **90**
substrate   **40–1**
supernova   265

Index

surface area
    nanoparticles   117–18
    and rate of diffusion   14
    and reaction rate   152, 154
    and terminal velocity   205

syrup production   47, 48

systems of the body   11

T

temperature
    changes in chemical reactions   160–61
    and enzyme reactions   39–40, 41–42
    in greenhouses   27, 29
    measuring changes in   161–62
    and rate of photosynthesis   24
    and rate of reaction   152, 153–54
    and resistance of thermistors   235–36
    and smart alloys   109–10

tension forces   192

terminal velocity   **204–05**

theoretical yield   **138–39**

therapeutic stem cell transplantation   **72**

thermal decomposition   163

thermistors   **235–36**
    symbol for   227

thermosetting materials   **114–15**

thermosoftening materials   **114–15**

thinking distance   **202–03**

Thomson, J. J., 'plum-pudding' model   249

time period   **238**

tinstone, extracting tin from   135–36

tissues   **10**, 12

total momentum, conservation of   218–20

U

unbalanced forces   192–93

undifferentiated cells   **71**

universal indicators   171

uranium   251, 252
    effects of exposure to   260
    and nuclear fission   261

V

valency   **96**, **100**

validity   **32**

velocity   **196**
    and momentum   218
    terminal   204–05

velocity–time graphs   200–02, 204–05

Virchow, Rudolph   5

voltage *see* potential difference

W

washing powders   48–49

water
    electrolysis of   181
    for photosynthesis   21

watts (W)   216–17

weight   204–05, 207

white dwarf stars   264–65

work   **211**
    done against friction   211–12
    energy transfer and power   216–17
    and force   211
    and gravitational potential energy   212–13
    and kinetic energy   213
    and transfer of energy   211

X

X-chromosomes   65–6

Y

Y-chromosomes   65–6

yeast cells   8

yield of chemical reactions   138–40

Z

zygotes   64

Periodic Table

Group 1	Group 2											Group 3	Group 4	Group 5	Group 6	Group 7	Group 0
						1 **H** hydrogen 1											4 **He** helium 2
7 **Li** lithium 3	9 **Be** beryllium 4											11 **B** boron 5	12 **C** carbon 6	14 **N** nitrogen 7	16 **O** oxygen 8	19 **F** fluorine 9	20 **Ne** neon 10
23 **Na** sodium 11	24 **Mg** magnesium 12											27 **Al** aluminium 13	28 **Si** silicon 14	31 **P** phosphorus 15	32 **S** sulfur 16	35.5 **Cl** chlorine 17	40 **Ar** argon 18
39 **K** potassium 19	40 **Ca** calcium 20	45 **Sc** scandium 21	48 **Ti** titanium 22	51 **V** vanadium 23	52 **Cr** chromium 24	55 **Mn** manganese 25	56 **Fe** iron 26	59 **Co** cobalt 27	59 **Ni** nickel 28	63.5 **Cu** copper 29	64 **Zn** zinc 30	70 **Ga** gallium 31	73 **Ge** germanium 32	75 **As** arsenic 33	79 **Se** selenium 34	80 **Br** bromine 35	84 **Kr** krypton 36
85 **Rb** rubidium 37	88 **Sr** strontium 38	89 **Y** yttrium 39	91 **Zr** zirconium 40	93 **Nb** niobium 41	96 **Mo** molybdenum 42	[98] **Tc** technetium 43	101 **Ru** ruthenium 44	103 **Rh** rhodium 45	106 **Pd** palladium 46	108 **Ag** silver 47	112 **Cd** cadmium 48	115 **In** indium 49	119 **Sn** tin 50	122 **Sb** antimony 51	128 **Te** tellurium 52	127 **I** iodine 53	131 **Xe** xenon 54
133 **Cs** caesium 55	137 **Ba** barium 56	139 **La** lanthanum 57	178 **Hf** hafnium 72	181 **Ta** tantalum 73	184 **W** tungsten 74	186 **Re** rhenium 75	190 **Os** osmium 76	192 **Ir** iridium 77	195 **Pt** platinum 78	197 **Au** gold 79	201 **Hg** mercury 80	204 **Tl** thallium 81	207 **Pb** lead 82	209 **Bi** bismuth 83	[209] **Po** polonium 84	[210] **At** astatine 85	[222] **Rn** radon 86
[223] **Fr** francium 87	[226] **Ra** radium 88	[227] **Ac*** actinium 89	[261] **Rf** rutherfordium 104	[262] **Db** dubnium 105	[266] **Sg** seaborgium 106	[264] **Bh** bohrium 107	[277] **Hs** hassium 108	[268] **Mt** meitnerium 109	[271] **Ds** darmstadtium 110	[272] **Rg** roentgenium 111							

Key

relative atomic mass
atomic symbol
atomic name
proton number

Elements with atomic numbers 112–116 have been reported but not fully authenticated

*The lanthanides (atomic numbers 58–71) and the actinides (atomic numbers 90–103) have been omitted. **Cu** and **Cl** have not been rounded to the nearest whole number.

AQA GCSE Additional Science